Plasmonic Sensors and their Applications

Plasmonic Sensors and their Applications

Edited by

Adil Denizli

WILEY-VCH

Editor

Adil Denizli
Department of Chemistry
Hacettepe University
Ankara
Turkey

Cover

Cover Design: Wiley
Cover Image: © Creations/Shutterstock

Library of Congress Card No.: applied for

British Library Cataloguing-in-Publication Data
A catalogue record for this book is available from the British Library.

Bibliographic information published by the Deutsche Nationalbibliothek
The Deutsche Nationalbibliothek lists this publication in the Deutsche Nationalbibliografie; detailed bibliographic data are available on the Internet at <http://dnb.d-nb.de>.

Print ISBN: 978-3-527-34847-3
ePDF ISBN: 978-3-527-83033-6
ePub ISBN: 978-3-527-83035-0
oBook ISBN: 978-3-527-83034-3

Typesetting Straive, Chennai, India
Printing and Binding CPI Group (UK) Ltd, Croydon, CR0 4YY

Printed on acid-free paper

C109776_150721

Contents

Preface

I welcome the publication of this book titled *Plasmonic Sensors and Their Applications*. In recent years, plasmonic sensors have been employed for various applications from medical diagnosis, environmental monitoring, pharmaceutical analysis, food quality detection to defense, and security fields. The development and progress of the plasmonic sensors cover chemistry, physics, material science, nanotechnology, and engineering. A huge body of information on plasmonic sensors and applications is already existed and continuing to create more reliable, selective, sensitive, and low-cost sensors for a variety of applications although a complicated and time-consuming production procedure.

This book contains 13 chapters, which contain plasmonic sensors prepared by different methods and used for various applications. In the first chapter, following the mention of the fundamentals of plasmonic sensors, new trends in plasmonic sensors for the applications in disease diagnosis are extensively reviewed with future perspectives. In Chapter 2, nanosensors based on localized surface plasmon resonance are highlighted. The historical and theoretical background, fabrication of metal nanostructures, improving detection limit, and integration of sensors with other molecular identification techniques are discussed. Highly sensitive and selective plasmonic-sensing platforms in medical and environmental applications are comprehensively evaluated in Chapter 3. The next chapter, Chapter 4, concentrates on the detection of chemical and biological warfare agents using plasmonic sensors with recent studies. Chapter 5 includes plasmonic-sensing platforms based on molecularly imprinted polymers for medical applications. Chapter 6 summarizes the performance and analytical features of the magnetoplasmonic sensors. In Chapter 7, overview of vitamin detection using plasmonic sensors can be found. Proteomic applications of plasmonic sensors are reviewed in Chapter 8. Cancer cell recognition via plasmonic sensor systems is given in Chapter 9. Plasmonic nanoparticles, which are prepared by different strategies for the ultrasensitive sensing platforms, are combined in Chapter 10. The next chapter, Chapter 11, gives details about the application of surface-enhanced Raman scattering sensors for chemical and biological sensing. Carbon nanomaterials as plasmonic sensors in biotechnological and biomedical applications are summarized in Chapter 12. Finally, surface plasmon resonance sensors based on molecularly imprinted polymers are highlighted in detail in Chapter 13.

I believe this book provides an overview and highlights some of the recent research including the extensively studied topics. I would like to deeply thank WILEY-VCH and all the contributors to the generation of this book possible. I hope this book will reach a broad range of readers.

Prof. Dr. Adil Denizli
Editor
Ankara, Turkey, 2021

1

Deciphering Plasmonic Modality to Address Challenges in Disease Diagnostics

Esma Derin[1,2], Özgecan Erdem[1], and Fatih Inci[1,2]

[1] UNAM-National Nanotechnology Research Center, Bilkent University, Ankara, Turkey
[2] Institute of Materials Science and Nanotechnology, Bilkent University, Ankara, Turkey

1.1 Introduction

Recent advances in health technologies have remarkable impact on health-care system. Advanced health technologies are however not affordable and available for resource-constraint settings. From global health perspective, recent improvements in health technologies need to reflect alternative solutions for overcoming such inequalities by developing new technologies and strategies in the manners of cost-effective, rapid, easy-to-use, and portable size, hereby minimizing the disparities between resource-rich and limited settings [1, 2]. In this regard, biosensing arena has enormous applications in diverse fields, spanning from biomedicine to agriculture; for instance, monitoring vital information for diseases detection or determining the presence of contaminants in water and soil [2]. The concentrations of analytes can be accurately determined based on a linear association between signal intensity and analyte concentrations via an analytical biosensor. In the field, four types of biosensors, such as optical (mostly plasmonic sensors), electrochemical, piezoelectric, and magnetic biosensor, are mainly employed for analytical measurements [3]. Optical biosensors (e.g. surface plasmon resonance (SPR), localized SPR (LSPR), surface-enhanced Raman scattering (SERS), plasmon-enhanced fluorescence (PEF), surface-enhanced infrared absorption spectroscopy (SEIAS), etc.) are some of the mostly applied biosensors in health-care biomedicine [4, 5]. Especially, SPR-stemmed platforms have been benefitted in a vast majority of bio-analytical analyses since this technique enables rapid observation of bio-analytes in specimens, at the same time provides high sensitivity and selectivity in comparison to currently available instruments [2, 6]. These sensors basically monitor minute alterations in spectral properties of the plasmon by acting as a transducer of the sensing signal [5]. The sensing mechanism is constituted by recognizing and capturing the target analytes through bioreceptor which are immobilized on the metal surface. Then local refractive index increases due to capturing and SPR signals are shifted [7]. Their plasmonic fashion can be designed by considering the material and plasmonic features, like surface-based strategies that support SPP mode or nanoparticle-based

Plasmonic Sensors and their Applications, First Edition. Edited by Adil Denizli.

modality that employs localized surface plasmon resonance (LSPR). The intensity and peak position of the SPR have been influenced by the size, shape, and composition of the nanostructures and also the surrounding environment's dielectric properties. [5]. To touch upon the fundamentals of plasmonic sensors, they are stemmed from surface plasmon polariton (SPPs) or plasmonics, which are basically defined as the collective oscillations of free electrons at the metal surface [8]. Plasmonics deal with the electromagnetic (EM) wave and free electron interactions through excitation on conductors, such as metals, semimetals or semiconductors [9]. The excited electrons leads to collective oscillation with the close frequency to EM wave [9].

The history of SPP modality takes back more than a century, however, the improvements in the field have not accelerated until notable leaps in the nanotechnology (e.g. nanoscale fabrication techniques) to achieve appropriate sized structures for the further discoveries [10]. For instance, importantly note here, Kretschmann and Otto are two pioneers to create Surface Plasmon Resonance (SPR) through coupling strategies. Afterwards, SPR-based sensor was leveraged by Liedberg et al. through the antibody and antigen relations, which can be identified as a critical direction for biosensing platforms [11]. Today, SPR has been considered as one of the most powerful biosensing platforms, especially in analytical chemistry and medicine [7].

1.2 Surface Plasmon Polaritons

SP is the propagating light waves at the conductor surface via trapping due to interactions with free electrons [12]. Per these interactions, collective oscillations with light waves result in a resonance as a response of free electrons [12]. The formation of the SP is carried out by the resonant interactions of surface charge oscillations and electromagnetic field of light, and these interactions enhance the dominancy of SPs simultaneously [12]. The definition of the SPP can be simplified as an electromagnetic wave, and more specifically, it is transverse magnetic (TM)-polarized optical surface wave that propagates directly along a surface between dielectric and a metal surface (Figure 1.1a) [13, 14]. In addition, metal surfaces have crucial charge density wave, which constitutes SPPs with the combination with electromagnetic fields that are maximum at the interface on the contrast through both media since they are decreased exponentially [14]. The amplitude of the SPP is influenced inversely (exponential decay) with the distance of each medium from the interface [13]. In addition to the metal surface, SPPs can be obtained from different metal structures, such as thin films, stripes, differently sized and shaped nanoparticles or differently patterns (e.g. holes, slits, grooves, gaps, or corrugations, etc.) [14]. SPPs or plasmonics are mostly utilized by the fields of medical diagnostics, biosensing, spectroscopy, nanophotonic, imaging, or circuitry due to their substantial properties, including energy asymptotes in dispersion curves, resonances, field enhancement and localization, high surface and bulk sensitivities, and subwavelength confinements [8, 14]. Another point is that SPPs have subwavelength property and field confinement since their ability is overweighted to traditional optical elements (e.g. lenses, spatial light modulators) by considering ability to spatial field modulations at nanoscale [8]. Majorly, Maxwell equations between interface of conductor (e.g. metals) and a dielectric layer are crucial for the investigation of physical properties of SPPs [15].

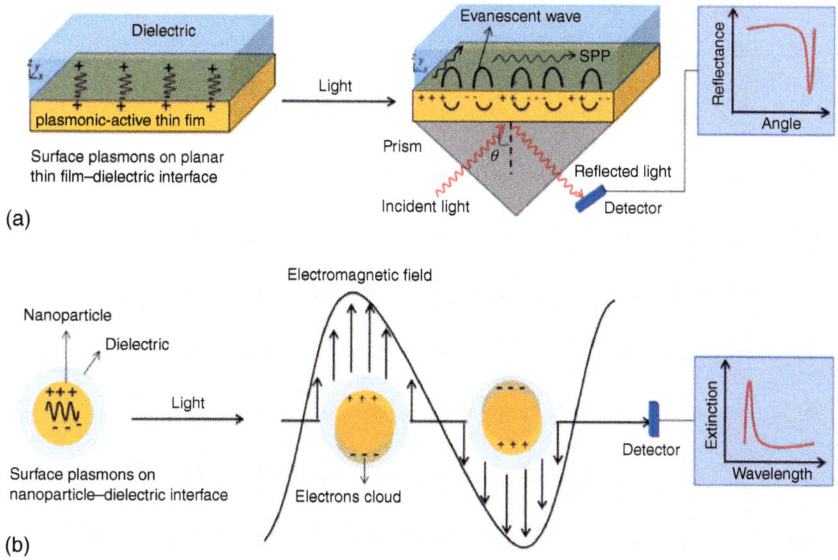

Figure 1.1 (a) SPP propagation is illustrated through thin film with the surface charges at the metal and dielectric interface with the excitation, and the spectra is obtained after surface plasmon excitation. (b) SPP is generated through nanoparticle surface to achieve localized surface plasmon resonance (LSPR), and the spectra is obtained after surface plasmon excitation. Source: Reprinted with permission from Bhattarai et al. [19]. © MDPI.

The characteristics of the SP modes, which can be either localized SPs of individual particles or several propagating SPPs on flat and curved, single and multiple surfaces, are determined by the topology of the metal surface. In addition, SPP modes of complex particle arrays and metal nanostructures are also determined. Based on this unique property of the SPP-based waveguides, the bandwidth of information can be transported by plasmonic waveguides through conventional (dielectric-based) photonics. The interest in plasmon-based nanophotonics is increased substantially [16]. When the SPP interacts with the metal, its energy dissipates. Free electron scattering in the metal, which is bounded with an ideal dielectric, creates loss due to absorption via inter-band transitions at a short enough wavelength [14]. For the SPP, this loss is fundamental. Operating wavelength selection can be selected carefully to avoid absorption via inter-band transitions or advanced fabrication techniques can decrease the free-electron scattering, and however, the both cannot be eliminated [14]. As aforementioned, surface of the metal interface has influence on the SPP, and moreover, the roughness creates additional loss since SPPs are scattered into bulk waves [14]. The main drawbacks of this loss are limiting practical applications of SPPs when it is excessive [14].

1.2.1 Excitation of the SPP

The excitation of the surface plasmon is mostly related to the conservation of the energy and momentum of photon in the incident light on a metal–dielectric interface. The excitation of the SPPs requires the momentum and energy matching to the incident photon's and plasmon

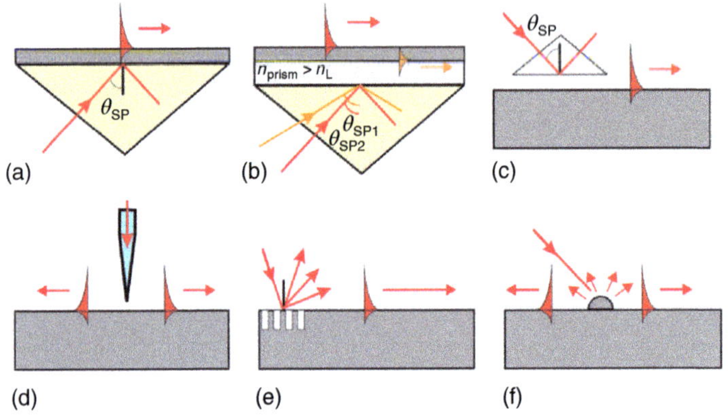

Figure 1.2 The common configuration of SPP excitations is depicted. (a) Kretschmann configuration, (b) two-layer Kretschmann configuration, (c) Otto configuration, (d) excitation with an SNOM probe, (e) grating diffraction, and (f) diffraction on surface features. Source: Reprinted with permission from Zayats et al. [24]. © 2005, Elsevier.

modes in order to obtain charge-coupled oscillations [17]. Special techniques are the primary requirement for the excitation of three-dimensional light beams in order to employ phase-matching [15]. The excitation of surface charges route can be explained by the presence of metal and dielectric interfaces and incoming p-polarized wave, which is transverse mode of the parallel electric field vector to the interface with an angle. The incident wave reaches the interface and split into two waves that are propagated in different directions [18].

Effective plasmon generation is achieved with the optical coupling element integration into the system. Prism, grating, and waveguide coupling methods (Figure 1.2) are the widely used light coupling techniques in comparison to waveguide, photonic crystal, and fiber-optic based coupling. These applications utilize attenuated total reflection (ATR), light diffraction or evanescent wave coupling from waveguide modes [17].

1.3 Surface Plasmon Resonance (SPR)

SPR is an optical biosensor that relies on the refractive index change of sensor surface, denoting label-free and in real-time detection [2, 20]. The plasmonics can be defined basically interaction of light with metals or metallic nanostructures; hence, this mechanism combines photonics and electronics to measure optical properties, e.g. spectra and refractive index changes at the nanoscale [3]. The interests in SPR biosensors have been increased enormously over the years [21]. The application of the SPR mostly focuses on clinical diagnostics, biological and pharmaceutical analysis, food quality and safety evaluation since it enables to monitor molecular interactions and quantify biomarkers, such as proteins, DNA or whole cells [22].

SPP or SP is the main requirement to induce SPR in the interface between the metal (e.g. gold or silver) and dielectric materials based on light excitation [22]. The generation of SPP in the form of EM wave occurs after the interactions of the incident light with metal since, collective oscillation of the free electrons is induced by the photons in the conduction band [23].

SPPs can be only sustained by p-polarized electromagnetic or transverse magnetic wave at an interface of metal and dielectric medium. The sign of the dielectric constant has to be opposite for instance gold, silver, copper, or aluminum, which have negative real and positive imaginary dielectric constants, and therefore, they can be used for the SPP generation [23]. These indicated metals have both pros and cons; for instance, the negative real dielectric constant of silver is the largest, which creates higher sensitivity against to refractive index (RI) changes. However, its chemical stability is lower due to easy oxidation in air. In contrast to silver, RI sensitivity of gold is lower, yet it holds higher stability and chemical versatility, and hence, the functionalization of the sensor surface can be carried out much easier.

1.4 Localized Surface Plasmon Resonance (LPSR)

Localized surface plasmon resonance (LSPR) is an optical phenomenon produced as a result of interactions between the incoming light and surface electrons in a conduction band through a light wave trapped in conductive nanoparticles smaller than the wavelength of light. This phenomenon is reliant on the size, geometry, dielectric environment, composition, and particle–particle separation distance of the nanoparticles [25]. Since metal nanostructures interact with a beam of light, some of the incoming photons are absorbed, and the rest are scattered in different directions. When LSPR is stimulated, these absorptions and scattering events increase greatly. In metal nanostructures, LSPR is most easily detected by an optical spectroscopic method, and this measurement is usually based on the extinction or scattering events [26]. The differences between SPR and LSPR are illustrated in Figure 1.3.

In addition to gold and silver, which are the most commonly used plasmonic materials, the other metals, such as copper [27] and aluminum [28, 29] also exhibit plasmon resonance features [30]. Mostly, the physical properties of metal particles change considerably when the size of particles is around nanoscale, and also, smaller than the wavelength of

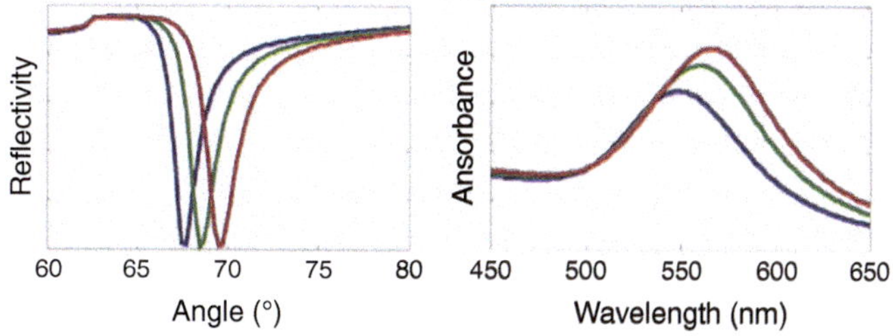

Figure 1.3 The schematic represents the basic principle and the difference between SPR and LSPR. Source: Reprinted with permission from Jatschka et al. [37]. © Elsevier.

light used to illuminate them [31]. In recent years, biosensors based on LSPR (majorly stemmed from metallic nanoparticles) have begun to draw attention for the label-free biosensing approaches due to its easy and colorimetric sensing features, as well as portability and its ability to interface with multiplexed devices [30, 31]. LSPR biosensors can also be easily integrated into miniaturized devices for point-of-care (POC) applications in order to save cost and reach the assay at different settings [32–34]. Integrating full-automation to these sensors also helps minimizing inter-personal errors for the measurements. Such integrations enable rapid acceleration for biosensor deployment into the health-care settings by promoting wider POC applications, such as bedside diagnosis, personalized medicine, and wearable devices [35]. Mostly, two main methods are used in LSPR biosensors: direct and indirect strategies. The first strategy tracks the shifts in the LSPR absorption peak due to the refractive index changes upon binding of the target molecule. This direct analysis requires less time and cost, but it has a limited sensitivity. The latter strategy is reliant on a sandwich analysis, where LSPR is used to stimulate the labels. As the light is on, metal nanostructures produce LSPR and are used to capture light near the surfaces [36].

On the other hand, there are also some obstacles. Since LSPR-based strategies are mostly dependent on the changes in the refractive index at the close vicinity of nanoparticles, a large number of molecules need to be localized around the particles in order to create a plasmonic shift. Strategies, such as adjusting the size and shape of the nanoparticle material, could be utilized to overcome this limit [30]. Another obstacle could be the reproducibility of sensor surface comprised of nanoparticles, ultimately limiting their utility and expansion to hurdle the real-world problems in clinical use.

From an application perspective, sensitive and selective detection of cancer biomarkers is of great importance in the early diagnosis of this disease. In a study, for instance, a LSPR lab-on-a-chip was designed to detect human alpha fetoprotein and prostate-specific antigen, which are cancer markers [38]. The microfluidic chip, which was developed by combining plasmonic, microfluidics, nanofabrication, and surface chemistry, accommodated 32 detection areas distributed across 8 independent microfluidic channels. The relevant markers could be detected quickly at a low concentration of $500\,\mathrm{pg\,ml^{-1}}$ in a complex medium containing human serum, and the chip could be used multiple times. As another example, extracellular vesicles are abundant in various biological fluids, such as blood, saliva, urine, and extracellular matrix. Toxic signals derived from extracellular vesicles can spread on tissues adjacent to the damaged area in some diseases, including brain tumors and neurodegenerative disorders. In this regard, extracellular vesicles that can be used clinically for liquid biopsy, needs to be better characterized. An LSPR biosensor containing self-assembly gold nanoislands (SAM-AuNIs) was used to detect and differentiate SH-SY5Y from microvesicles isolated from A-549 cells [39]. Blood serum, lung cancer cell, and urine samples obtained from the mouse model were used as biological samples. Exosomes have been shown to produce a discernible response in the LSPR biosensor compared to microvesicles. According to these results, there was a different biophysical interaction between exosomes and microvesicles with SAM AuNIs.

In addition to their single-mode measurements, LSPR sensors can be integrated with different modalities. For instance, a dual-mode plasmonic biosensor that combines plasmonic photothermal effect and LSPR sensing transduction has been developed as a promising alternative method for the diagnosis of COVID-19 disease [40]. On this sensor, SARS-CoV-2-specific sequences can be detected precisely using 2D AuNIs functionalized with complementary DNA receptors. When the thermoplasmonic heat was illuminated at the same AuNI chip at plasmonic resonance frequencies, a more sensitive sensing performance was provided.

1.5 Raman Spectroscopy and SERS

Raman spectroscopy is a method that measures the frequency shifts of the inelastic diffuse light from the sample when the photons hit a molecule and produce a diffused photon [41]. The photons of the laser light are absorbed successively by the sample, and the wavenumbers of re-sent photons are shifted up or down compared to the original monochrome waves (termed as the Raman effect). The resultant shift provides information about vibration, rotation, and other low wavelength transitions in molecules [42]. Near-IR (NIR), visible, or UV range monochromatic light is usually utilized for the Raman effect, which defines the photons to be adapted to virtual energy states, or energy stock generated due to the interaction of light with vibration modes associated with chemical bonds in the sample. Discrete vibration modes of the polarizable molecules are analyzed with such changes in energy, thereby obtaining a qualitative measurement of the biochemical composition [43]. Raman spectroscopy is a powerful analytical technique used in many areas, including detection of illegal drugs [44–46], toxic substances in the environment [47–49], and chemical [50–52] and biological warfare agents [53, 54], as well as *ex vivo* and *in vivo* applications of tissue diagnosis [55–58], and biomedical applications in which *in vitro* drug–cell interaction studies [59–62] are performed.

Resonance Raman effects that provide 10^2–10^6 enhancement, and Surface Enhanced Raman spectroscopy (SERS) resulting in up to 10^8 or both used together, which can provide up to 10^{16} enhancement, are used to increase the Raman signal [63]. SERS phenomenon is based on reduced Raman scattering when an analyte is adsorbed onto metal surface. The differences between SERS and Raman technique are represented in Figure 1.4. From the time when the discovery of the SERS, many researchers have begun to apply this method for molecular-level analysis, taking advantages of SERS, including high sensitivity, unique molecular fingerprint, and narrow spectral bandwidth for multiplex detection [64]. Known as an ultra-sensitive method that can detect even single molecules, SERS has long been considered a powerful tool, including the analysis of biomarkers that have been present in trace amounts. It offers an exceptional "signature" spectrum profile with very narrow peaks, capable of detecting multiple analytes simultaneously [65].

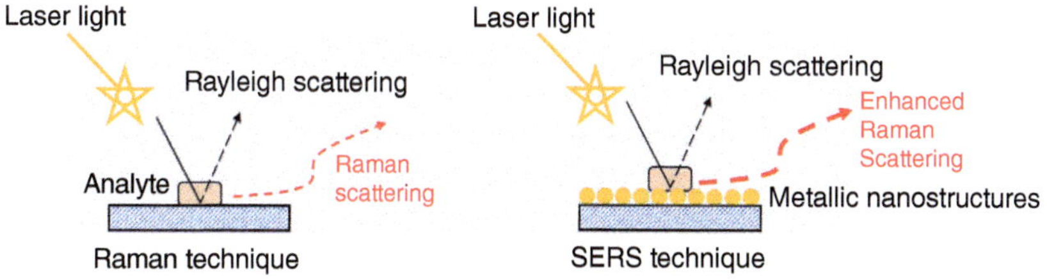

Figure 1.4 Basic principles of Raman and SERS technique. Source: Adapted and redrawn according to Zheng and He [66]. © John Wiley & Sons.

As a couple of examples in recent literature, a SERS sensor was fabricated using an Ag nanorod array by combining molecular signatures in the form of special hairpin to detect lung cancer-related miRNA biomarkers. With a portable-sized sensor, three different miRNAs (miRNA-21, miRNA-486 and miRNA-375) related to lung cancer were detected qualitatively and quantitatively [65]. As another example, a new paper-based, surface-enhanced SERS detection platform was developed to detect two key cytokines related to atherosclerosis, causing many cardiovascular and cerebrovascular diseases [67]. Two key cytokines, i.e., Interleukin 10 and monocyte chemoattractant protein 1, play key roles in the progression of this disease at different levels, and they are used for monitoring status and early diagnosis. A nanoporous networking membrane as a substrate and SERS nanotags as a signal reading probe were designed as a sandwich strategy, thereby enabling precise and specific identification of cytokine targets in human serum. In another study, ultrasonic surface-assisted SERS biosensor of the target-bound, acute myocardial infarction-related miRNA (miR-133a) was developed for the detection of disease-related biomarkers [68]. Bimetallic probes with high stability and a strong surface plasmon resonance effect were captured with a duplex connector to perform signal amplification after synthesis with a controllable silver and gold ratio through a galvanic replacement method. In this way, the target miR-133a could be detected in a wide linear range with high selectivity compared to other miRNAs expressed in human heart. The multiplex detection of biomarkers of Alzheimer's disease is of great importance for early diagnosis and personalized treatment of the disease. As the last example here, different Raman dye coded polyA aptamer–AuNPs conjugates were employed as SERS agents for simultaneous detection of Ap (1–42) oligomers and Tau protein [69]. Here, specific protein-aptamer binding mediated aggregation of AuNPs and the accompanying plasmonic coupling effect enabled to detect protein biomarkers within 15 minutes.

1.6 Whispering Gallery Mode (WGM)

Basically, WGM sensors are resonating micro- or nano-structures that provide high quality factors (Q). Where the changes in Q or shear resonance wavelength is used to evaluate surrounding milieu or binding events on the WGM resonator's surface. Optical WGMs are a family of electromagnetic modes built in a resonator with axial symmetry. WGMs create resonances at certain frequencies that depend on the geometry of the resonator, the refractive indices of both the resonator and the surrounding environment, and also the polarization of the modes [70]. WGM resonators can be fabricated in different morphologies with specific spectral properties, such as narrow line width, high stability, and adjustability [71].

WGM-based sensors have been used to detect biological molecules; for instance, a platform monitoring the shifts in the WGM resonance frequency were fabricated to measure enzymatic oxidation of glucose (Figure 1.5) [72]. The platform was modified with glucose oxidase and gold nanoparticles. Throughout the enzymatic reaction catalyzed by glucose oxidase, electrons were transferred to gold nanoparticles, and the optical signals produced

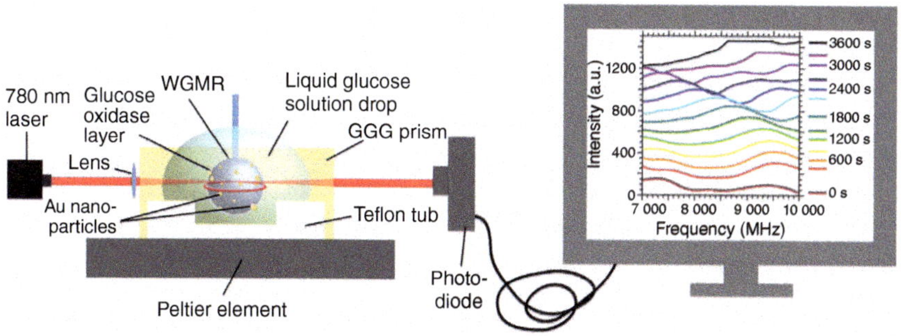

Figure 1.5 Schematic representation of WGM-based glucose sensor. Source: Reprinted with permission from Brice et al. [72]. © 2020, Elsevier.

by WGM resonators consisting of standard telecommunication fiber optics dissolved in a hydrogen flame were evaluated. As a result of tests that were performed at various glucose concentrations, the WGM-resonance frequency shift rate increased significantly at higher glucose concentrations.

In another study, a WGM resonator-based on fluorescence imaging has been reported to detect CA-125, i.e., an ovarian cancer biomarker [73]. Measurements have been extended using a simplified approach to initiate WGM resonances through excitation light coupled with the Dove prism. The mod structure in each resonator emerges thanks to the improved phase matching, thereby providing significant improvements in signal-to-noise. In addition, fluorescence imaging of the WGM resonances enabled for repeatable detection of biomarkers in complex biological fluids.

1.7 Fiber Cables Sensors

In recent years, the applications of fiber optic sensors in modern medical technologies and devices have been leveraged. Since the first generation of probes for *in vivo* pressure detection has been commercialized, the research is underway to develop new generation of fiber optic systems that have been significantly improved over the other sensing technologies, such as micro- and nanoelectromechanical systems [74]. Usually, a fiber optic probe is functionalized using biorecognition elements that can be selectively linked with target molecules. In the system, the response occurs based on a change in the local refractive index caused by the target. The biorecognition elements generally include various protein and nucleic acid-based molecules. The label-free detection methods using optical fibers enable to reach very low target detection limits. Gold, silver, magnetic nanoparticles and nanostructures with different shapes and sizes can be used as energy concentrators to expand the detection limits of optical fibers and obtain high-precision biosensing probes [75]. Fiber optic biosensors reliant on SPR [76–78], long-period grating [79–81] and fiber Bragg grating have been employed in a variety of fields and they provide rapid and precise detection [82].

As an example, fiber optic-based SPR type platform was utilized to detect acetylcholine, a pivotal neurotransmitter involved in the regulation of behavioral activities in human [83]. Dysfunction in acetylcholine regulation, for instance, has been linked to a variety of neurological disorders, including Alzheimer's disease. In this study, the sensing probe consisted of multiple layers of silver metal and tantalum-v-oxide nanoflakes functionalized with acetylcholinesterase on the uncoated core of an optical fiber. Once the sensing probe was exposed to acetylcholine solutions, the RI changed, and accordingly, the sensor provided a detection limit down to 38 nM of acetylcholine. In another study, a black phosphorus fiber optic biosensor was developed for the ultrasensitive detection of human neuron-specific enolase (a cancer biomarker) [84]. Bio-functionalized black phosphorus nanosheets by poly-L-lysine were exploited by integrating them into a largely curved fiber grid. After the nanosheets were synthesized by a liquid ultrasonication exfoliation, they were deposited on the fiber device by a layer-by-layer method. The anti-NSE immobilized BP-TFG biosensor was able to detect small cell lung cancer with a detection limit of 1.0 $pg\,ml^{-1}$. A new non-invasive measuring probe based on the fiber Bragg grid (FBG) was designed as a hybrid multi-channel fiber optic sensor system [85]. The probe specifically monitored body temperature, breathing rate, and heart rate, and it was capable of processing signals coming up to 128 people continuously (Figure 1.6).

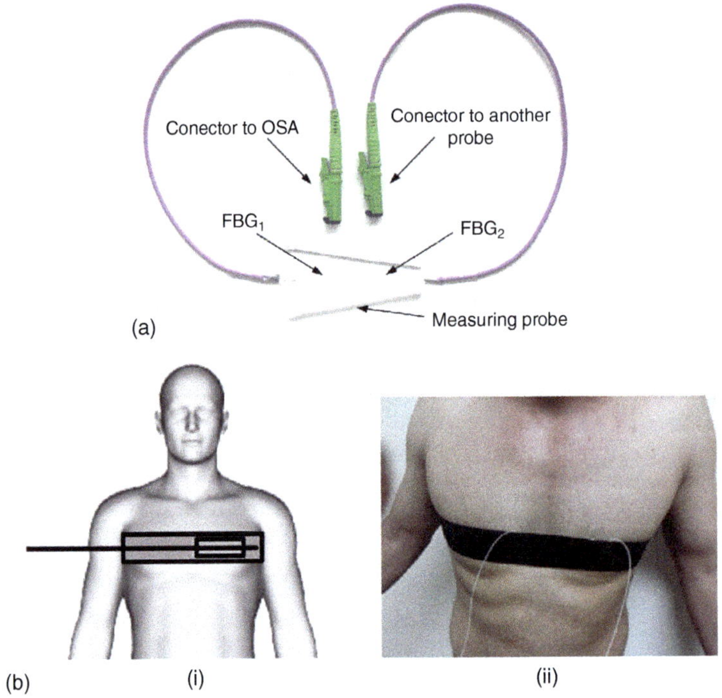

Figure 1.6 (a) The fiber-optic probe. (b) (i) A schematic diagram of the experimental set up. (ii) Experimental set up to acquire vital signals from a human subject using the probe embedded in a thoracic elastic strap. Source: Reprinted with permission from Fajkus et al. [85]. © MDPI.

1.8 New Trends in Plasmonic Sensors for the Applications in Disease Diagnosis

1.8.1 Mobile Phone-Integrated Platforms

The utilization of mobile phones in daily life mediates substantial opportunity for biomedical applications as a mobile device, potentially accelerating to reach health-care. The unique properties of the mobile phones (e.g., powerful CPUs, touch screen displays, advanced connectivity, high pixel-count, sensitive cameras, and integrated light sources) leverage their applicability to the biosensor realm [86]. These features also enable applications at resource-constrained settings, where the dedicated instruments and laboratory conditions do not exist [87]. As a consequence, integrating mobile phones with optical detection platforms creates a crucial niche that has evolved rapidly and denoted diverse biosensing approaches in the fields of immunodiagnostic assays, lateral flow assays, microscopic imaging, flow cytometry, colorimetric detection, photonic crystal, and SPR [86]. Talking over the examples, an enzyme-mediated LSPR strategy was employed to measure the RI changes around gold nanorods (AuNRs) while interacting with serum myoglobin, i.e., a biomarker for acute myocardial infarction (AIM) [88]. To eliminate complex readout systems, smartphone was turned into plasmonic immunoassay reader, which was designed as an ambient light sensor (ALS), and measured transmitted light intensity of AuNRs. The synthesis of AuNRs was carried out through the preparation of the seeded gold, which was used to create a sandwich immunoassay while exhibiting plasmonic properties [88]. The sensing strategy was based on hydrogen peroxide (H_2O_2) generation from glucose and gluconic acid catalysis since H_2O_2 caused a blue shift in the SPR spectrum of AuNRs. Then, blue shift was positively correlated with the target concentrations. Two pieces of the reader were 3D-printed, one of them fixed on a smart phone to supply a stable light source powered by two batteries, and the other one provided a host to microwells (Figure 1.7) [88]. The linear detection range of this

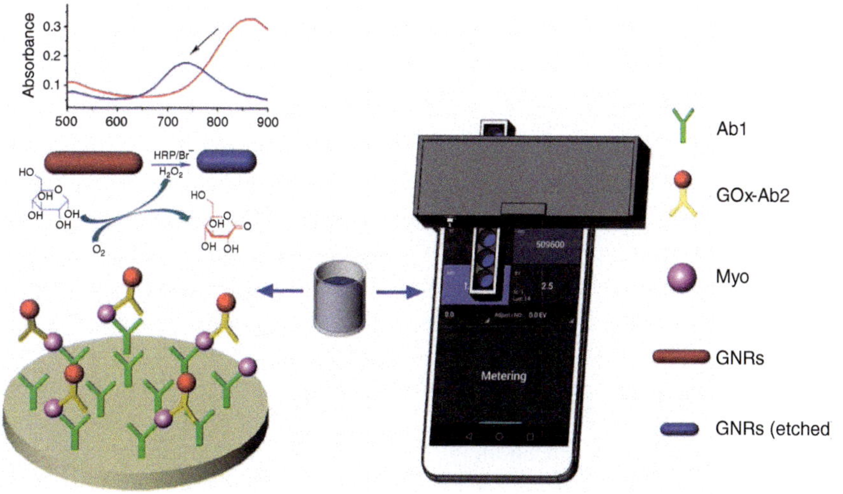

Figure 1.7 Schematic illustration of the AuNR-based plasmonic immunoassay for Myo detection. Source: Reprinted with permission from Yang et al. [88]. © Springer Nature.

platform was 0.1–1000 ng ml^{-1}, and LOD value for myoglobin detection was 0.057 ng ml^{-1}. This platform has presented high potential for biomedical applications due to higher sensitivity in comparison to conventional ELISA. Simply to put smartphone-based plasmonic immunoassay reader has enabled an easy access when the sources are limited [88].

1.8.2 Smart Material Integration

Recent studies have introduced micro/nano-size photonic/plasmonic structures, including photonic crystals (PC) or SPR-based metal nanostructures [89]. The PCs have ability to manipulate light in nontraditional ways because of photonic band structure concepts [90]. These structures are periodic arrangements of dielectric materials that interact with the light to provide reflection at specific wavelengths [91, 92]. The structure of PCs can be formed naturally or created rationally using different fabrication methods [91, 92]. In nature, well-known examples are present, such as *Morpho rhetenor* butterfly peacock, *Eupholus magnificus* insect, sea mouse, and opal. In an artificial way, PC structures can be fabricated in different dimensions (e.g., 1D, 2D, or 3D) with diverse materials including silicon, glass, polymers, colloids, and silk [91]. Among them, the unique and crucial properties (e.g., optical transparency, mechanical robustness, biocompatiblity, biodegradability, and facile functionalization) of silk fibroin provide important developments in photonics and biomedical device application [93]. One of the recent studies introduced robust, free-standing, 3D, PC fabrication in the form of inverse opal by using silk fibroin and created different lattice constants [94]. The methodology can be simply defined as a template that was constituted via placing differently sized (PMMA) submicrometric spheres on silicon substrate in the face-centered cubic (fcc) conformation and pouring silk fibroin extract. After solidifying of the silk solution, silicon substrate was detached; immersed in acetone to dissolve the PMMA spheres; and finally, amorphous free-standing inverse silk opal structure were obtained [94]. Moreover, as an example of hybrid PC-plasmonic platform, a recent study demonstrated an assay approach that was integrating plasmonic nanoparticle tags, imaging-based optical biosensor, and microfluidic device [95]. The integration of the microfluidic device reduced the assay time for the binding analytes to the biorecognition elements through diffusion. In this study, PC behaved as an active transducer for rapid imaging. Plasmonic AuNP created high-contrast digital resolution sensing of analytes on the LSPR spectra by overlapping with the resonance reflection of the PC [95]. AuNPs and immobilized PC were conjugated with antibodies that have an affinity to target analytes. When the AuNP-antibody conjugates capture HIV-1 capsid antigen (target analyte), they were applied through a microfluidic device, antibodies that are immobilized on PC captured AuNP-HIV-1 conjugates. [95]. The design of the sensing system utilized an absorbing paper pad to control single-pass flow rate. The assay was completed within 35 minutes, and LOD value was 1 pg ml^{-1}. This approach would be applied as a POC device for infectious disease diagnostics and early stage disease monitoring by the virtue of ultrasensitive and ultrafast biomolecule detection [95].

A recent study focused on to create an imaging-based plasmonic biosensor combining AuNPs and gold nanohole arrays (AuNHAs) in order to achieve highly sensitive detection of human C-reactive protein (CRP) for acute inflammatory diseases [96]. The detection principle integrates a heat-map generation based on the single NPs position. The capture antibodies, which were immobilized in AuNHAs, reacted with the target biomarkers and second antibody that was immobilized on the NPs to recognize the biomarker (Figure 1.8a), hence sandwich assay approach was achieved. After the binding, AuNPs created intensity dips (i.e., red spots) on a captured image due to strong local suppression in the transmission (Figure 1.8b). CRP could be detected as low as 27 $pg\,ml^{-1}$ and this enabled to reach to the clinically relevant concentrations at four orders of magnitude. Single NP-labeled proteins could also detected via the digital quantification and localization of individual AuNPs.

Moreover, combining SPR strategy with NPs provides notable improvements in sensing manners [97]. In particular, optical, chemical, electronic, and catalytic properties of colloidal plasmonic nanoparticles (PNPs) have created paramount interest in plasmonics, sensing, catalysis, biomedical imaging, diagnostics, and therapeutics [98]. The applications of AuNPs are highly overweighted while comparing them with the other PNPs in the biomedical applications owing to chemical/biological inertness and low cytotoxicity, versatile, and straightforward surface functionalization (e.g., oligonucleotides, proteins, or antibodies) [98]. On the other hand, signal enhancement is crucial approach for biosensors as demonstrated in recent studies. For instance, a plasmonic patch, which was fabricated through absorbing plasmonic nanoparticles (different shapes were also investigated) into PDMS elastomer, was introduced to enhance the signal (Figure 1.9) [99]. The idea behind this fluorescence enhancement technique was to add a plasmonic patch onto diverse fluorescent surfaces in order to enable large and uniform fluorescence enhancement. In contrast to conventional plasmon enhancement techniques, this system does not require any protocol modification [99]. The mechanism of the enhancement was occurred via close proximity of the elastomeric film and fluorescent species on the surface. The enhancement of fluorescence was reached 100 times higher with the ~300-fold increase in the sensitivity due to the enhanced electromagnetic field. The sensing platform was applied with the acute kidney injury markers (e.g., KIM-1 and NGAL) to indicate diagnostic capability from urine samples [99].

As another example, the periodic patterns of the optical discs (e.g., compact discs [CDs], digital versatile discs [DVDs], and Blu-ray discs), which are commercially available, enable inexpensive (0.90–$1.50) and large area of active sensing for ultrasensitive biosensing applications [100, 101]. Their production scale is high to compensate for the demand in the market besides their potential usage for plasmonic biosensors. The quality of the optical discs has been evolved with time in terms of molding processes to obtain nanometer precision for their encoded structure [101]. Optical discs have been combined with several kinds of plasmonic sensor approaches, including SPR [100], LSPR [102], and SERS [101]. One of the recent studies demonstrated the detection of HIV-1 particles on a plasmonic metasurface DVD, exciting plasmonic Fano resonance modes to increase sensitivity for both multimodal and multiplex sensing. In this study, two or more molecular oscillations and small perturbation interferences created large spectral shifts in the resonance frequency as a response [103]. With the use of inherent periodic arrays of DVDs, the requirements of metasurface fabrication were remarkably reduced in terms of cost, time, and personnel efforts, compared to the conventional techniques, which have complex, lengthy, and highly expensive procedures. Overall, plastic-templated metasurfaces hold great potential for various applications (e.g., biochemical sensing, optoelectronics, and optical spectroscopy devices), and they

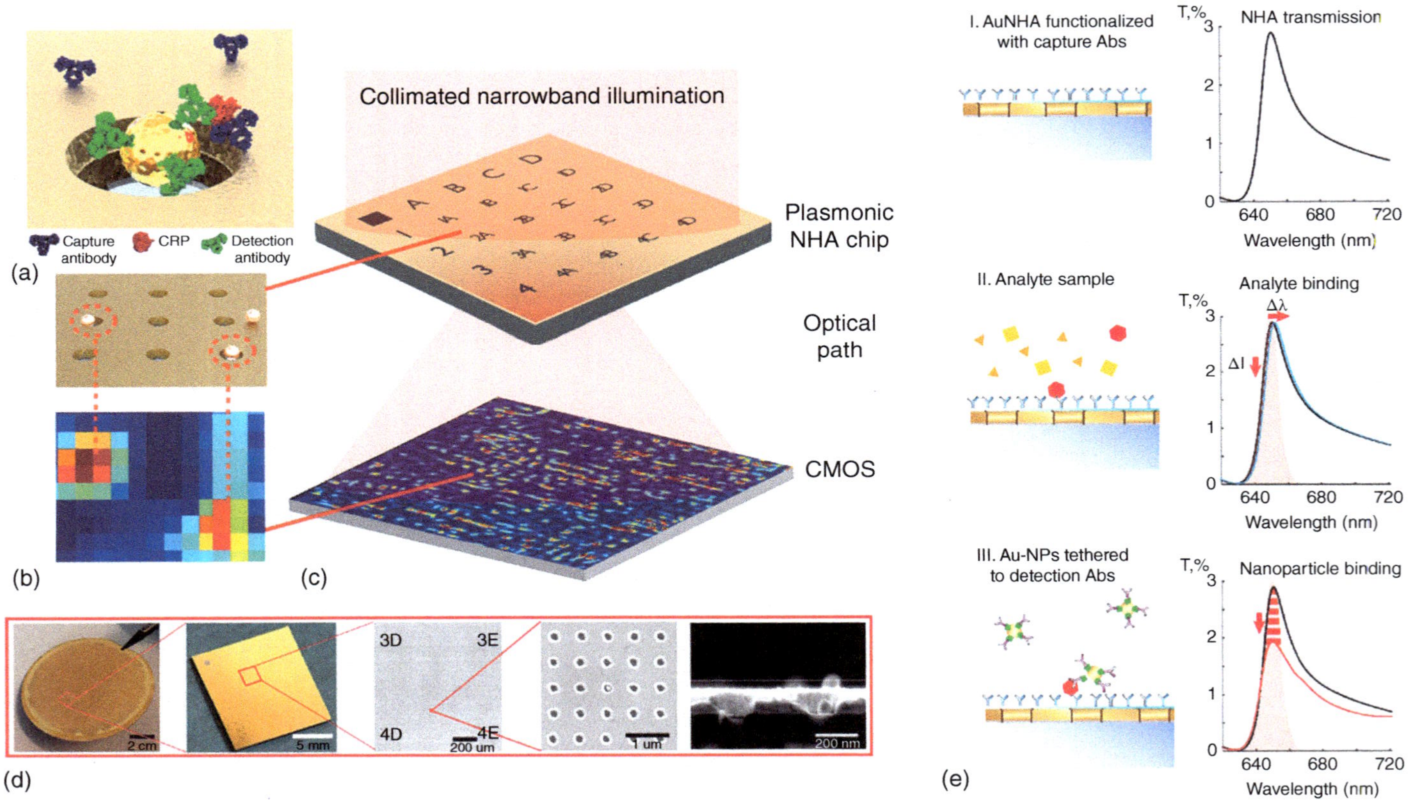

Figure 1.8 CRP detection via AuNP-enhanced plasmonic imager. (a) The schematic represents the capture of CRP (red) by an antibody (blue) on AuNHA and the detection of CRP by detection antibodies (green) tethered to AuNPs, respectively. (b) AuNPs create intensity dips (i.e. red spots) on captured image due to strong local suppression in the transmission. (c) Array labeling for accurate image processing and heatmap generation via CMOS camera for digital detection. (d) AuNHA wafer with 50 sensors. The plasmonic sensor surface is covered with AuNHAs. An image of microarray labels after patterning. The image of nanoholes (D = 200 nm, P = 600 nm) and the image of the nanohole containing AuNP are represented in the order from left to right. (e) (i) Transmission spectrum of the AuNHAs. (ii) peak position shift after the analyte binding, and (iii) local suppression after the binding of AuNPs to the AuNHA surface. Source: Reprinted with permission from Belushkin et al. [96]. © 2018, American Chemical Society.

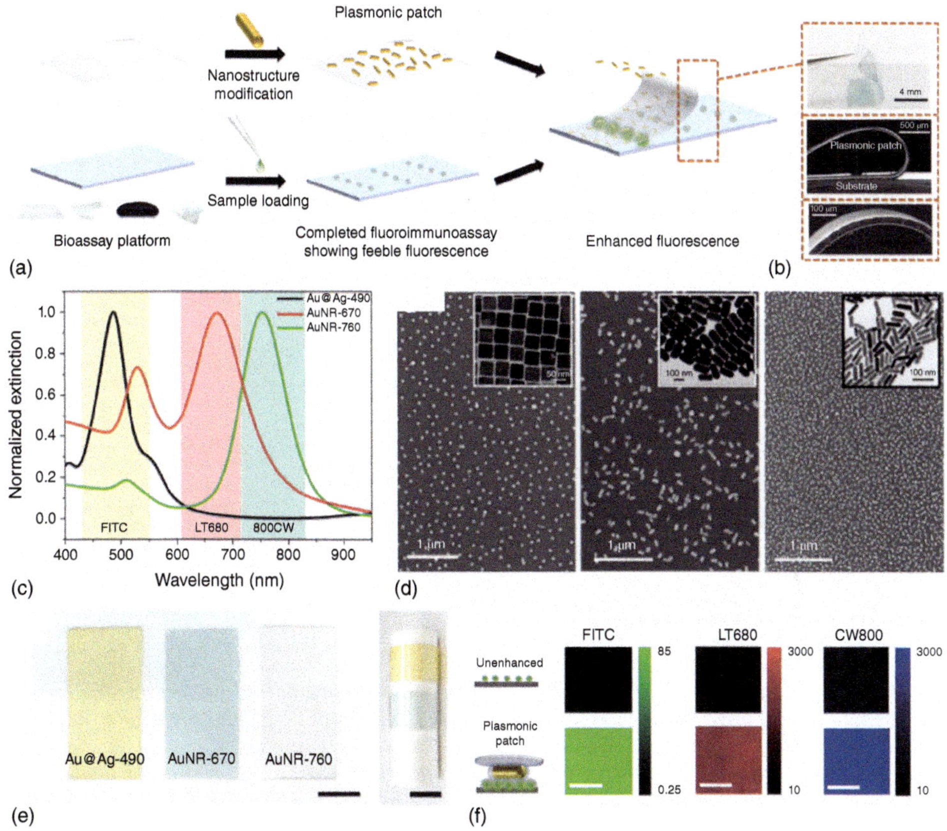

Figure 1.9 (a) The diagram of the plasmonic patch fabrication and application procedure for fluoroimmunoassays. (b) The image of the plasmonic patch transfer to assay surface, and an SEM image of the patch and its properties (e.g. thickness as 30 μm, flexibility, conformability). (c) Normalized extinction spectra of aqueous solutions for the differently shaped plasmonic nanostructures. (d) Characterization of the differently shaped AuNPs via an SEM imaging of plasmonic patch surface to demonstrate uniform distribution. (e) The patch image after modification with differently shaped nanostructures (scale bar represents 1 cm). (F) The fluorescence map of three fluorophores that were absorbed onto silicon wafer in the presence of plasmonic patch (the scale bar on green represents 10 μm; and the scale bar on red and blue represents 1 mm). Source: Reprinted with permission from Luan et al. [99].© Springer Nature.

present portability, cost-effectivity, and disposability properties [103]. In addition, plasmonic grating structures were utilized to estimate the thicknesses of biolayers formed on a plasmonic sensor by measuring entangled Fabry–Perot cavities (EFPC), leading to dual-mode sensing (SPP and LSPP) in order to eliminate background RI variations (Figure 1.10) [102]. This work was also applied to measure the size of exosomes, which is the useful information for neurodegenerative and cardiovascular disorders, and cancer. From the assay cost perspective, the fabrication was established by a large-area photolithography instead of e-beam lithography, and portable spectrometer was adapted to visible wavelengths, minimizing the need for complex and expensive instrumentation and setup.

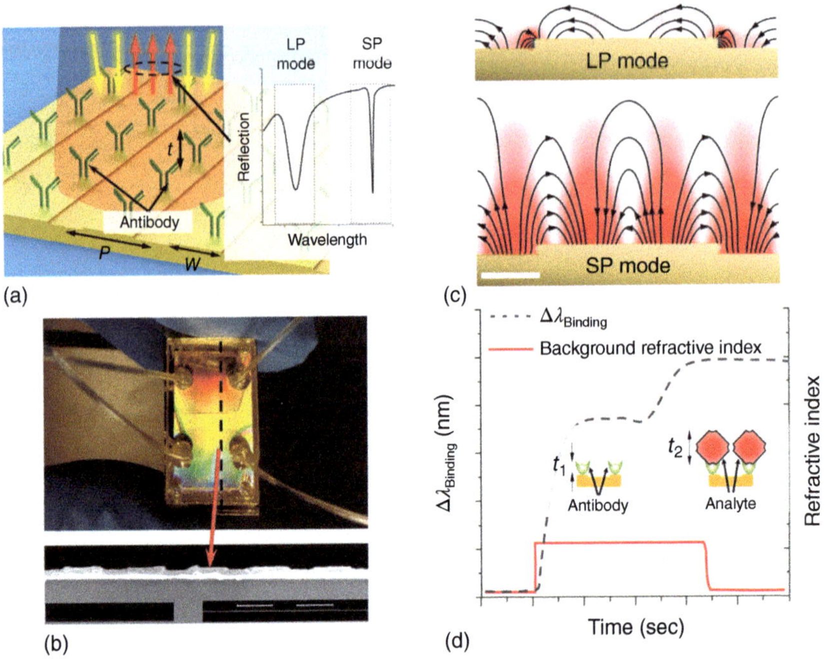

Figure 1.10 Schematic represents the working principle of plasmonic Fabry–Perot cavities. (a) The schematic illustrates the surface after the immobilization of antibodies to capture cellular entities and a typical reflection spectrum is depicted. (b) The image of the two-channel microfluidic chip, and an SEM image shows the cross-section of the sensor. (c) SPP and LSPP electric field distribution and field lines are demonstrated (scale bar 200 nm). (d) The typical data after multistep detection is presented and the data indicates the resonance shift data with respect to surface layer formation, bulk RI value, and the thickness of formed layers on the sensor surface. Source: Reprinted with permission from Mataji-Kojouri et al. [102]. © 2020, American Chemical Society.

1.8.3 Naked-Eye Detection

Technological advancements in the platforms-based on naked-eye detection are increasing promptly in the fields of health, environmental chemistry, food and beverage, bio-defense, and fermentation industry. Colorimetric sensors are important tools in diagnosing of various diseases, such as diabetes, cancer, Alzheimer, Parkinson, bacterial infections, depression, and infertility. These devices offer advantages as a real-time, high-precision, specific, and cost-effective alternative that are capable of analyzing results with naked-eye, simple integration with smartphones or color detection devices for the quantitative results [104]. A visually distinguishable color change is observed in the instant detection of the analyte on a colorimetric sensor. Metallic nanoparticles, such as Au, Ag, Cu, are widely employed in visual detection due to their optical properties [105]. Plasmonic colorimetric diagnostic platforms are the most suitable methods deployed into POC diagnosis. As a detection platform, their effortlessness in producing a signal output based on color change is advantageous [32]. The main theme in colorimetric sensors is how to transform the interactions of complex stimuli, such as molecule, temperature, pH value into visible color changes that can be seen by the naked-eye [106]. Smart-phones can be integrated with biosensors to develop POC devices that the end user can use at remote settings easily. Along with the technological developments, camera quality on smart phones have been updated, and nowadays, they are one of the perfect tools for optical detection through monitoring the primary color bands, namely red, green, and blue [107].

Paper-based lateral flow assays (LFA) have been largely utilized in this regard. Typically, biomolecules, such as labeled antibodies are used to capture and detect a biomolecule through colorimetric or fluorescence reading. At the end of the test period, a color band is analyzed for qualitative and quantitative reading [108]. An example of interest in LFAs was implemented on a paper or nitrocellulose membrane [109]. The filtered solution then migrated to the conjugate pad through capillary forces, where tracers were released, such as the gold nanoparticle conjugated with the receptor. As another example, aptamer-based LFA was designed for rapid detection of cortisol in sweat, an important parameter to monitor physiological stress [110]. For this purpose, cortisol-specific aptamers were conjugated to AuNPs. In the presence of cortisol molecules, they interacted with the aptamer–AuNP complex and they were desorbed from the surface of AuNPs. Free gold nanoparticles were then captured by a reaction with the immobilized cysteamine in the test region of the LFA strip. Thus, cortisol was visually perceived within minutes. Figure 1.11 details the colorimetric

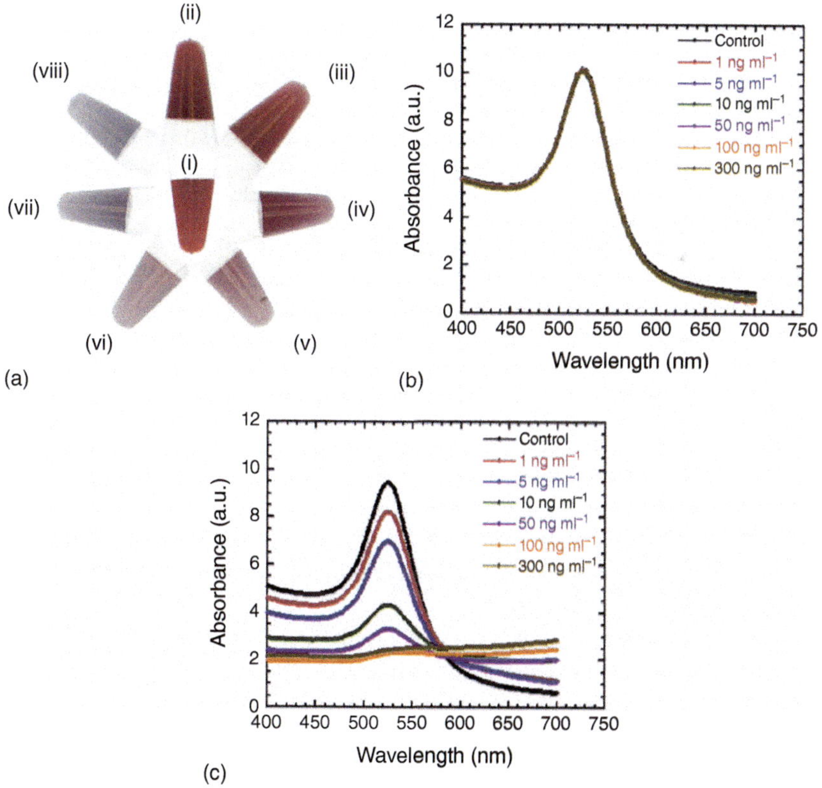

Figure 1.11 (a) Colorimetric detection of cortisol in artificial sweat (ii–viii:0–300 ng ml^{-1}). Optical spectra of mixture of DNA-AuNP and artificial sweat solutions with various cortisol concentrations: (b) without NaCl; (c) with NaCl. Source: Reprinted with permission from Dalirirad and Steckl [110]. © 2019, Elsevier.

detection of cortisol in artificial sweat. As another application, quantum dots (QDs) were integrated with an LFA to detect glutathione, which is a biothiol abundant in cells and plays a key role in many biological processes [111]. Herein, QDs- functionalized with bovine serum albumin were modified as signal reporter on the test line. When a binding occurs between Ag^+ and the QDs, the fluorescence level was effectively extinguished by electron transfer from QDs to Ag^+. In the presence of glutathione, Ag^+ preferred to react with glutathione by forming an Ag^+-S bond due to its stronger affinity. As a result, glutathione measurement was provided efficaciously using the fluorescence of the quantum dots. Furthermore, a fluorescent and colorimetric method was developed for multiplexed monitoring of cancer cells using graphene oxide-based aptameric nanosensors in microfluidic paper analytical devices [112]. Mesoporous silica nanoparticles-coated with quantum dots labeled as aptamers in a flexible single-stranded state was adsorbed on the graphene oxide surface. Here, three different cancer cells were identified simultaneously with a single excitation light, and the changes in colors were easily observed with the naked-eye.

1.9 Outcomes and Conclusion

In this chapter, we summarize the fundamentals of plasmonic modalities along with their applications in disease diagnostics. On the course of dramatic transition of biosensors from laboratory-based diagnostic assays to POC devices, there are important assets needed to be revisited before their implementation to the clinical settings, which include portability, cost, short assay time, less equipment need, quantitative results, acceptable LOD levels, reliability, and easy-to-use to the end-user, especially for an untrained personnel [95]. Especially, the rare number/level of biotargets is still an obstacle, guiding researchers to develop highly sensitive plasmonic tools [95, 113]. Considering the adaptation of plasmonic sensing to the POC settings, sensitivity could be improved via different strategies: (i) sensing based on target-induced local refractive index changes, (ii) colorimetric sensing based on LSPR coupling, and (iii) amplification of detection sensitivity based on nanoparticle growth [113]. As elaborated above, a variety of approaches, using large-area periodic nano-array patterns, such as nanosphere array, nano-disc array, or nano-triangle array instead of planar metal film would further improve refractive sensitivity since they support both a stronger local EM field and higher sensing area [5]. Furthermore, coupling plasmonic nanoparticles with Au films in the sandwich form would enhance the SPR shifts, compared to the measurements on a single Au film [5]. Inherent periodic arrays on plastic templates and low-cost fabrication techniques enable to fabricate sensitive plasmonic metasurfaces for not only monitoring the presence of biotargets, and also, measuring their sizes to help in biological investigations for research and clinical use.

In summary, plasmonic modality is a powerful technique capable of measuring nano-sized biological entities from biospecimens. With further developments described above and also with others such as integrating antifouling strategies, their sensitivity and reliability fashions would be improved remarkably. Although algorithms reliant on Artificial Intelligence and Machine Learning have not been elaborated here, plasmonic sensors would further benefit from these approaches, especially in terms of signal processing, accelerating assay time, and analysis for their proper utilization in the POC settings.

References

1 Howitt, P., Darzi, A., Yang, G.Z. et al. (2012). *The Lancet* 380 (9840): 507–535.
2 Firdous, S., Anwar, S., and Rafya, R. (2018). *Laser Physics Letters* 15 (6): 29–34.
3 Liu, J., Jalali, M., Mahshid, S., and Wachsmann-Hogiu, S. (2020). *The Analyst* 145 (2): 364–384.
4 Damborský, P., Švitel, J., and Katrlík, J. (2016). *Essays in Biochemistry* 60 (1): 91–100.
5 Li, M., Cushing, S.K., and Wu, N. (2015). *The Analyst* 140 (2): 386–406.
6 Inci, F., Celik, U., Turken, B. et al. (2015). *Biochemistry and Biophysics Reports* 2: 115–122.
7 Guo, X. (2012). *Journal of Biophotonics* 5 (7): 483–501.
8 Li, L. (2013). *Manipulation of Near Field Propagation and Far Field Radiation of Surface Plasmon Polariton*. Springer.
9 Nickelson, L. (2019). *Electromagnetic Theory and Plasmonics for Engineers*. Springer.
10 Barnes, W.L. (2006). *Journal of Optics A: Pure and Applied Optics* 8: 4.
11 Han, X., Liu, K., and Sun, C. (2019). *Materials* 12: 9.
12 Barnes, W.L., Dereux, A., and Ebbesen, T.W. (2003). *Nature* 424 (6950): 824–830.
13 Maradudin, A.A. (2014). *Modern Plasmonics*, vol. 4. Elsevier B.V.
14 Berini, P. and De Leon, I. (2012). *Nature Photonics* 6 (1): 16–24.
15 A. S.Maier, Plasmonics: Fundamentals and Applicaitons 2007.
16 Han, Z. and Bozhevolnyi, S.I. (2014). *Modern Plasmonics*, vol. 4. Elsevier B.V.
17 Tokel, O., Inci, F., and Demirci, U. (2014). *Chemical Reviews*.
18 Zhang, J., Zhang, L., and Xu, W. (2012). *Journal of Physics D: Applied Physics* 45: 11.
19 Bhattarai, J.K., Uddin Maruf, M.H., and Stine, K.J. (2020). *Processes* 8 (1): 1–20.
20 Inci, F., Saylan, Y., Kojouri, A.M. et al. (2020). *Applied Materials Today* 18: 100478.
21 Hoa, X.D., Kirk, A.G., and Tabrizian, M. (2007). *Biosensors and Bioelectronics* 23 (2): 151–160.
22 Qu, J.H., Dillen, A., Saeys, W. et al. (2020). *Analytica Chimica Acta* 1104: 10–27.
23 Mauriz, E., Dey, P., and Lechuga, L.M. (2019). *The Analyst* 144 (24): 7105–7129.
24 Zayats, A.V., Smolyaninov, I.I., and Maradudin, A.A. (2005). *Physics Reports* 408 (3–4): 131–314.
25 Petryayeva, E. and Krull, U.J. (2011). *Analytica Chimica Acta* 706 (1): 8–24.
26 Sepúlveda, B., Angelomé, P.C., Lechuga, L.M., and Liz-Marzán, L.M. (2009). *Nano Today* 4 (3): 244–251.
27 Liu, P., Wang, H., Li, X. et al. (2015). *RSC Advances* 5 (97): 79738–79745.
28 Zoric, I., Zach, M., Kasemo, B., and Langhammer, C. (2011). *ACS Nano* 5 (4): 2535–2546.
29 Wang, B., Singh, S.C., Lu, H., and Guo, C. (2020). *Plasmonics* 15 (3): 609–621.
30 Unser, S., Bruzas, I., He, J., and Sagle, L. (2015). *Sensors* 15 (7): 15684–15716.
31 Cao, J., Sun, T., and Grattan, K.T.V. (2014). *Sensors and Actuators B: Chemical* 195: 332–351.
32 Li, Z., Leustean, L., Inci, F. et al. (2019). *Biotechnology Advances* 37: 8, 107440.
33 Inci, F., Filippini, C., Baday, M. et al. (2015). *Proceedings of the National Academy of Sciences* 112 (32): E4354–E4363.
34 Inci, F., Karaaslan, M.G., Mataji-Kojouri, A. et al. (2020). *Applied Materials Today* 20: 100709.
35 Wang, Y., Zhou, J., and Li, J. (2017). *Small Methods* 1 (11): 1700197.

36 Sun, L.L., Leo, Y.S., Zhou, X. et al. (2020). *Materials Science for Energy Technologies* 3: 274–281.

37 Jatschka, J., Dathe, A., Csáki, A. et al. (2016). *Sensing and Bio-Sensing Research* 7: 62–70.

38 Acimovic, S.S., Ortega, M.A., Sanz, V. et al. (2014). *Nano Letters* 14 (5): 2636–2641.

39 Thakur, A., Qiu, G., Siu-Pang, N.G. et al. (2017). *Biosensors and Bioelectronics* 94: 400–407.

40 Qiu, G., Gai, Z., Tao, Y. et al. (2020). *ACS Nano* 14 (5): 5268–5277.

41 Rostron, P., Gaber, S., and Gaber, D. (2016). *Laser* 21: 24.

42 Wang, C., Zeng, L., Li, Z., and Li, D. (2017). *Journal of Raman Spectroscopy* 48 (8): 1040–1055.

43 Butler, H.J., Ashton, L., Bird, B. et al. (2016). *Nature Protocols* 11 (4): 664–687.

44 Haddad, A., Green, O., and Lombardi, J.R. (2019). *Spectroscopy Letters* 52 (8): 462–472.

45 Zhao, H., Hasi, W., Bao, L. et al. (2017). *Chinese Journal of Chemistry* 35 (10): 1522–1528.

46 de Oliveira Penido, C.A.F., Pacheco, M.T.T., Lednev, I.K., and Silveira, L. Jr. (2016). *Journal of Raman Spectroscopy* 47 (1): 28–38.

47 Liu, D., Wan, J., Liu, Z. et al. (2020). *Lasers in Medical Science*: 1–8.

48 Yang, S., Chen, Q., Shi, M. et al. (2020). *Nanomaterials* 10 (4): 770.

49 Bai, S., Serien, D., Hu, A., and Sugioka, K. (2018). *Advanced Functional Materials* 28: 23, 1706262.

50 Hu, G., Xiong, W., Luo, H. et al. (2018). *Applied Spectroscopy* 72 (1): 151–158.

51 Wiktelius, D., Ahlinder, L., Larsson, A. et al. (2018). *Talanta* 186: 622–627.

52 Tyndall, N.F., Stievater, T.H., Kozak, D.A. et al. (2018). *Optics Letters* 43 (19): 4803–4806.

53 Pearman, W.F. and Fountain, A.W. (2006). *Applied Spectroscopy* 60 (4): 356–365.

54 Yan, F. and Vo-Dinh, T. (2007). *Sensors and Actuators B: Chemical* 121 (1): 61–66.

55 Sinjab, F., Kong, K., Gibson, G. et al. (2016). *Biomedical Optics Express* 7 (8): 2993–3006.

56 Lin, K., Zheng, W., Lim, C.M., and Huang, Z. (2017). *Theranostics* 7 (14): 3517.

57 Haifler, M., Pence, I., Sun, Y. et al. (2018). *Journal of Biophotonics* 11: 6, e201700188.

58 Chen, H., Li, X., Broderick, N. et al. (2018). *Journal of Biophotonics* 11: 9, e201800016.

59 Farhane, Z., Bonnier, F., and Byrne, H.J. (2018). *Journal of Biophotonics* 11: 1, e201700112.

60 Zhang, H., Xiao, L., Li, Q. et al. (2018). *Biomicrofluidics* 12: 2, 24119.

61 da Silva, E.C.O., dos Santos, F.M., Ribeiro, A.R.B. et al. (2019). *The Analyst* 144 (5): 1622–1631.

62 Byrne, H.J., Bonnier, F., and Farhane, Z. (2019). *Journal of Biophotonics* 12: 3, e201800328.

63 Moore, T.J., Moody, A.S., Payne, T.D. et al. (2018). *Biosensors* 8 (2): 46.

64 Lee, M., Oh, K., Choi, H.-K. et al. (2018). *ACS Sensors* 3 (1): 151–159.

65 Song, C.Y., Yang, Y.J., Yang, B.Y. et al. (2016). *Nanoscale* 8 (39): 17365–17373.

66 Zheng, J. and He, L. (2014). *Comprehensive Reviews in Food Science and Food Safety* 13 (3): 317–328.

67 Li, C., Liu, Y., Zhou, X., and Wang, Y. (2020). *Journal of Materials Chemistry B* 8 (16): 3582–3589.

68 Sun, Y. and Li, T. (2018). *Analytical Chemistry* 90 (19): 11614–11621.

69 Zhang, X., Liu, S., Song, X. et al. (2019). *ACS Sensors* 4 (8): 2140–2149.

70 Kim, E., Baaske, M.D., and Vollmer, F. (2017). *Lab on a Chip* 17 (7): 1190–1205.

71 Nunzi Conti, G., Berneschi, S., and Soria, S. (2016). *Biosensors* 6 (3): 28.

72 Brice, I., Grundsteins, K., Atvars, A. et al. (2020). *Sensors and Actuators B: Chemical*: 128004.

73 Huckabay, H.A., Wildgen, S.M., and Dunn, R.C. (2013). *Biosensors and Bioelectronics* 45: 223–229.

74 Tosi, D., Poeggel, S., Iordachita, I., and Schena, E. (2018). Fiber optic sensors for biomedical applications. In: *Opto-Mechanical Fiber Optic Sensors* (ed. H. Alemohammad). Butterworth-Heinemann.

75 Tabassum, S. and Kumar, R. (2020). *Advanced Materials Technologies* 5 (5): 1900792.

76 Prakashan, V.P., George, G., Sanu, M.S. et al. (2020). *Applied Surface Science* 507: 144957.

77 Lu, Y., Li, H., Qian, X. et al. (2020). *Optical Fiber Technology* 56: 102187.

78 Shah, K., Sharma, N.K., and Sajal, V. (2018). *Optical and Quantum Electronics* 50 (6): 265.

79 Esposito, F., Sansone, L., Taddei, C. et al. (2018). *Sensors and Actuators B: Chemical* 274: 517–526.

80 Hromadka, J., Tokay, B., Correia, R. et al. (2018). *Sensors and Actuators B: Chemical* 255: 2483–2494.

81 Partridge, M., James, S.W., and Tatam, R.P. (2016). *Journal of Lightwave Technology* 34 (19): 4506–4510.

82 Kaushik, S., Pandey, A., Tiwari, U.K., and Sinha, R.K. (2018). *Optical Fiber Technology* 46: 95–103.

83 Kant, R. and Gupta, B.D. (2018). *Journal of Lightwave Technology* 36 (18): 4018–4024.

84 Zhou, L., Liu, C., Sun, Z. et al. (2019). *Biosensors and Bioelectronics* 137: 140–147.

85 Fajkus, M., Nedoma, J., Martinek, R. et al. (2017). *Sensors* 17 (1): 111.

86 Guner, H., Ozgur, E., Kokturk, G. et al. (2017). *Sensors and Actuators, B: Chemical* 239: 571–577.

87 Liu, Y., Liu, Q., Chen, S. et al. (2015). *Scientific Reports* 5: 1–9.

88 Yang, Q., Cai, R., Xiao, W. et al. (2018). *Nanoscale Research Letters* 13: 262.

89 Kim, S., Mitropoulos, A.N., Spitzberg, J.D. et al. (2013). *Optics Express* 21 (7): 8897.

90 Lu, T., Peng, W., Zhu, S., and Zhang, D. (2016). *Nanotechnology* 27: 050201.

91 Inan, H., Poyraz, M., Inci, F. et al. (2017). *Chemical Society Reviews* 46 (2): 366–388.

92 Sukhoivanov, I.A. and Gurvey, I.V. (2009). *Photonic Crystals: Physics and Practical Modeling*. Springer.

93 Lin, D., Tao, H., Trevino, J. et al. (2012). *Advanced Materials* 24 (45): 6088–6093.

94 Kim, S., Mitropoulos, A.N., Spitzberg, J.D. et al. (2012). *Nature Photonics* 6 (12): 818–823.

95 Che, C., Li, N., Long, K.D. et al. (2019). *Lab on a Chip* 19 (23): 3943–3953.

96 Belushkin, A., Yesilkoy, F., and Altug, H. (2018). *ACS Nano* 12 (5): 4453–4461.

97 Canovi, M., Lucchetti, J., Stravalaci, M. et al. (2012). *Sensors (Switzerland)* 12 (12): 16420–16432.

98 Park, J.E., Kim, M., Hwang, J.H., and Nam, J.M. (2017). *Small Methods* 1: 3.

99 Luan, J., Morrissey, J.J., Wang, Z. et al. (2018). *Light: Science and Applications* 7: 1.

100 Choi, B., Dou, X., Fang, Y. et al. (2016). *Physical Chemistry Chemical Physics* 18 (37): 26078–26087.

101 Radu, A.I., Ussembayev, Y.Y., Jahn, M. et al. (2016). *RSC Advances* 6 (50): 44163–44169.

102 Mataji-Kojouri, A., Ozen, M.O., Shahabadi, M. et al. (2020). *ACS Nano* 14: 8518–8527.

103 Ahmed, R., Ozen, M.O., Karaaslan, M.G. et al. (2020). *Advanced Materials* 32 (19): 1–11.

104 Shende, P., Prabhakar, B., and Patil, A. (2019). *TrAC, Trends in Analytical Chemistry* 121: 115687.

105 A. Piriya, V.S., Joseph, P., K. Daniel, S.C.G. et al. (2017). *Materials Science and Engineering: C* 78: 1231–1245.
106 Liu, D., Fang, L., Zhou, F. et al. (2018). *Advanced Functional Materials* 28: 18, 1707392.
107 Purohit, B., Kumar, A., Mahato, K., and Chandra, P. (2020). *Current Opinion in Biomedical Engineering* 13: 42–50.
108 Carrell, C., Kava, A., Nguyen, M. et al. (2019). *Microelectronic Engineering* 206: 45–54.
109 Nguyen, V.-T., Song, S., Park, S., and Joo, C. (2020). *Biosensors and Bioelectronics* 152: 112015.
110 Dalirirad, S. and Steckl, A.J. (2019). *Sensors and Actuators B: Chemical* 283: 79–86.
111 Liu, J., Ji, D., Meng, H. et al. (2018). *Sensors and Actuators B: Chemical* 262: 486–492.
112 Liang, L., Su, M., Li, L. et al. (2016). *Sensors and Actuators B: Chemical* 229: 347–354.
113 Guo, L., Jackman, J.A., Yang, H.H. et al. (2015). *Nano Today* 10 (2): 213–239.

2

Nanosensors Based on Localized Surface Plasmon Resonance

Deniz Umut Yildirim[1,2], *Amir Ghobadi*[1,2], *and Ekmel Ozbay*[1,2,3,4]

[1]*NANOTAM-Nanotechnology Research Center, Bilkent University, Ankara, Turkey*
[2]*Department of Electrical and Electronics Engineering, Bilkent University, Ankara, Turkey*
[3]*Department of Physics, Bilkent University, Ankara, Turkey*
[4]*UNAM-Institute of Materials Science and Nanotechnology, Bilkent University, Ankara, Turkey*

2.1 Historical and Theoretical Background

Noble metal nanoparticles, such as gold, silver, copper, and aluminum nanoparticles, have been in use since the fourth century CE with the invention of the Lycurgus Cup. The vibrant colors of these cups were obtained by mixing small amounts of gold and silver into the glass as decorative pigments, thereby creating nanoparticles embedded in the glass.

As our understanding about nanoparticles and their interaction with light became deeper owing to the spectroscopic techniques, it was understood that these nanoparticles are able to resonantly couple the incoming photons to the collective oscillations of free electrons on the surface of the nanoparticles, and this phenomenon is called localized surface plasmon resonance (LSPR). LSPRs of noble metals, such as gold and silver, are of a very localized nature, and therefore, they have a high quality factor. Nanosensors based on LSPRs operate on the principle that the spectral position of the resonances of metallic nanoparticles changes upon changes in the dielectric environment within the electromagnetic near field [1].

The optical properties of nanoparticles are already well explained in several textbooks [2, 3]. Here, we will consider the optical polarizability of metallic nanospheres, the radius of which are r and which are covered with a dielectric film of thickness d, as described in Figure 2.1a. If the size of the nanosphere is many times smaller, compared to the wavelength of the incident radiation, the polarizability, $\alpha(\lambda)$, of this nanosphere can be expressed as follows [4]:

$$\alpha(\lambda) = 4\pi\epsilon_3(\lambda)(r+d)^3 \frac{\epsilon_2(\lambda)\epsilon_a(\lambda) - \epsilon_3(\lambda)\epsilon_b(\lambda)}{\epsilon_2(\lambda)\epsilon_a(\lambda) + 2\epsilon_3(\lambda)\epsilon_b(\lambda)} \tag{2.1}$$

Plasmonic Sensors and their Applications, First Edition. Edited by Adil Denizli.

where λ is the wavelength, ε_1, ε_2, and ε_3 are the dielectric permittivity functions of the metallic nanosphere, dielectric film, and the ambient medium, respectively. The parameters ε_a and ε_b are described as follows:

$$\epsilon_a = \epsilon_1 (3 - 2P) + 2\epsilon_2 P \tag{2.2}$$

$$\epsilon_b = \epsilon_1 P + \epsilon_2 (3 - P) \tag{2.3}$$

$$P = 1 - \left(\frac{r}{r+d} \right)^3 \tag{2.4}$$

The factor P is the ratio of the dielectric shell volume to the total particle volume. The polarizability, $\alpha(\lambda)$, reaches a maximum when the denominator is minimum, which is the basis of an excited LSPR. The scattering and absorption spectra of the nanospheres can then be determined from the polarizability, $\alpha(\lambda)$.

Figure 2.1b shows the scattering efficiency of gold nanospheres of 40 nm in diameter covered with a constant permittivity dielectric thin film (ε_2 = 2.25), in which the

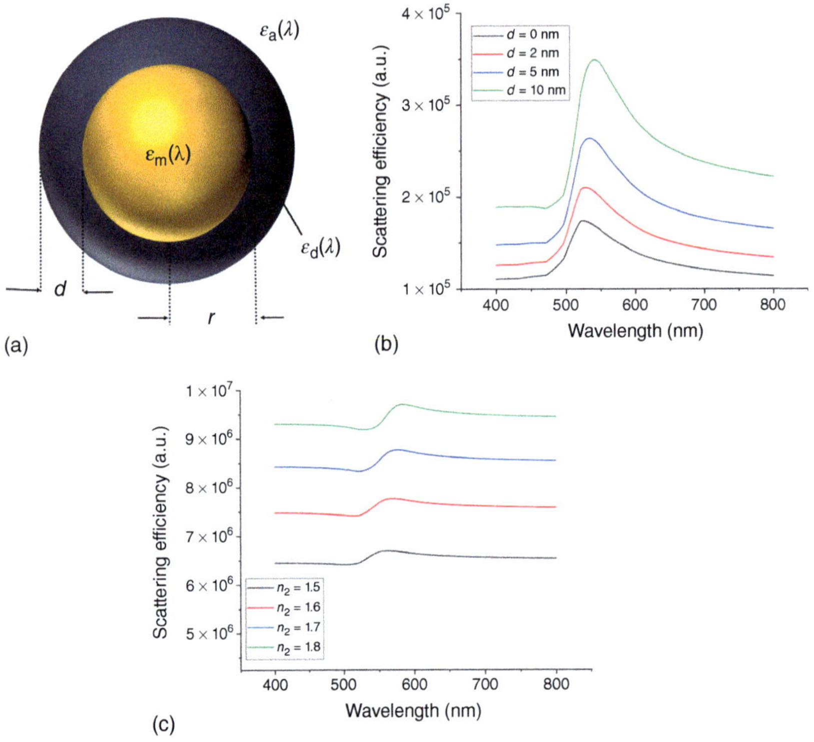

Figure 2.1 Optical scattering efficiency of a metallic nanosphere covered with a thin dielectric film. (a): Geometry of a gold nanosphere, covered with a thin metallic film, (b): Calculated scattering efficiency of a gold nanosphere with varying thickness of dielectric films with n_2 = 1.5, (c): Calculated scattering efficiency of a gold nanosphere with varying n_2 and a constant thickness of dielectric films, d = 100 nm.

thicknesses of the dielectric film are 2, 5, and 10 nm. Figure 2.1c shows the scattering efficiency of the same gold nanospheres that are covered with a constant thickness dielectric film ($d = 100$ nm) with its refractive index, $n_2 = \sqrt{\epsilon_2}$ swept from 1.5 to 1.8, with 0.1 step increments.

From Figure 2.1b and c; it is clear that adjusting P and ε_2 tunes the spectral position of the resonance, which makes measuring the thickness and the optical properties of the thin film, i.e. refractive index sensing, possible. Scrutinizing Eq. (2.1) shows that LSPRs are achievable in any metal, alloy, or semiconductor with a large negative dielectric permittivity, ε_1', and a small imaginary dielectric constant, ε_1'' Because the imaginary part limits the magnitude of the field enhancement at the LSPR conditions, due to ohmic losses and damping [4], noble metals such as gold and silver have been the most commonly used materials in the LSPR-based sensing. Silver is capable of producing sharper and more intense LSPRs than gold. For example, while gold nanospheres of 50–60 nm diameter were observed to have a refractive index sensitivity of 60 nm per refractive index unit (RIU), this metric for silver nanospheres of the same size was 160 nm per RIU [5, 6]. Nevertheless, silver is prone to oxidation [7, 8], so the chemical stability of gold enabled it to dominate over the former in the biosensing applications [9]. Nevertheless, for ultraviolet (UV) applications, both silver and gold are strongly absorptive, which makes aluminum the most preferred choice of material in that spectral range [10] because of the fact that it has a strong interband transition centered narrowly around 1.5 eV. However, above this energy, it acts very much Drude-like [11]. Copper, due to being an abundant metal in the Earth's crust, is also worth considering. The problem associated with it, however, is that it can easily form a copper oxide layers on its surface, which drastically affects its plasmonic properties by broadening and damping the LSPR. Removal of this oxide layer is shown to restore the narrow resonance linewidth [12].

The nanoparticles can indeed absorb and scatter light, i.e. cause extinction, so intensely that single nanoparticles can easily be observed using dark-field (optical scattering) microscopy [5, 13]. In one of the pioneering experiments on colorimetric sensing, Mock and colleagues observed a change in the color of silver nanoparticles from blue to green, as shown in Figure 2.2, upon the addition of oil, which indicates a red shift in the scattering spectra of the nanoparticles, in accordance with Figure 2.1c. [5]. Simple colorimetric detection has

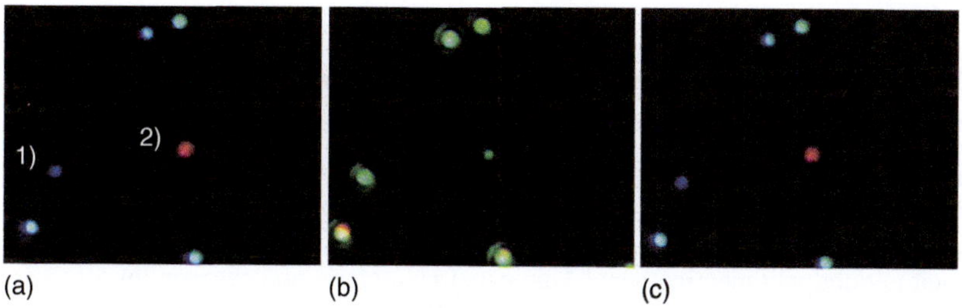

Figure 2.2 Typical field of silver nanoparticles immobilized on SiO_2 wafer and imaged under dark-field illumination with 100× objective. Color images taken with Nikon Coolpix 995. (a): Before oil, (b): With 1.44 index oil spread over particles, (c): After the removal of 1.44 index oil with the described protocol. Source: Reprinted with permission from [5]. Copyright (2003) American Chemical Society.

indeed been one of the most basic, yet effective methods of identification [14–19], due to offering a simple readout by the naked eye.

Van Duyne and colleagues observed the measurable shift of LSPR λ_{max}. They showed that silver nanoparticles can be used to detect the formation of monolayers on the surface of nanoparticles, when these nanoparticles are in different solvents, such as nitrogen, methanol, 1-propanol, chloroform, and benzene [20]. This was done by scrutinizing the scattering spectrum of the nanoparticles in different solvents, as shown in Figure 2.3a.

Concentrating on the peak wavelengths, λ_{max} of the scattering spectra of Figure 2.3a, the red shift of them can be easily observed and when plotted, this increase in the peak wavelength was shown to be linear, as shown in Figure 2.3b. This behavior is consistent with the approximate description of LSPR spectral shift in the literature, as given by Eq. (2.5) [21],

$$\Delta\lambda_{max} = m\Delta n\left[1-\exp\left(\frac{-2d}{l_d}\right)\right] \tag{2.5}$$

where m is the bulk refractive index sensitivity factor (in nm per RIU), $\Delta n = n_{adsorbate} - n_{medium}$ is the refractive index change in the surrounding medium, induced by the adsorbate, d is the effective thickness of the adsorbate layer (in nm), and l_d is the electromagnetic field decay length (in nm), which is approximated by an exponential decay. This thickness-dependent LSPR sensing was further confirmed in the experiment of Whitney and colleagues [22]. LSPR λ_{max} shift of silver nanoparticles was scrutinized for the deposition of Al_2O_3, the thickness of which varied from 0.1 to 66 nm. At first, LSPR band shift followed a steep linear slope, but leveled off as the thickness of the Al_2O_3 film increased, confirming the qualitatively expected behavior of Eq. (2.5). In the literature, it is quite common for d to be much larger than l_d, for example, in those cases when sensors are exposed to water,

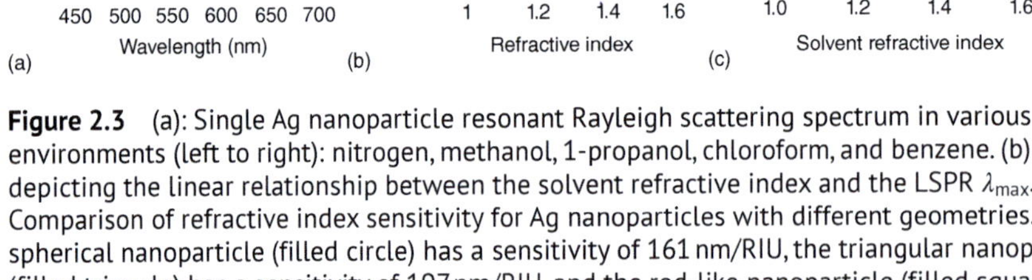

Figure 2.3 (a): Single Ag nanoparticle resonant Rayleigh scattering spectrum in various solvent environments (left to right): nitrogen, methanol, 1-propanol, chloroform, and benzene. (b): Plot depicting the linear relationship between the solvent refractive index and the LSPR λ_{max}. (c): Comparison of refractive index sensitivity for Ag nanoparticles with different geometries. The spherical nanoparticle (filled circle) has a sensitivity of 161 nm/RIU, the triangular nanoparticle (filled triangle) has a sensitivity of 197 nm/RIU, and the rod-like nanoparticle (filled square) has a sensitivity of 235 nm/RIU. Source: Adapted with permission from [20]. Copyright (2003) American Chemical Society.

isopropanol, or PMMA [23]. In such cases, it is common to calculate m from the formula, $m = \Delta\lambda_{max}/\Delta n$ [24–27].

Importantly, LSPR spectral shape and the peak wavelength, λ_{max}, were found in the literature to be strongly dependent on the size, shape, orientation of the nanoparticles [28], which is observed by comparing the different sensitivities of spherical, triangular, and rod-like nanoparticles, as shown in Figure 2.3c.

The LSPRs can then be tuned during the fabrication process, by controlling the above-mentioned parameters by a variety of chemical syntheses [29] and lithographic methods [30, 31]. These methods will be discussed in detail in the fabrication of the metal nanostructures section. By this way, tunability throughout the visible range was demonstrated [32], as shown in Figure 2.4. Silver nanoparticles with increasing size and aspect ratios were shown to cover the entire spectral range from 380 to 6 μm [33].

As opposed to fluorophores, which are commonly used for fluorescent labeling and are not straightforward or desirable to use in many situations [34], plasmonic nanoparticles also do not bleach or blink, which is a significant advantage over the former. By this way, molecular binding events can be observed for arbitrarily long periods of time [10]. The operating principle of this sensing scheme is demonstrated in Figure 2.5 [9]. When a sample solution involves analyte molecules, they react with the receptor molecules that are immobilized on the surface, resulting in an increase in the thickness of the surface dielectric layer, d. Since any molecule that causes a refractive index change in the environment can trigger shifts in the LSPR λ_{max}, an important aspect of biosensor design then becomes ensuring selectivity to the analyte of interest, which is generally achieved by functionalizing the metallic surface so that only the target analyte is bound to the surface [1, 19]. This

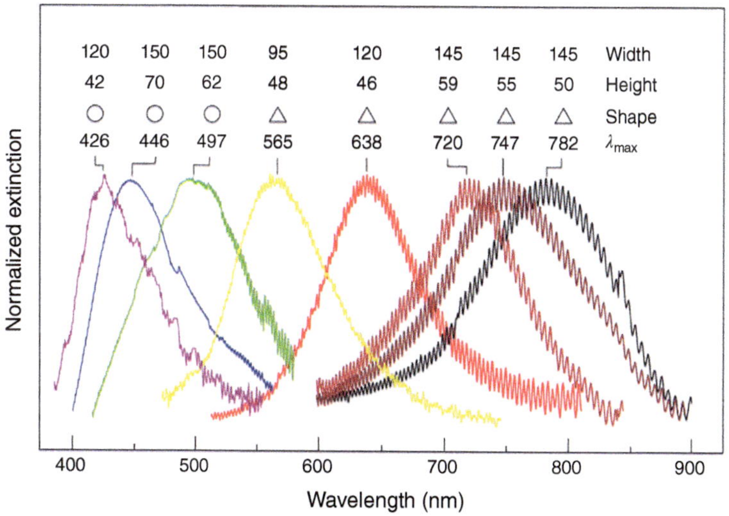

Figure 2.4 Effect of size and shape on LSPR extinction spectrum for silver nanoprisms and nanodiscs formed by nanosphere lithography (NSL). The high-frequency signal on the spectra is an interference pattern from the reflection at the front and back surfaces of the mica. Source: Reprinted with permission from [32]. Copyright (2001) Materials Research Society.

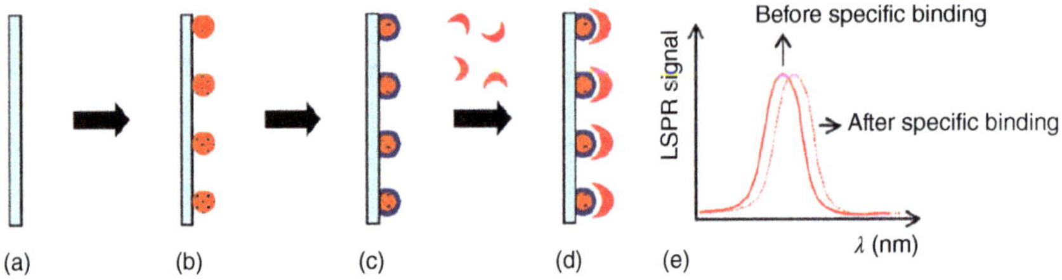

Figure 2.5 Schematic representation of the preparation and response of LSPR biosensors based on refractive index changes. (a) A substrate is chosen, (b) metal nanoparticles are attached to it by means of chemical linkers or prepared by nanolithography, (c) the metal particles are modified with the sensor moiety, (d) the analyte attaches from solution specifically onto the recognition function adsorbed onto the particles, causing a change in the refractive index around the particle resulting in an LSPR shift (e). Source: Reprinted with permission from [9]. Copyright (2009) Elsevier Ltd.

topic is to be discussed in detail in the upcoming parts of this chapter. As an example of the observation of the binding events, Yonzon and colleagues investigated the real-time binding event of Concanavalin A to mannose-functionalized nanoparticles [35].

Compared to sensors that are on propagating surface plasmons, i.e. surface plasmon polaritons (SPPs), LSPR sensors demonstrated greater spatial resolution in both lateral and normal directions, due to the smaller characteristic decay length of the electromagnetic field of the LSPR sensors (≈5–6 nm) [21] compared to that of SPP sensors (≈200 nm) [36] and single nanoparticles that can achieve the ultimate lateral spatial resolution [10, 35]. One other key difference that LSPR-based sensors offer over SPP-based sensors is that the latter requires a grating or prism elements to couple incident radiation to the propagating modes, i.e. match their momentum. This makes the former type of sensors more compact, simpler to use, and easier to miniaturize [37]. Schultz and colleagues demonstrated that a single, 80 nm diameter silver nanosphere scatters 445-nm blue light with a scattering cross section of 3×10^{-2}, which is a millionfold improvement compared to the fluorescence cross section of a fluorescein molecule and a thousandfold greater than the cross section of a similarly sized nanosphere filled with fluorescein to the self-quenching limit [13]. LSPR-based biosensors can also be used for the characterization of proteins. To give an example, Haes and colleagues developed an LSPR-based spectroscopy device, which consists of NSL-synthesized Ag triangles, to monitor the interaction between the antigen, amyloid-β-derived diffusible ligands (ADDLs), which are thought to be an important Alzheimer's disease pathogen and specific anti-ADDL antibodies. They were able to measure the concentration of ADDLs as small as 100 fM [38].

These properties all mean that, low-cost, label-free, highly sensitive point-of-care devices that can perform real-time identification of even molecular monolayers, via spectral changes, are realizable and have received enormous research efforts over the past two decades [39–47]. The remainder of this chapter discusses the following: fabrication of these nanostructures, improving the sensitivity and the limit of detection, integrating LSPR-based assays with other molecular identification methods, such as metal-enhanced fluorescence, surface-enhanced Raman spectroscopy (SERS), and mass spectroscopy; and finally, the practical issues and future prospects for LSPR-based sensors.

2.2 Fabrication of Metal Nanostructures

Metallic nanostructures can be fabricated by using top-down or bottom-up fabrication methods. While the former includes using lithographic techniques that were discussed briefly previously and are to be discussed in detail in this section; the latter includes chemical reactions that give rise to the nanomaterials, the composition, size, and shape of which are determined by the process conditions.

Regarding the bottom-up techniques, there are four commonly employed methods to synthesize metallic nanoparticles: reduction of metal salt precursors, electrochemical synthesis, reduction of organic ligands in organometallic precursors, and metal vapor chemistry [48]. Among these methods of producing metallic nanoparticles, the chemical reduction reaction of a metal salt in the presence of a stabilizer is considered as one of the simplest and has been used to produce Ag (Silver), Au (Gold), Pd (Palladium), Cu (Copper), Pt (Platinum), Os (Osmium), Rh (Rhodium), and Ir (Iridium) nanoparticles [49–52]. Seeded growth, during which metal atoms are clustered to establish small metallic nanoparticles nuclei, called "seed" particles, has proven to be much more versatile in terms of size and shape control, thus enabling control over the LSPR spectra of the nanoparticles. Figure 2.6 shows an example of various nanoparticle morphologies and the LSPR spectra corresponding to them.

The control over the nanoparticle properties is because stabilizing agents are commonly introduced in the reaction to control the particle growth and aggregation. They can then induce anisotropic growth in certain directions, as well as change the growth habit of different facets in the metal surface. Different stabilizing agents such as surfactants, polymers, or organic ligands can influence the size, shape, and morphologies of the final nanoparticles [48]. In an aqueous solution, the biofunctionalization of these nanoparticles poses a challenge, so they are usually assembled on solid surfaces for LSPR-biosensing applications [9]. These synthesized nanoparticles in solution can then be transferred to

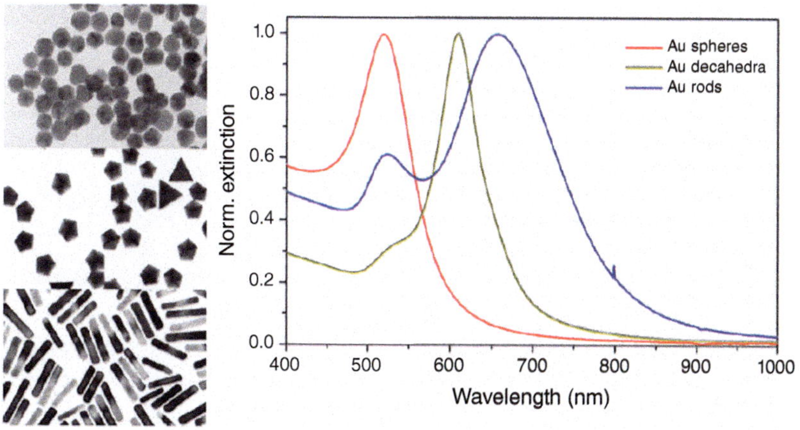

Figure 2.6 Transmission electron micrographs and UV–Visible extinction spectra of gold nanoparticle colloids with various geometries, as indicated. Source: Reprinted with permission from [9]. Copyright (2009) Elsevier Ltd.

oxide-bearing substrates, such as silica or quartz, by using organic adhesion layers of thiol-terminated silanes [19, 53].

Surface-linked metal nanostructures are mainly fabricated by using top-down lithographic approaches, which are ubiquitous in realizing nanostructures to be used in LSPR sensing. Lithography techniques such as electron-beam lithography (EBL) [30, 54, 55] and focused ion beam (FIB) [56] allow the fabrication of metallic nanostructures with the desired size and shape, but their upscaling, reproducibility, large-area compatibility, and mass production are inherently limited [23, 57, 58]. For example, Zhu and coworkers fabricated a silver rhombic nanoparticle array, as in Figure 2.7a, via the EBL method. They

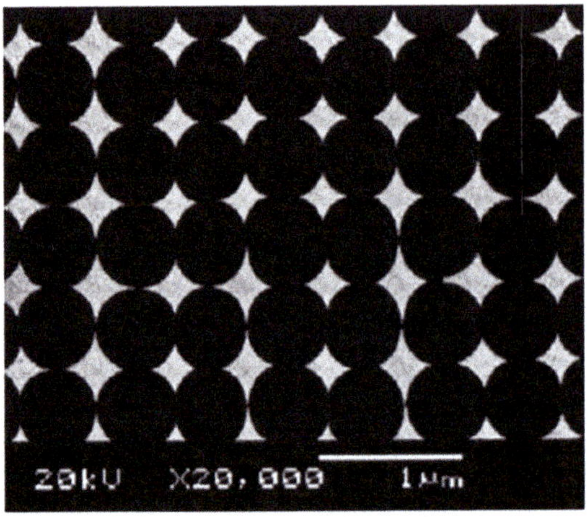

(a)

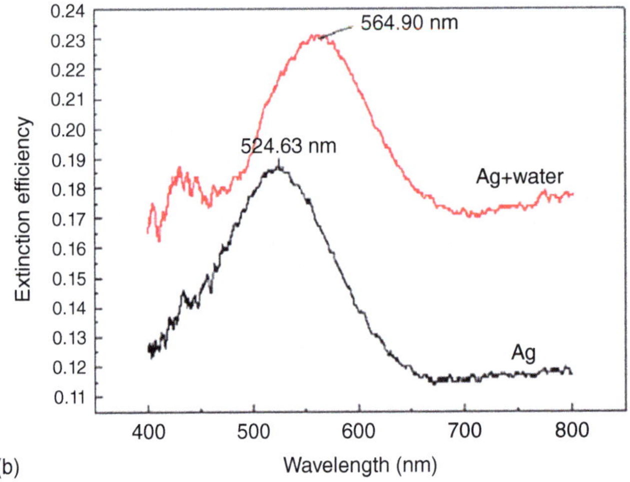

(b)

Figure 2.7 (a) Representative scanning electron microscope image of substrate. (b) Extinction spectra of the triangular metallic nanoparticles-based biochip with air and water on its surface. Source: Reprinted with permission from [59]. Copyright (2008) Elsevier B.V.

obtained an experimental refractive index sensitivity of 266.2 nm/RIU by making the necessary multiplication to the sensitivity value of 121 nm/RIU for triangular nanoparticles, which is demonstrated in Figure 2.7b [59].

In an effort to improve the large-scale compatibility of lithographic methods, other methods are also pursued and researched in the literature. The most commonly used method under this category is the one called nanosphere lithography [20, 28, 60]. NSL allows controlling the height of the produced nanoprisms by the vapor-deposition time and rate, while the lateral width and the interspacing of the particles are controlled by changing the diameter of the nanospheres in the mask [10]. Other examples of large-scale compatible methods include diblock copolymer micelle nanolithography [61], nanoimprint lithography [62] or soft interference lithography [63, 64], deep ultraviolet lithography [65].

There also exist some lithography-free fabrication routes for the realization of metallic nanostructures. In the method known as oblique-angle deposition [23, 58, 66], line-of-sight coating of physical vapor deposition systems and shadowing are combined to realize disordered and densely packed nanorods. The characteristic of the structures produced by this method is the extremely small gaps that result in large field-enhancement factors and high sensitivity, but because of the inherent randomness in the size of the nanostructures, their resonance bandwidth is also broadened. Chemical processes and postannealing of fabricated films are the other two methods in realizing disordered and randomness, but their resonance bandwidth is also very broad, and the processes are usually slow and require multiple fabrication steps with the precise control of parameters [67–71].

2.3 Improving Detection Limit of LSPR Sensors

In practical LSPR-based nanobiosensors, the size of the relevant molecules is typically in the 1–10 nm range, but having a significant increase in the *effective* thickness, d of the adsorbate layer on nanoparticle surface necessitates a large number of these molecules to coat the surface, so, usually, increase in d is in the order of 0.1–1 nm because the molecules do not form a densely packed structure. Therefore, optimizing the detection limit and improving the sensitivity, m of Eq. (2.5), of these biosensors are essential to probe such small increases in thickness.

For this aim, optimizing m and l_d via careful optimization and selection of nanoparticle size, shape, and composition is one of the fundamental ways to increase the sensitivity of LSPR sensors, because of their aforementioned major influence on plasmon characteristics, such as the decay length (l_d), LSPR strength, field-enhancement value, and the linewidth of resonances. In the literature, a variety of different shapes, such as holes in thin metal films [72], nanoprisms [73], nanorings [74], nanorices [29], nanostars [75], nanocubes [76], etc. were explored to investigate their refractive index sensitivity. In general, particles with higher aspect ratios and sizes produce red-shifted LSPRs [77], while the former feature also allows for higher sensitivity values [78]. Having sharp nanoparticle features allow for hot spots with higher local electromagnetic field enhancement factors, which in turn increase the sensitivity to the local refractive index [79]. Anisotropic shapes are also shown to increase sensitivity, at the cost of broader plasmonic reasonances [80]. It is also critical to consider that the addition of functionalization/molecular recognition elements will decrease some part of the interaction of the LSPR-induced near-field and the analyte, thus decreasing the sensitivity compared to what would happen in the absence of them [19]. In such a case, Eq. (2.5) can be modified to Eq. (2.6):

$$\Delta\lambda_{\max} = m\Delta n \exp\left(\frac{-2d_l}{l_d}\right)\left[1-\exp\left(\frac{-2d}{l_d}\right)\right] \tag{2.6}$$

where d_l is the thickness of the molecular recognition of element. The effect of this phenomenon was observed by Haes and coworkers, and it was found that the initial m value before the modification by hexadecanethiol was 196 nm/RIU, but it decreased to 159 nm/RIU after the functionalization [79]. The sensitivity of LSPR nanosensors is projected to improve as nanoparticle designs and nanofabrication methods advance.

Although most LSPR spectroscopy has been carried out with large clusters of nanoparticles, single-nanoparticle spectroscopy represents the absolute detection limit, and there is a huge amount of research efforts in the literature to push the sensitivity to the single-molecule detection limit [81, 82]. This method of sensing offers narrower plasmonic resonances (due to the reduced coupling of LSPRs of different nanoparticles) higher spatial resolution, miniaturized detection volume, prevention of random events for single nanoparticles, ability to monitor the spectral shifts of many nanoparticles in parallel, which is crucial for multiplexing, enabling more targeted binding, and reducing nonspecific interactions.

Nevertheless, measuring the extinction spectrum of a single nanoparticle limits the overall signal, so lower signal-to-noise ratios can be the limiting factors. Furthermore, using and probing a single nanoparticle are difficult in practical point-of-care settings, because it requires trained personnel for handling small volumes. The binding kinetics would also be very small due to the small surface area associated with a single nanoparticle. The readers are referred to an excellent review article on this topic [83] for further information.

In addition to controlling the plasmonic properties of individual nanoparticles, three other methods are outlined by Van Duyne and colleagues to enhance the sensitivity of the nanoparticles to the single-molecule limit, by increasing the LSPR shift [10, 84]. The first one is that large molecules, such as proteins and macromolecules, cause a larger peak shift because of having a larger overlap with the decaying electromagnetic field outside the nanoparticle [22, 79, 85]. Quantitatively, an increase in d corresponds to a larger $\Delta\lambda_{\max}$, from Eq. (2.5).

The second method is based on chromophores, which absorb visible light to give objects their distinct colors. When these molecules are adsorbed onto the nanoparticles, the spectral overlap between the absorption bands of chromophores and the LSPR causes enhanced spectral shifts, compared to nonoverlapping conditions.

This method, to exemplify, is used to detect the binding event between camphor (a small molecule) and the cytochrome P450cam protein, CYP101, on self-assembled monolayer (SAM) functionalized silver nanoparticles, fabricated with NSL. CYP101 has an absorption band at around 417 nm, which blue-shifts to 391 nm upon binding with camphor, as shown in Figure 2.8a–d. In the experiment, it was observed that the set of nanoparticles, the LSPR

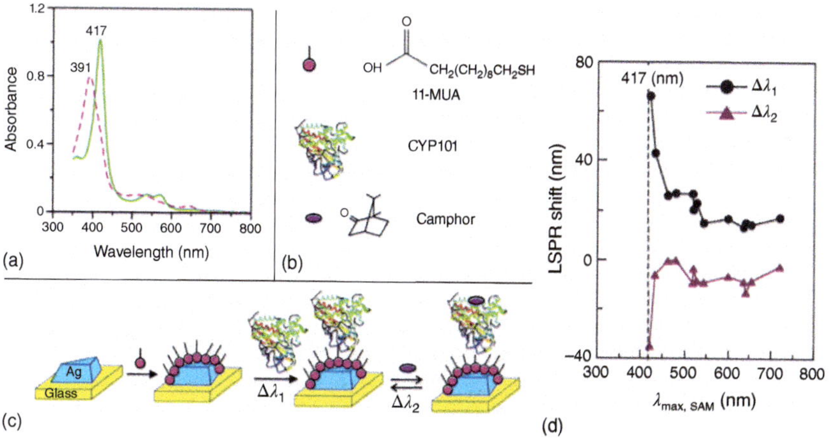

Figure 2.8 (a) UV–Visible absorption spectra of CYP101(Fe^{3+}) (green solid line) with a Soret band at 417 nm (low spin) and camphor-bound CYP101(Fe^{3+}) (pink dashed line) with a Soret band at 391 nm (high spin). (b) Schematic notations of 11-MUA (SAM), CYP101, and camphor. (c) Schematic representation of functionalizing the nanoparticles, CYP101 protein immobilized Ag nanobiosensor, followed by binding of camphor. The Ag nanoparticles are fabricated using nanosphere lithography on a glass substrate. (d) plots of LSPR shifts. The vertical black dotted line denotes the molecular resonance of Fe^{3+}CYP101 at 417 nm. Source: Reprinted with permission from [86]. Copyright (2006) American Chemical Society.

peak positions of which are closer to the CYP101 resonance band, experienced larger LSPR shifts, compared to the nanoparticles, the LSPR positions of which are distant to the CYP101 resonances. While an amplified LSPR red shift as large as 67 nm was observed for protein binding, and a 37-nm blue shift for camphor binding for the former set of nanoparticles, it was, on average, 19 nm red shift for protein binding and 6 nm blue shift for camphor binding for the latter set of nanoparticles. This result is demonstrated in Figure 2.8d, where $\Delta\lambda_1 = \lambda_{\text{max,CYP101}} - \lambda_{\text{max,SAM}}$ denotes the LSPR peak spectral position shift upon binding of CYP101 on SAM-functionalized Ag nanoparticles, whereas $\Delta\lambda_2 = \lambda_{\text{max,CYP101 - CAM}} - \lambda_{\text{max,CYP101}}$ is due to LSPR shift on binding camphor. Positive values indicate red shift, while negative values indicate blue shift [86].

Finally, the third method is decreasing the gap size between individual nanoparticles, which is shown to increase the near-field local electromagnetic field enhancement [58], thus increasing the sensitivity [23], at the cost of resulting in broader (in the sense of full-width-at-half-maximum of the resonances, FWHM) resonances and more strenuous lithography steps [85]. It is shown that nanoparticles, the separation of which are less than 2.5 times their radii, show coupling of LSPRs and pronounced spectral shifts [87]. The fact that larger particles can support higher sensitivities at the cost of broadened plasmonic resonances necessitates accounting for both factors in a general form and defining a metric for the biosensors. A more general term, called Figure of Merit (FoM), has been defined and used extensively for comparing different nanostructures and is defined as in Eq. (2.7):

$$\text{FoM} = \frac{m(\text{nm}/\text{RIU})}{\text{FWHM}(\text{nm})} \tag{2.7}$$

Researchers could employ appropriate combinations of these methods in realizing highly sensitive methods, with the goal of obtaining single-molecule sensitivity without requiring sophisticated or bulky experimentation.

2.4 Integration of LSPR with Other Molecular Identification Techniques

The enhancements in the local electromagnetic field around metallic nanoparticles and the increased absorption of incident electromagnetic radiation make the integration of LSPR with other molecular identification techniques possible and very attractive. In this section, we will outline three such techniques, namely surface-enhanced Raman spectroscopy, laser desorption ionization mass spectrometry, and metal-enhanced fluorescence, and how LSPR-based sensing techniques can be integrated with them.

2.4.1 Metal-Enhanced Fluorescence

Due to the widespread use of fluorescence-based sensing [88–93], the study of fluorescence near metal nanostructures has attracted enormous research interest, in order to improve the capabilities of fluorescence technology [94] and consequently give rise to the spectroscopic technique called plasmon-enhanced fluorescence spectroscopy [95]. While the emission of nearby fluorophores is generally quenched near planar metallic films, due to image theory [96], the situation is more complicated near metallic nanostructures. Depending on the parameters for the metallic nanostructure/fluorophore system, fluorescence quenching (metal-quenched fluorescence, MQF) [97], enhancement (metal-enhanced fluorescence, MEF) [98], or both [99] can be observed. More effective bioassays necessitate an enhancement of fluorescence, which strongly depends on the size and shape of the metal nanoparticles, dipole orientation of the fluorophores with respect to the surface, the distance between fluorophores and metal nanoparticles as well as the spectral overlap between the LSPR and the emission spectrum of the emitter [100, 101].

Distance dependence of the fluorescence intensity is studied extensively in the literature, by using DNA spacers [102], layer-by-layer (LbL) technique [103], and deposition of SiO_2 as a spacer [104]. In general, quenching is observed when the emitter is too close (within a few nanometers) to the fluorophore, while separation larger than quenching distance leads to enhancement [105], with a maximum occurring at an optimum distance [99]. Akbay and colleagues investigated the effect of nanoparticle–fluorophore distance on fluorescence intensity, by controlling the distance by the LbL technique. Distance-dependent intrinsic fluorescence of proteins in the UV spectral region was investigated in the vicinity of aluminum nanoparticles, and the results are shown in Figure 2.9. The largest enhancement of ninefold is observed for the BSA-adsorbed poly(styrene sulfonate) (PSS) and poly(allylamine hydrochloride) (PAH) nanocomposite with the probe distance of 9.2 nm from the Al surface. Approximately sixfold and sevenfold increases in fluorescence intensities of goat and rabbit IgG on metallic nanocomposites were observed with the probe distance of 8.2 nm, respectively [103].

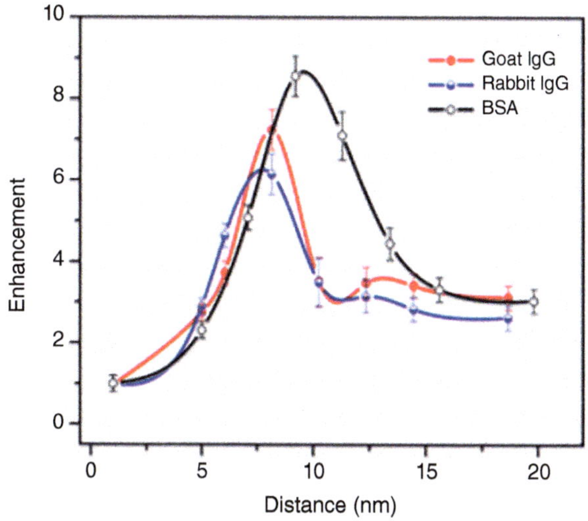

Figure 2.9 The effect of the probe distance on the fluorescence enhancement of proteins, bovine serum albumin (BSA), goat and rabbit immunoglobulins (IgG). Source: Reprinted with permission from [103]. Copyright (2012) American Chemical Society.

The general consensus in the literature regarding the distance-dependent quenching/emission is the relative contributions of nonradiative energy transfer (or the radiative rate suppression), and surface plasmon-coupled emission determines whether the fluorescence signal will be quenched or enhanced. At shorter distances, quenching occurs because the energy of the emitter nonradiatively couples to the plasmons of the nanoparticle, while around the optimum intermediate distances, the cooperative process between the emitter and the plasmons dominates, thus enhancing fluorescence. At distances much larger than the optimum enhancement distance, plasmonic fields decay considerably to be able to couple to the fluorophores [100].

Carefully arranged nanoantennas allow for 100–1000-fold enhancement in fluorescence intensity [106–109]. The hot spots between two nanoparticles are shown to produce a larger fluorescence enhancement than the sum of the effects of two noninteracting, largely separated nanoparticles, as Bek and colleagues showed that in the presence of a dimer, a 272% increase in fluorescence intensity was observed. The enhancement was observed to be very sensitive to the placement of the fluorescent molecule, because the largest enhancement of 272% was recorded when the molecule was located in a line connecting the center of two gold nanoparticles. When the molecule was located slightly off-axis, by moving it with an atomic force microscopy (AFM) tip, this enhancement was less pronounced, only 127%, which is shown in Figure 2.10. This constitutes further evidence for the fact that the intensity of the local electric field around the fluorophore is highly influential in determining the enhancement factor [107]. The spectral overlap between the emission spectrum of fluorophores and the LSPR spectrum of the metal nanoparticles is another important factor in determining the fluorescence enhancement [100]. Tam and colleagues found that the fluorescence enhancement of adjacent indocyanine green (ICG) dye molecules is the highest when their emission wavelength is closest to the LSPR frequency of gold nanoparticles

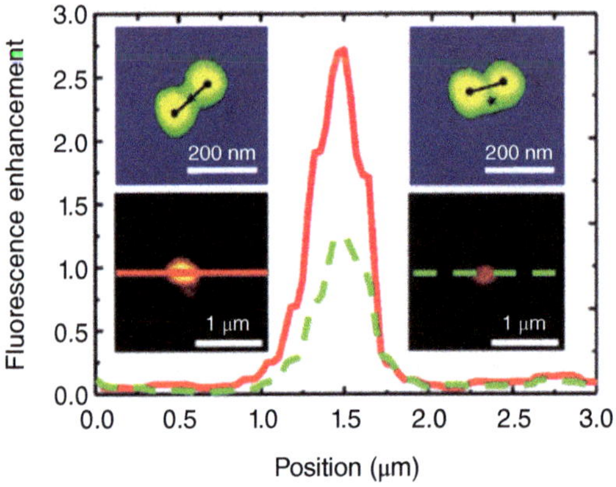

Figure 2.10 The fluorescence enhancement is very sensitive to the exact placement of the fluorescent sphere (FS) between the Au Nanoparticles (AuNPs). The graph shows the fluorescence intensities obtained with two different FS positions. The AFM height images are shown in the upper two insets, and the corresponding fluorescence images are shown in the lower two insets. The exact positions of the AuNPs and FS are indicated by solid black dots and asterisks, respectively. The axis connecting the two AuNP centers is also indicated. In the left column, the FS is sandwiched in the hot spot right between the two AuNPs, which leads to a strong fluorescence enhancement (solid red line). In the right column, the FS is not in the hot spot, and consequently, the fluorescence enhancement is less pronounced (dashed green line). Source: Reprinted with permission from [107]. Copyright (2008) American Chemical Society.

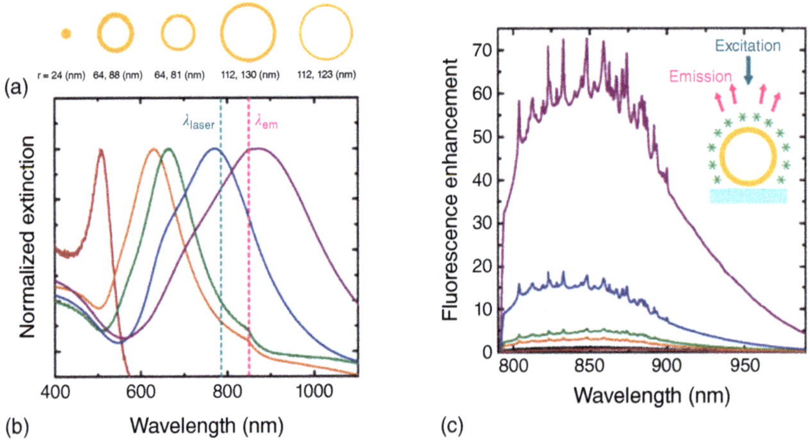

Figure 2.11 (a) Schematic of gold nanoparticles used as fluorescence enhancement substrates, arranged from short to long plasmon resonance wavelength. One Au colloid and four nanoshells of various $[r_1, r_2]$ were used. (b) Normalized extinction measurements from nanoparticle substrates corresponding to (a) in air prior to HSA and ICG deposition. The laser excitation is at 785 nm (λ_{laser}), and the emission wavelength of ICG attached to HSA is 850 nm (λ_{em}). (c) Corresponding fluorescence emission from ICG conjugated to the nanoshell substrates adjusted for surface area available for fluorophore conjugation and normalized to the fluorescence from a control sample with no nanoparticles (black). Inset schematic illustrates experimental geometry. Source: Reprinted with permission from [110]. Copyright (2007) American Chemical Society.

[110]. This is shown in Figure 2.11, where the nanoparticles of size parameters outlined in Figure 2.11a had the LSPR spectrum in Figure 2.11b. Although the size of the nanospheres is influential in determining the fluorescence enhancement, due to larger scattering cross section of the particles, it is not the sole factor in doing so. This observation was made based on the measurement that nanoparticles with similar sizes (more specifically, the ones with $[r_1, r_2] = [112, 130]$ and $[112, 123]$) differed remarkably in their enhancement values when their LSPR λ_{max} values were tuned to the fluorophore excitation versus emission wavelengths, in which the latter case ($[r_1, r_2] = [112, 123]$) showed almost three times more enhancement compared to the former ($[r_1, r_2] = [112, 130]$), as shown in Figure 2.11c. Chen [111] and Thomas [112] found more specifically that the optimum LSPR wavelength should be between the dye emission and absorption maxima.

The combination of the well-developed fluorescence spectroscopy has benefited immensely from the field of plasmonics. There is still some work to be done for a better understanding of the theory that governs the cross talk between the metallic nanoparticles and emitters. The end goal is to realize ultrasensitive biosensors for the fields that include but are not limited to biology, chemistry, and environmental control.

2.4.2 Surface-Enhanced Raman Spectroscopy

Raman spectroscopy is a highly specific molecular identification technique whose basis of operation is the unique molecular vibrational energy levels and the resultant fingerprints. In Raman scattering, which essentially is an elastic scattering of photons, the collected photons (elastically scattered photons are filtered out by a notch filter) are collected and dispersed into a detector. These photons are observed to either lose energy (Stokes shift) or gain energy (anti-Stokes shift). Historically, signal obtained by spontaneous Raman scattering was very weak, which limited its applications. Surface-enhanced Raman spectroscopy was discovered around the 1970s due to enabling huge increases in the measured signal (by up to 10^{14}–10^{15} for single molecules). The reason for this, in the view of electromagnetic theory, is that Raman intensities are proportional to the electric field, so they benefit from the large enhancements in the electromagnetic field surrounding the metals during LSPRs. [113, 114]. As a result, the enhancement is maximum when nanoparticle physical dimensions are arranged so that λ_{max} of the LSPR modes of these nanoparticles falls between the excitation wavelength and the wavelength of the scattered photon [115], and the optimum condition is outlined to be such that λ_{max} is equal to the excitation wavelength (in absolute wavenumbers) minus one-half of the Stokes shift of the band [10].

The dependence of SERS signal intensity on LSPR wavelength is studied systematically in the literature, in the work of Zhang and colleagues [116]. First, SERS-active silver film over nanosphere (AgFON) substrates were fabricated. Varying the diameter of nanospheres gave the opportunity to match the λ_{max} of the LSPRs of nanospheres to the optimum wavenumber that yields the maximum SERS signal intensity. Nanospheres with 390, 510, and 600 nm diameter values showed their LSPR peaks (which are dips in their reflection spectra) at 531.5, 676.5, and 753.1 nm; respectively, as in Figure 2.12a. The aim of the experiment was to maximize the SERS signal intensity of the $1003\,cm^{-1}$ benzenethiol band, acquired with 750-nm laser illumination. According to the qualitative formula above, this corresponds to the optimal LSPR λ_{max} of 779 nm (excitation wavelength [in absolute

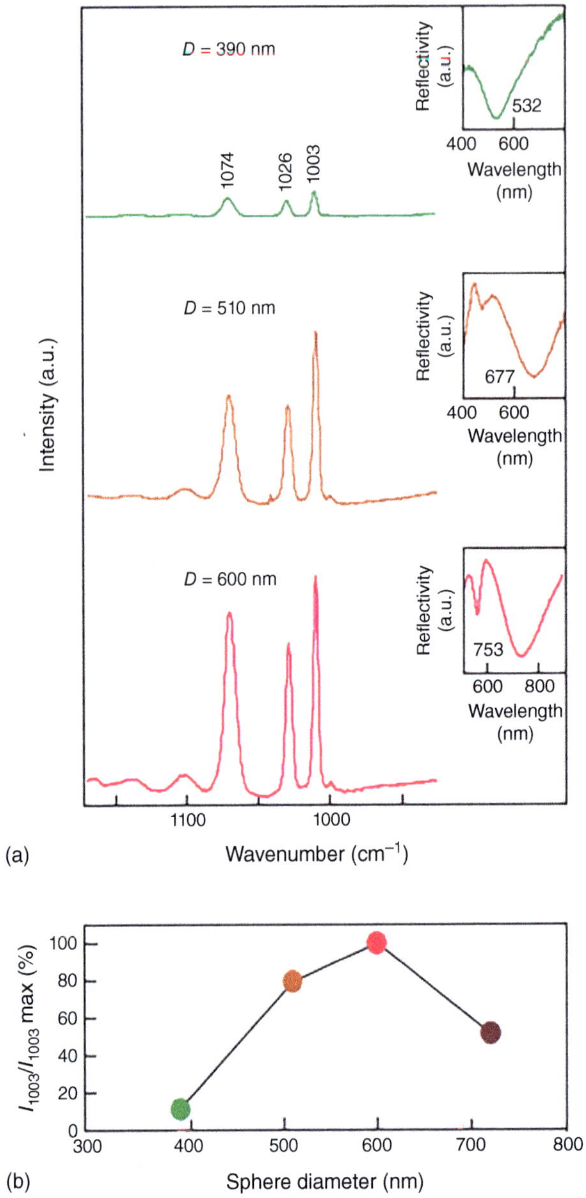

Figure 2.12 Tuning the LSPR to maximize the SERS signal. (a) SERS spectrum of benzenethiol on AgFONs with varying nanosphere diameters and corresponding resonances: at 532 nm, sphere diameter *D* = 390 nm (green), at 677 nm, *D* = 510 nm (orange), and at 753 nm, *D* = 600 nm (red). The reflection spectrum is shown in the insets, with minimal reflection corresponding to maximum LSPR-induced absorbance and scattering. (b) Relative peak heights of the 1003 cm^{-1} SERS peak (in-plane ring deformation mode) as a function of the nanosphere diameter. Source: Reprinted with permission from [10]. Copyright (1969), Springer Nature. Original study is [116].

wavenumbers] minus one-half of the Stokes shift of the band). Therefore, the particles of 600 nm in diameter were expected and were confirmed to yield the largest SERS signal intensity, as shown in in Figure 2.12b.

LSPRs, in summary, can greatly aid SERS measurements, in the sense that LSPR can be used for quantification and SERS can be used for identification. This approach enables LSPR spectroscopy to differentiate between isotopically labeled molecules, which would not be possible based on mere refractive index sensing via measuring LSPR shifts or observing colorimetry [117].

2.4.3 Matrix-Assisted Laser Desorption Ionization Mass Spectroscopy

Mass spectroscopy is another platform that can benefit from integration with LSPR-based spectroscopic devices. The working principle of laser desorption ionization mass spectroscopy is that a laser is used to ablate and desorb the sample. The analyte molecules are then ionized and accelerated into the mass spectrometer. Sometimes, for more effective and easier ionization, the analyte may be embedded in a solid or liquid, laser energy absorbing material, called the "matrix." This particular technique is known as matrix-assisted laser desorption ionization (MALDI) [118]. Chen and colleagues demonstrated plasmonic-enhanced, direct laser desorption/ionization mass spectroscopy by using gold nanorods, fabricated by electrodepositing gold into the pores of a porous alumina template. When the LSPRs of these gold nanorods, which were observed to peak around 520 nm, were excited by using a frequency-doubled Nd:YAG laser of 532 nm, the desorption and ionization were found to be favored compared to a continuous Au film. Due to two main reasons, this increase was attributed to the plasmonic-enhanced absorption. Firstly, the ion abundance was more intense for the frequency-doubled case, because the laser wavelength is closer to the LSPR λ_{max} compared to the laser wavelength in the frequency-tripled case, 355 nm. Secondly, the mass spectrometry intensity was found to depend on the length of the nanorods and the polarization state of the incident light, which also influences the LSPR properties. This enhancement was attributed to the enhanced optical absorption of 532 nm light due to its overlap with the LSPRs [119]. More up-to-date literature shows that plasmon-induced hot carriers promote the desorption of molecules [120], which is supported by recent experimental pieces of evidence [121, 122].

Overall, the distinct developments in mass spectroscopy and plasmonic nanoparticles for sensing need to be integrated in the future for the further development of this technology.

2.5 Practical Issues

The promise and applications of LSPR-based sensors, as well as their combination with other molecular identification methods, are well established. In this section, we will outline some considerations for the realization of rapid, high-throughput, reliable, and well-characterized sensing platforms that consist of uniform nanoparticles fabricated on stable substrates [10]. First of all, the nanoparticles themselves have to be uniformly fabricated in terms of their size, shape, and composition. Uniform nanoparticles can be reliably produced by the aforementioned lithographic methods. Techniques such as electrophoresis can be used to separate and purify nanoparticles based on their size, shape, and charge [123]. Optical damage threshold is another important consideration in the sense that sensors should ensure stable operation under high temperatures that may be influenced by the laser fluence. The ohmic losses of the metallic nanoparticles that are exacerbated when their LSPRs are excited imply that significant damage can occur under low laser intensities, so the magnitude of near-field enhancement should be balanced with Joule losses to minimize heating and possible irreversible damage to the nanoparticles [124, 125]. Regarding the fabrication, an important issue that requires attention is the poor adhesion of noble metal nanoparticles to most inorganic substrates. This can result in morphological changes compared to the desired shapes and the consequent change in optical performance. To minimize the adverse effects of this issue, metallic (Cr, Ni, Ti) adhesion/coupling layers can be deposited prior to the noble metals, [126] overcoating layers such as diamond [127], or amorphous carbonated silicon thin films [128] can be used, or nanoparticles can be embedded into dielectric matrices [129].

Although LSPRs can themselves act as sensors by measuring $\Delta\lambda_{max}$, as discussed up to this point, it is inherently a nonspecific technique because the LSPR shift is obtained by any refractive index around the metal. However, if the molecule to be detected is not known, then LSPR-based detection schemes are not effective, and labeling is required as a part of the experiment, or LSPRs need to be combined with the other identification methods that are discussed previously. Specificity, or selectivity, is achieved by chemically modifying the metal interface, by employing biological recognition elements, such as antibodies together with hydrophilic SAMs and blocking agents; as well as utilizing chromophore-coupled plasmonic nanoparticles. One other advantage of forming SAMs is that it improves the affinity between some target analytes, such as glucose, and the bare metal surface [9, 10].

Thiolated organic compounds are able to form strong covalent bonds and consequent SAMs with metals, which make them one of the most important and widely used head groups to stabilize and "cap" the metallic nanoparticles and make the anchoring of other functionalizers possible. It is worth mentioning that the surface chemistry of gold makes establishing sulfur bonds between gold atoms and thiols relatively easy, which is another reason behind gold's domination in the LSPR-based bioassay devices market [1, 19].

Nevertheless, thiol-based SAMs suffer from time-dependent, long-term structural rearrangement to ordered and stable structures [130], thermal desorption [131] and photooxidation [132], so research is conducted to come up with alternative surface chemistries. Another such chemistry includes depositing conformal layers of alumina or titania on top metallic nanoparticles. The deposition method can be atomic layer deposition (ALD) or solution-phase sol–gel deposition. These conformal layers can then protect nanoparticles from annealing, as well as offering new chemical functionalities. To illustrate, alumina may significantly reduce the proneness of nanoparticles to oxidation and may aid in the selective adsorption of polar compounds. Regarding the former benefit, Zhang and colleagues utilized ALD- deposited alumina-functionalized AgFON substrates for the quantitative detection of anthrax biomarker calcium dipicolinate. Compared to their earlier experiment of the same group in 2005 [116], the result of which is shown in Figure 2.12, the authors found a twofold increase in the limit of detection, as well as yielding a stable SERS intensity for over nine months [133]. Readers may refer to the following excellent review articles for further information on improving the selectivity of LSPR-based assays in complex solutions: Refs. [19, 134]

Another fundamental necessity for LSPR-based biosensing platforms to reach their full potential and become more widespread in the sensing market is interfacing with existing multiplexed point-of-care instruments with high throughput, reducing the requirement for sample volumes, and eliminating the necessity for labeling [9, 10, 42, 80, 134–136].

One of the initial efforts toward reaching label-free, high-throughput LSPR assays with parallel detection capability is the work of Endo and colleagues [137]. In this work, a multiarray nanochip composed of 300 sensing spots was used for the detection of six different proteins, i.e. antigen–antibody reactions. To do this, gold-deposited silica nanospheres were situated on top a thin gold film. SAM formation was performed by using 4,4′-dithiodibutyric acid (DDA). Each spot was separated by 1 mm to avoid cross-contamination. Figure 2.13a shows the photograph of the multiarray sensor, with a uniform color throughout, indicating a uniform LSPR spectrum. The antibodies were immobilized on the sensor by using a nanoliter dispensing system. Different concentrations of specific antigens were dispersed onto the multiarray sensor surface (each antibody was dispensed onto 50 spots) and the change in LSPR λ_{max} was recorded for each of the 300 sensing spots, to observe the specificity and measure the limit of detection. Figure 2.13b,c shows the change in absorbance upon the application of varying concentrations of immunoglobulin A (IgA) to six different types of antibodies. In this experiment, they found the limit of detection to be 100 pg ml^{-1}, and the sensor response scaled linearly with concentrations up to 1 μg ml^{-1}.

Microfluidics-based biosensing has also attracted enormous attention from the research community because of its high-throughput, low-cost mass fabrication [138]. Integration of microfluidics with LSPR-based nanoparticles combines advantages of both technologies and allows for significantly lowered sample/reagent volume, miniaturized device footprints that may accelerate immunoreaction times and allow for shortening the assay times [139]. Synthesis of nanoparticles in the microfluidic environment is also shown to improve the uniformity of the nanoparticle size, as found by SadAbadi and colleagues. They found that in situ synthesis results in 8% size variation, while for synthesis in the macroenvironment, this variation was 67% [140]. In a recent experiment, Chen and colleagues realized a multiarrayed LSPR-based microfluidic optical sensing platform that consists of 480 nanoplasmonic sensing spots and demonstrated parallel multiplex immunoessay of six cytokines in a complex serum on a single chip. The limit of detection values reached as low as 5–20 pg ml^{-1} from a 1 μl serum sample. They also reported that the entire experimentation time, which consisted of washing of samples and reagents and 10fold replicated multianalyte detection for each sample using the entire biosensor arrays, was completed within 40 minutes [141].

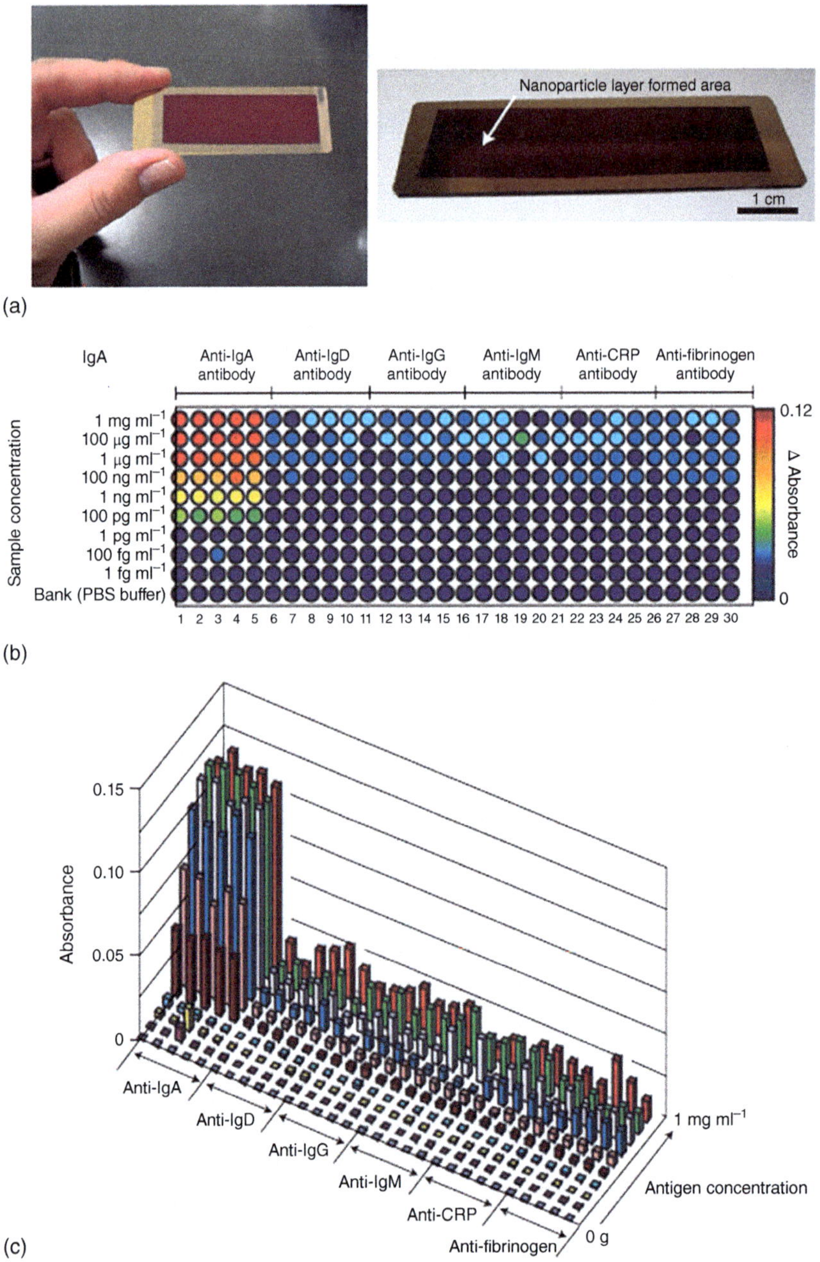

Figure 2.13 Uniform biochip for multiplexed LSPR detection. (a) Photograph of the LS PR-based nanochip with dimensions 20 × 60 mm. The nanochip structure consists of silica nanospheres deposited on a flat gold film and coated with a gold overlayer. (b, c) Absorbance measurements at each spot of the multiarray nanochip for binding of immunoglobulin A (IgA) to six different types of antibodies. (b) The antibodies were immobilized on the chip, resulting in a total of 300 spots separated by 1 mm (to prevent cross-contamination). (c) Different concentrations of antigen were incubated for 30 minutes and LSPR absorption spectra were then acquired with a fiber-coupled ultraviolet–visible spectrometer in a reflection geometry. Source: Reprinted with permission from [10]. Copyright (1969), Springer Nature. Original study is [137].

2.6 Conclusions and Future Prospects

Although LSPR-based biosensors, a fundamental physical understanding of increasing their performance, integration of LSPR phenomenon with other molecular identification platforms, and interfacing with multiplexed platforms have received an enormous amount of research efforts, clear challenges and limitations still exist, and the number of commercial products and devices is still rather limited (an overview of current LSPR-related companies and the principle of operation of their devices can be found in Ref. [142]).

One of the main drawbacks of LSPR sensors is their reproducibility and monodispersity in the size and shape of them, as well as low stability. Regarding the first two issues, the advancing nanofabrication, microscopy, and characterization techniques are expected to make the nanoparticles support higher-quality LSPRs and approach single-molecule detection limit. The stability problem indeed, as mentioned previously, explains the discrepancy that silver is optically a better material than gold, yet the latter dominates the LSPR-based devices. Hybrid LSPR structures that employ dielectric overcoatings pose a solution to this issue [143].

One other key challenge is to improve the sensitivity and decrease the limit of detection of the existing nanoparticles. In addition to the strategies outlined in this chapter, enzyme-mediated and biomolecule conformationally mediated amplification techniques are among two methods to achieve this purpose, and the recent literature shows that combining more than one amplification method can yield significantly low limits of detection [134, 144].

Another consideration that is discussed in this chapter is the challenge of functionalization and making the LSPR-based assays selective. In addition to our improved understanding and application of the self-assembled molecules, which combine attracting the molecule of interest while repelling the others, biological scaffolds emerged as highly promising in this regard [145, 146]. Although they appear as more selective, they would make the devices more susceptible to denaturation and decrease portability, i.e. make handling more difficult [134].

The most important advantages of LSPR-based assays, in addition to their remarkable optical properties, are their miniaturization, compatibility with microfluidics and in-flow assays, which is the reason behind the recent research efforts toward multiplexed and microfluidic LSPR sensing platforms. Another recent current trend in LSPR-based biosensing is making the sensors more approachable to the untrained personnel and allowing for a smartphone-based optical readout [80, 147–149]. Moreover, LSPR-based sensing is currently being explored beyond the detection of solution-based binding reactions, by demonstrating the ability to sense inert gasses such as He, Ar, and N_2 [150]. Single-particle measurements are also expected to be researched and pursued more heavily in the future, because of enhancing the detection sensitivities by eliminating averaging effects and offering ultimate spatial resolution limits that reveal the actual events that occur on their surfaces [151].

In conclusion, a combined and multidisciplinary research effort is needed to enable LSPR-based assays to reach their full potential and establish a competent technology for the current, commercially available devices. Future point-of-care devices will most probably require interfacing robust, sensitive, and selective LSPR nanoparticles with cell lysing and separation platforms in an integrated multiplex design, if they are to be used for true lab-on-a-chip applications [152, 153].

References

1 Stefan A. Maier. Plasmonics: Fundamentals and Applications. 2007. doi: https://doi.org/10.1007/0-387-37825-1.

2 Bohren, C.F. and Huffman, D.R. (2008). *Absorption and Scattering of Light by Small Particles*, Wiley Science Series. Wiley https://books.google.com.tr/books?id=ib3EMXXIRXUC.

3 Quinten, M. (2010). *Optical Properties of Nanoparticle Systems: Mie and beyond*. John Wiley & Sons.

4 Neeves, A.E. and Birnboim, M.H. (1989). Composite structures for the enhancement of nonlinear-optical susceptibility. *Journal of the Optical Society of America B: Optical Physics* 6 (4): 787–796. https://doi.org/10.1364/JOSAB.6.000787. http://josab.osa.org/abstract.cfm?URI=josab-6-4-787.

5 Mock, J.J., Smith, D.R., and Schultz, S. (2003). Local refractive index dependence of Plasmon resonance spectra from individual nanoparticles. *Nano Letters* 3 (4): 485–491. https://doi.org/10.1021/nl0340475.

6 Sun, Y. and Xia, Y. (2002). Increased sensitivity of surface Plasmon resonance of gold Nanoshells compared to that of gold solid colloids in response to environmental changes. *Analytical Chemistry* 74 (20): 5297–5305. https://doi.org/10.1021/ac0258352.

7 Fornasiero, D. and Grieser, F. (1987). Analysis of the visible absorption and SERS excitation spectra of silver sols. *The Journal of Chemical Physics* 87 (5): 3213–3217. https://doi.org/10.1063/1.453009.

8 Haes, A.J., Zou, S., Zhao, J. et al. (2006). Localized surface Plasmon resonance spectroscopy near molecular resonances. *Journal of the American Chemical Society* 128 (33): 10905–10914. https://doi.org/10.1021/ja063575q.

9 Sepúlveda, B., Angelomé, P.C., Lechuga, L.M., and Liz-Marzán, L.M. (2009). LSPR-based nanobiosensors. *Nano Today* 4 (3): 244–251. https://doi.org/10.1016/j.nantod.2009.04.001. http://www.sciencedirect.com/science/article/pii/S1748013209000280.

10 Anker, J.N., Hall, W.P., Lyandres, O. et al. (2008). Biosensing with plasmonic nanosensors. *Nature Materials* 7 (6): 442–453. https://doi.org/10.1038/nmat2162. http://www.ncbi.nlm.nih.gov/pubmed/18497851.

11 Han, X., Liu, K., and Sun, C. (2019). Plasmonics for biosensing. *Materials* 12 (9) https://doi.org/10.3390/ma12091411.

12 Chan, G.H., Zhao, J., Hicks, E.M. et al. (2007). Plasmonic properties of copper nanoparticles fabricated by Nanosphere lithography. *Nano Letters* 7 (7): 1947–1952. https://doi.org/10.1021/nl070648a.

13 Schultz, S., Smith, D.R., Mock, J.J., and Schultz, D.A. (2000). Single-target molecule detection with nonbleaching multicolor optical immunolabels. *Proceedings of the National Academy of Sciences* 97 (3): 996–1001. https://doi.org/10.1073/pnas.97.3.996.

14 Alzahrani, E. (2020). Colorimetric detection based on localized surface Plasmon resonance optical characteristics for sensing of mercury using green-synthesized silver nanoparticles. *Journal of Analytical Methods in Chemistry* 2020: 6026312. https://doi.org/10.1155/2020/6026312.

15 Englebienne, P. (1998). Use of colloidal gold surface plasmon resonance peak shift to infer affinity constants from the interactions between protein antigens and antibodies specific

for single or multiple epitopes. *Analyst* 123 (7): 1599–1603. https://doi.org/10.1039/A804010I.

16 Li, T., Sun, J., Liu, J. et al. (2020). Constructing boronate-bridged core-satellite gold nanoassembly and its application in high sensitive colorimetric detection of benzoyl peroxide residues in food matrices. *Chinese Chemical Letters* 31 (2): 439–442. https://doi.org/10.1016/j.cclet.2019.08.051. http://www.sciencedirect.com/science/article/pii/S1001841719305248.

17 Liu, J. and Lu, Y. (2003). A colorimetric Lead biosensor using DNAzyme- directed assembly of gold nanoparticles. *Journal of the American Chemical Society* 125 (22): 6642–6643. https://doi.org/10.1021/ja034775u.

18 Mirkin, C.A., Letsinger, R.L., Mucic, R.C., and Storhoff, J.J. (1996). A DNA-based method for rationally assembling nanoparticles into macroscopic materials. *Nature* 382 (6592): 607–609. https://doi.org/10.1038/382607a0.

19 Szunerits, S. and Boukherroub, R. (2012). Sensing using localised surface plasmon resonance sensors. *Chemical Communications* 48 (72): 8999–9010. https://doi.org/10.1039/C2CC33266C.

20 McFarland, A.D. and Van Duyne, R.P. (2003). Single silver nanoparticles as real-time optical sensors with Zeptomole sensitivity. *Nano Letters* 3 (8): 1057–1062. https://doi.org/10.1021/nl034372s.

21 Haes, A.J. and VanDuyne, R.P. (2002). A nanoscale optical biosensor: sensitivity and selectivity of an approach based on the localized surface Plasmon resonance spectroscopy of triangular silver nanoparticles. *Journal of the American Chemical Society* 124 (35): 10596–10604. https://doi.org/10.1021/ja020393x.

22 Whitney, A.V., Elam, J.W., Zou, S. et al. (2005). Localized surface Plasmon resonance Nanosensor: a high-resolution distance-dependence study using atomic layer deposition. *The Journal of Physical Chemistry B* 109 (43): 20522–20528. https://doi.org/10.1021/jp0540656.

23 Soydan, M.C., Ghobadi, A., Yildirim, D.U. et al. (2020a). Lithography-free random bismuth nanostructures for full solar Spectrum harvesting and mid-infrared sensing. *Advanced Optical Materials* 8 (4): 1901203. https://doi.org/10.1002/adom.201901203.

24 Li, R., Wu, D., Liu, Y. et al. (2017b). Infrared Plasmonic refractive index sensor with ultra-high figure of merit based on the optimized all-metal grating. *Nanoscale Research Letters* 12 (1) https://doi.org/10.1186/s11671-016-1773-2.

25 Luo, S., Zhao, J., Zuo, D., and Wang, X. (2016). Perfect narrow band absorber for sensing applications. *Optics Express* 24 (9): 9288. https://doi.org/10.1364/oe.24.009288.

26 Wu, D., Li, R., Liu, Y. et al. (2017). Ultra-narrow band perfect absorber and its application as Plasmonic sensor in the visible region. *Nanoscale Research Letters* 12 https://doi.org/10.1186/s11671-017-2203-9.

27 Yong, Z., Zhang, S., Gong, C., and He, S. (2016). Narrow band perfect absorber for maximum localized magnetic and electric field enhancement and sensing applications. *Scientific Reports* 6 https://doi.org/10.1038/srep24063.

28 Jensen, T.R., Duval, M.L., Kelly, K. et al. (1999). Nanosphere lithography: effect of the external dielectric medium on the surface Plasmon resonance Spectrum of a periodic Array of silver nanoparticles. *The Journal of Physical Chemistry B* 103 (45): 9846–9853. https://doi.org/10.1021/jp9926802.

29 Wang, H., Brandl, D.W., Le, F. et al. (2006). Nanorice: a Hybrid Plasmonic Nanostructure. *Nano Letters* 6 (4): 827–832. https://doi.org/10.1021/nl060209w.

30 Barbillon, G., Bijeon, J.-L., Plain, J. et al. (2007). Electron beam lithography designed chemical nanosensors based on localized surface plasmon resonance. *Surface Science* 601 (21): 5057–5061. https://doi.org/10.1016/j.susc.2007.09.005. http://www.sciencedirect.com/science/article/pii/S0039602807009181.

31 Haynes, C.L. and Van Duyne, R.P. (2001). Nanosphere lithography: a versatile nanofabrication tool for studies of size-dependent nanoparticle optics. *The Journal of Physical Chemistry B* 105 (24): 5599–5611. https://doi.org/10.1021/jp010657m.

32 Haes, A.J., Haynes, C.L., and Van Duyne, R.P. (2000). Nanosphere lithography: self-assembled photonic and magnetic materials. *MRS Proceedings* 636: D4.8.1. https://doi.org/10.1557/PROC-636-D4.8.1. https://www.cambridge.org/core/article/nanosphere-lithography-selfassembled-photonic-and-magnetic-materials/A5319AF6EDF8A84A64B3E6B4DC25A861.

33 Jensen, T.R., Malinsky, M.D., Haynes, C.L., and Van Duyne, R.P. (2000). Nanosphere lithography: tunable localized surface Plasmon resonance spectra of silver nanoparticles. *The Journal of Physical Chemistry B* 104 (45): 10549–10556. https://doi.org/10.1021/jp002435e.

34 Sanders, G.H.W. and Manz, A. (2000). Chip-based microsystems for genomic and proteomic analysis. *TrAC Trends in Analytical Chemistry* 19 (6): 364–378. https://doi.org/10.1016/S0165-9936(00)00011-X. http://www.sciencedirect.com/science/article/pii/S016599360000011X.

35 Yonzon, C.R., Jeoung, E., Zou, S. et al. (2004). A comparative analysis of localized and propagating surface Plasmon resonance sensors: the binding of Concanavalin a to a monosaccharide functionalized self-assembled monolayer. *Journal of the American Chemical Society* 126 (39): 12669–12676. https://doi.org/10.1021/ja047118q.

36 Jung, L.S., Campbell, C.T., Chinowsky, T.M. et al. (1998). Quantitative interpretation of the response of surface Plasmon resonance sensors to adsorbed films. *Langmuir* 14 (19): 5636–5648. https://doi.org/10.1021/la971228b.

37 Willets, K.A. and Van Duyne, R.P. (2007). Localized surface Plasmon resonance spectroscopy and sensing. *Annual Review of Physical Chemistry* 58 (1): 267–297. https://doi.org/10.1146/annurev.physchem.58.032806.104607.

38 Haes, A.J., Chang, L., Klein, W.L., and Van Duyne, R.P. (2005). Detection of a biomarker for Alzheimer's disease from synthetic and clinical samples using a nanoscale optical biosensor. *Journal of the American Chemical Society* 127 (7): 2264–2271. https://doi.org/10.1021/ja044087q.

39 Brolo, A.G. (2012). Plasmonics for future biosensors. *Nature Photonics* 6 (11): 709–713. https://doi.org/10.1038/nphoton.2012.266.

40 Howes, P.D., Rana, S., and Stevens, M.M. (2014). Plasmonic nanomaterials for biodiagnostics. *Chemical Society Reviews* 43 (11): 3835–3853. https://doi.org/10.1039/c3cs60346f.

41 Joshi, G.K., Deitz-Mcelyea, S., Johnson, M. et al. (2014). Highly specific plasmonic biosensors for ultrasensitive MicroRNA detection in plasma from pancreatic cancer patients. *Nano Letters* 14 (12): 6955–6963. https://doi.org/10.1021/nl503220s.

42 Lim, W.Q. and Gao, Z. (2016). Plasmonic nanoparticles in biomedicine. *Nano Today* 11 (2): 168–188. https://doi.org/10.1016/j.nantod.2016.02.002.

43 Mayer, K.M., Lee, S., Liao, H. et al. (2008). A label-free immunoassay based upon localized surface Plasmon resonance of gold Nanorods. *ACS Nano* 2 (4): 687–692. https://doi.org/10.1021/nn7003734.

44 Nath, N. and Chilkoti, A. (2004). Label-free biosensing by surface Plasmon resonance of nanoparticles on glass: optimization of nanoparticle size. *Analytical Chemistry* 76 (18): 5370–5378. https://doi.org/10.1021/ac049741z.

45 Saha, K., Agasti, S.S., Kim, C. et al. (2012). Gold nanoparticles in chemical and biological sensing. *Chemical Reviews* 112 (5): 2739–2779. https://doi.org/10.1021/cr2001178.

46 Shi, L., Jing, C., Ma, W. et al. (2013). Plasmon resonance scattering spectroscopy at the single-nanoparticle level: real-time monitoring of a click reaction. *Angewandte Chemie, International Edition* 52 (23): 6011–6014. https://doi.org/10.1002/anie.201301930.

47 Zhang, J., Wang, L., Pan, D. et al. (2008). Visual cocaine detection with gold nanoparticles and rationally engineered aptamer structures. *Small* 4 (8): 1196–1200. https://doi.org/10.1002/smll.200800057.

48 Khalil, M., Jan, B.M., Tong, C.W., and Berawi, M.A. (2017). Advanced nanomaterials in oil and gas industry: design, application and challenges. *Applied Energy* 191: 287–310. https://doi.org/10.1016/j.apenergy.2017.01.074.

49 Ayyappan, S., Gopalan, R.S., Subbanna, G.N., and Rao, C.N.R. (1997). Nanoparticles of Ag, Au, Pd, and Cu produced by alcohol reduction of the salts. *Journal of Materials Research* 12 (2): 398–401. https://doi.org/10.1557/JMR.1997.0057.

50 Jana, N.R., Gearheart, L., and Murphy, C.J. (2001). Evidence for seed-mediated nucleation in the chemical reduction of gold salts to gold nanoparticles. *Chemistry of Materials* 13 (7): 2313–2322. https://doi.org/10.1021/cm000662n.

51 Sanyal, U. and Jagirdar, B.R. (2012). Metal and alloy nanoparticles by amine-borane reduction of metal salts by solid-phase synthesis: atom economy and green process. *Inorganic Chemistry* 51 (23): 13023–13033. https://doi.org/10.1021/ic3021436.

52 Wostek-Wojciechowska, D., Jeszka, J.K., Amiens, C. et al. (2005). The solid-state synthesis of metal nanoparticles from organometallic precursors. *Journal of Colloid and Interface Science* 287 (1): 107–113. https://doi.org/10.1016/j.jcis.2005.01.109. http://www.sciencedirect.com/science/article/pii/S0021979705001311.

53 Truong, P.L., Kim, B.W., and Sim, S.J. (2012). Rational aspect ratio and suitable antibody coverage of gold nanorod for ultra-sensitive detection of a cancer biomarker. *Lab on a Chip* 12 (6): 1102–1109. https://doi.org/10.1039/C2LC20588B.

54 Cinel, N.A., Bütün, S., and Özbay, E. (2012). Electron beam lithography designed silver nano-disks used as label free nano-biosensors based on localized surface plasmon resonance. *Optics Express* 20 (3): 2587–2597. https://doi.org/10.1364/OE.20.002587. http://www.opticsexpress.org/abstract.cfm?URI=oe-20-3-2587.

55 Hicks, E.M., Zou, S., Schatz, G.C. et al. (2005). Controlling Plasmon line shapes through diffractive coupling in linear arrays of cylindrical nanoparticles fabricated by electron beam lithography. *Nano Letters* 5 (6): 1065–1070. https://doi.org/10.1021/nl0505492.

56 Brolo, A.G., Gordon, R., Leathem, B., and Kavanagh, K.L. (2004). Surface Plasmon sensor based on the enhanced light transmission through arrays of Nanoholes in gold films. *Langmuir* 20 (12): 4813–4815. https://doi.org/10.1021/la0493621.

57 Soydan, M.C., Ghobadi, A., Yildirim, D.U. et al. (2019). All ceramic-based metal-free ultra-broadband perfect absorber. *Plasmonics* https://doi.org/10.1007/s11468-019-00976-z.

58 Yildirim, D.U., Ghobadi, A., Soydan, M.C. et al. (2019a). Disordered and densely packed ITO Nanorods as an excellent lithography-free optical solar reflector Metasurface. *ACS Photonics* 6: 1812–1822. https://doi.org/10.1021/acsphotonics.9b00636.

59 Zhu, S., Li, F., Chunlei, D., and Yongqi, F. (2008). A localized surface plasmon resonance nanosensor based on rhombic Ag nanoparticle array. *Sensors and Actuators B: Chemical* 134 (1): 193–198. https://doi.org/10.1016/j.snb.2008.04.028. http://www.sciencedirect.com/science/article/pii/S092540050800302X.

60 Haynes, C.L., McFarland, A.D., Smith, M.T. et al. (2002). Angle-resolved Nanosphere lithography: manipulation of nanoparticle size, shape, and Interparticle spacing. *The Journal of Physical Chemistry B* 106 (8): 1898–1902. https://doi.org/10.1021/jp013570+.

61 Lohmueller, T., Bock, E., and Spatz, J.P. (2008). Synthesis of quasi-hexagonal ordered arrays of metallic nanoparticles with Tuneable particle size. *Advanced Materials* 20 (12): 2297–2302. https://doi.org/10.1002/adma.200702635.

62 Kim, S., Jung, J.-M., Choi, D.-G. et al. (2006). Patterned arrays of au rings for localized surface Plasmon resonance. *Langmuir* 22 (17): 7109–7112. https://doi.org/10.1021/la0605844.

63 Henzie, J., Lee, M.H., and Odom, T.W. (2007). Multiscale patterning of plasmonic metamaterials. *Nature Nanotechnology* 2 (9): 549–554. https://doi.org/10.1038/nnano.2007.252.

64 Li, H., Luo, X., Chunlei, D. et al. (2008). Ag dots array fabricated using laser interference technique for biosensing. *Sensors and Actuators B: Chemical* 134 (2): 940–944. https://doi.org/10.1016/j.snb.2008.06.039. http://www.sciencedirect.com/science/article/pii/S0925400508004577.

65 Yesilkoy, F., Terborg, R.A., Pello, J. et al. (2018). Phase-sensitive plasmonic biosensor using a portable and large field-of-view interferometric microarray imager. *Light: Science & Applications* 7 (2): 17152. https://doi.org/10.1038/lsa.2017.152.

66 Soydan, M.C., Ghobadi, A., Yildirim, D.U. et al. (2020b). Deep subwavelength light confinement in disordered bismuth Nanorods as a linearly thermal-tunable metamaterial. *Physica Status Solidi RRL: Rapid Research Letters*, n/a(n/a): 2000066. https://doi.org/10.1002/pssr.202000066. https://onlinelibrary.wiley.com/doi/abs/10.1002/pssr.202000066.

67 Bae, K., Kang, G., Cho, S.K. et al. (2015). Flexible thin-film black gold membranes with ultrabroadband plasmonic nanofocusing for efficient solar vapour generation. *Nature Communications* 6: 10103. https://doi.org/10.1038/ncomms10103.

68 Ghobadi, A., Dereshgi, S.A., Hajian, H. et al. (2017a). 97 percent light absorption in an ultrabroadband frequency range utilizing an ultrathin metal layer: randomly oriented, densely packed dielectric nanowires as an excellent light trapping scaffold. *Nanoscale* 9 (43): 16652–16660. https://doi.org/10.1039/c7nr04186a.

69 Ghobadi, A., Dereshgi, S.A., Hajian, H. et al. (2017b). Ultra-broadband, wide angle absorber utilizing metal insulator multilayers stack with a multi-thickness metal surface texture. *Scientific Reports* 7: 4755. https://doi.org/10.1038/s41598-017-04964-3.

70 Ghobadi, A., Hajian, H., Dereshgi, S.A. et al. (2017c). Disordered Nanohole patterns in metal-insulator multilayer for ultra-broadband light absorption: atomic layer deposition for lithography free highly repeatable large scale multilayer growth. *Scientific Reports* 7: 15079. https://doi.org/10.1038/s41598-017-15312-w.

71 Ghobadi, A., Hajian, H., Rashed, A.R. et al. (2018). Tuning the metal filling fraction in absorbers to maximize the absorption bandwidth. *Photonics Research* 6: 168–176. https://doi.org/10.1364/PRJ.6.000168.

72 Prikulis, J., Hanarp, P., Olofsson, L. et al. (2004). Optical spectroscopy of Nanometric holes in thin gold films. *Nano Letters* 4 (6): 1003–1007. https://doi.org/10.1021/nl0497171.

73 Sherry, L.J., Jin, R., Mirkin, C.A. et al. (2006). Localized surface Plasmon resonance spectroscopy of single silver triangular Nanoprisms. *Nano Letters* 6 (9): 2060–2065. https://doi.org/10.1021/nl061286u.

74 Aizpurua, J., Hanarp, P., Sutherland, D.S. et al. (2003). Optical properties of gold Nanorings. *Physical Review Letters* 90 (5): 57401. https://doi.org/10.1103/PhysRevLett.90.057401.

75 Nehl, C.L., Liao, H., and Hafner, J.H. (2006). Optical properties of star-shaped gold nanoparticles. *Nano Letters* 6 (4): 683–688. https://doi.org/10.1021/nl052409y.

76 Sherry, L.J., Chang, S.-H., Schatz, G.C. et al. (2005). Localized surface Plasmon resonance spectroscopy of single silver Nanocubes. *Nano Letters* 5 (10): 2034–2038. https://doi.org/10.1021/nl0515753.

77 Link, S. and El-Sayed, M.A. (1999). Spectral properties and relaxation dynamics of surface Plasmon electronic oscillations in gold and silver Nanodots and Nanorods. *The Journal of Physical Chemistry B* 103 (40): 8410–8426. https://doi.org/10.1021/jp9917648.

78 Chen, C.-D., Cheng, S.-F., Chau, L.-K., and Wang, C.R.C. (2007a). Sensing capability of the localized surface plasmon resonance of gold nanorods. *Biosensors and Bioelectronics* 22 (6): 926–932. https://doi.org/10.1016/j.bios.2006.03.021. http://www.sciencedirect.com/science/article/pii/S0956566306001825.

79 Haes, A.J., Zou, S., Schatz, G.C., and Van Duyne, R.P. (2004). Nanoscale optical biosensor: short range distance dependence of the localized surface plasmon resonance of noble metal nanoparticles. *Journal of Physical Chemistry B* 108 (22): 6961–6968. https://doi.org/10.1021/jp036261n.

80 Csaki, A., Stranik, O., and Fritzsche, W. (2018). Localized surface plasmon resonance based biosensing. *Expert Review of Molecular Diagnostics* 18 (3): 279–296. https://doi.org/10.1080/14737159.2018.1440208.

81 Ahijado-Guzmán, R., Prasad, J., Rosman, C. et al. (2014). Plasmonic nanosensors for simultaneous quantification of multiple protein–protein binding affinities. *Nano Letters* 14 (10): 5528–5532. https://doi.org/10.1021/nl501865p.

82 Ament, I., Prasad, J., Henkel, A. et al. (2012). Single unlabeled protein detection on individual Plasmonic nanoparticles. *Nano Letters* 12 (2): 1092–1095. https://doi.org/10.1021/nl204496g.

83 Dahlin, A.B. (2012). Size matters: problems and advantages associated with highly miniaturized sensors. *Sensors* 12 (3): 3018–3036. https://doi.org/10.3390/s120303018.

84 Long, Y.-T. and Jing, C. (2014). *Localized Surface Plasmon Resonance Based Nanobiosensors*. Springer.

85 Yildirim, D.U., Ghobadi, A., Soydan, M.C. et al. (2019b). Colorimetric and near- absolute polarization insensitive refractive-index sensing in all-dielectric guided-mode resonance based Metasurface. *The Journal of Physical Chemistry C* 123: 19125–19134. https://doi.org/10.1021/acs.jpcc.9b04748.

86 Zhao, J., Das, A., Zhang, X. et al. (2006). Resonance surface Plasmon spectroscopy: low molecular weight substrate binding to cytochrome P450. *Journal of the American Chemical Society* 128 (34): 11004–11005. https://doi.org/10.1021/ja0636082.

87 Reinhard, B.M., Sheikholeslami, S., Mastroianni, A. et al. (2007). Use of plasmon coupling to reveal the dynamics of DNA bending and cleavage by single EcoRV restriction enzymes. *Proceedings of the National Academy of Sciences* 104 (8): 2667–2672. https://doi.org/10.1073/pnas.0607826104.

88 Elf, J., Li, G.W., and Xie, X.S. (2007). Probing transcription factor dynamics at the single-molecule level in a living cell. *Science* 316 (5828): 1191–1194. https://doi.org/10.1126/science.1141967.

89 Huang, B., Wu, H., Bhaya, D. et al. (2007). Counting low-copy number proteins in a single cell. *Science* 315 (5808): 81–84. https://doi.org/10.1126/science.1133992.

90 Ji, X., Zheng, J., Xu, J. et al. (2005). (CdSe)ZnS quantum dots and organophosphorus hydrolase bioconjugate as biosensors for Detection of Paraoxon. *The Journal of Physical Chemistry B* 109 (9): 3793–3799. https://doi.org/10.1021/jp044928f.

91 Lang, M.J., Fordyce, P.M., Fordyce, A.M. et al. (2004). Simultaneous, coincident optical trapping and single- molecule fluorescence. *Nature Methods* 1 (2): 133–139. https://doi.org/10.1038/nmeth714.

92 Sako, Y., Minoghchi, S., and Yanagida, T. (2000). Single-molecule imaging of EGFR signalling on the surface of living cells. *Nature Cell Biology* 2 (3): 168–172. https://doi.org/10.1038/35004044.

93 Schuler, B., Lipman, E.A., and Eaton, W.A. (2002). Probing the free-energy surface for protein folding with single-molecule fluorescence spec-troscopy. *Nature* 419 (6908): 743–747. https://doi.org/10.1038/nature01060. http://www.ncbi.nlm.nih.gov/pubmed/12384704.

94 Lakowicz, J.R. (2006). Plasmonics in biology and Plasmon-controlled fluorescence. *Plasmonics* 1 (1): 5–33. https://doi.org/10.1007/s11468-005-9002-3.

95 Li, J.-F., Li, C.-Y., and Aroca, R.F. (2017a). Plasmon-enhanced fluo- rescence spectroscopy. *Chemical Society Reviews* 46 (13): 3962–3979. https://doi.org/10.1039/C7CS00169J.

96 Barnes, W.L. (1998). Fluorescence near interfaces: the role of photonic mode density. *Journal of Modern Optics* 45 (4): 661–699. https://doi.org/10.1080/09500349808230614.

97 Dulkeith, E., Ringler, M., Klar, T.A. et al. (2005). Gold nanoparticles quench fluorescence by phase induced radiative rate suppression. *Nano Letters* 5 (4): 585–589. https://doi.org/10.1021/nl0480969.

98 Kulakovich, O., Strekal, N., Yaroshevich, A. et al. (2002). Enhanced luminescence of CdSe quantum dots on gold colloids. *Nano Letters* 2 (12): 1449–1452. https://doi.org/10.1021/nl025819k.

99 Anger, P., Bharadwaj, P., and Novotny, L. (2006). Enhancement and quenching of single-molecule fluorescence. *Physical Review Letters* 96 (11): 113002. https://doi.org/10.1103/PhysRevLett.96.113002.

100 Kochuveedu, S.T. and Kim, D.H. (2014). Surface plasmon resonance mediated photoluminescence properties of nanostructured multicomponent fluorophore systems. *Nanoscale* 6 (10): 4966–4984. https://doi.org/10.1039/C4NR00241E.

101 Mishra, H., Mali, B.L., Karolin, J. et al. (2013). Experimental and theoretical study of the distance dependence of metal-enhanced fluorescence, phosphorescence and delayed fluorescence in a single system. *Physical Chemistry Chemical Physics* 15 (45): 19538–19544. https://doi.org/10.1039/C3CP50633A.

102 Zhou, Z., Li, T., Huang, H. et al. (2014). A dual amplification strategy for DNA detection combining bio-barcode assay and metal-enhanced fluorescence modality. *Chemical Communications* 50 (87): 13373–13376. https://doi.org/10.1039/C4CC05554C.

103 Akbay, N., Lakowicz, J.R., and Ray, K. (2012). Distance-dependent metal-enhanced intrinsic fluorescence of proteins using polyelectrolyte layer-by-layer assembly and aluminum nanoparticles. *The Journal of Physical Chemistry C* 116 (19): 10766–10773. https://doi.org/10.1021/jp2122714.

104 Sorokin, A.V., Zabolotskii, A.A., Pereverzev, N.V. et al. (2015). Metal-enhanced fluorescence of Pseudoisocyanine J-aggregates formed in layer-by-layer assembled films. *The Journal of Physical Chemistry C* 119 (5): 2743–2751. https://doi.org/10.1021/jp5102626.

105 Reineck, P., Gómez, D., Ng, S.H. et al. (2013). Distance and wavelength dependent quenching of molecular fluorescence by Au@SiO2 Core–Shell nanoparticles. *ACS Nano* 7 (8): 6636–6648. https://doi.org/10.1021/nn401775e.

106 Acuna, G.P., Möller, F.M., Holzmeister, P. et al. (2012). Fluorescence enhancement at docking sites of DNA-directed self- assembled Nanoantennas. *Science* 338 (6106): 506–510. https://doi.org/10.1126/science.1228638.

107 Bek, A., Jansen, R., Ringler, M. et al. (2008). Fluorescence enhancement in hot spots of AFM- designed gold nanoparticle sandwiches. *Nano Letters* 8 (2): 485–490. https://doi.org/10.1021/nl072602n.

108 Kinkhabwala, A., Yu, Z., Fan, S. et al. (2009). Large single-molecule fluorescence enhancements produced by a bowtie nanoantenna. *Nature Photonics* 3 (11): 654–657. https://doi.org/10.1038/nphoton.2009.187.

109 Punj, D., Mivelle, M., Moparthi, S.B. et al. (2013). A plasmonic 'antenna-in-box' platform for enhanced single-molecule analysis at micromolar concentrations. *Nature Nanotechnology* 8 (7): 512–516. https://doi.org/10.1038/nnano.2013.98.

110 Tam, F., Goodrich, G.P., Johnson, B.R., and Halas, N.J. (2007). Plasmonic enhancement of molecular fluorescence. *Nano Letters* 7 (2): 496–501. https://doi.org/10.1021/nl062901x.

111 Chen, Y., Munechika, K., and Ginger, D.S. (2007c). Dependence of fluorescence intensity on the spectral overlap between fluorophores and Plasmon resonant single silver nanoparticles. *Nano Letters* 7 (3): 690–696. https://doi.org/10.1021/nl062795z.

112 Thomas, M., Greffet, J.-J., Carminati, R., and Arias-Gonzalez, J.R. (2004). Single-molecule spontaneous emission close to absorbing nanostructures. *Applied Physics Letters* 85 (17): 3863–3865. https://doi.org/10.1063/1.1812592.

113 Nie, S. and Emory, S.R. (1997). Probing single molecules and single nanoparticles by surface-enhanced Raman scattering. *Science* 275 (5303): 1102–1106. https://doi.org/10.1126/science.275.5303.1102. https://science.sciencemag.org/content/275/5303/1102.

114 Sharma, B., Frontiera, R.R., Henry, A.-I. et al. (2012). SERS: materials, applications, and the future. *Materials Today* 15 (1): 16–25. https://doi.org/10.1016/S1369-7021(12)70017-2. http://www.sciencedirect.com/science/article/pii/S1369702112700172.

115 McFarland, A.D., Young, M.A., Dieringer, J.A., and Van Duyne, R.P. (2005). Wavelength-scanned surface-enhanced Raman excitation spectroscopy. *The Journal of Physical Chemistry B* 109 (22): 11279–11285. https://doi.org/10.1021/jp050508u.

116 Zhang, X., Young, M.A., Lyandres, O., and Van Duyne, R.P. (2005). Rapid detection of an anthrax biomarker by surface-enhanced Raman spectroscopy. *Journal of the American Chemical Society* 127 (12): 4484–4489. https://doi.org/10.1021/ja043623b.

117 Dieringer, J.A., Lettan, R.B., Scheidt, K.A., and VanDuyne, R.P. (2007). A frequency domain existence proof of single-molecule surface-enhanced Raman spectroscopy. *Journal of the American Chemical Society* 129 (51): 16249–16256. https://doi.org/10.1021/ja077243c.

118 Houck, M.M. and Siegel, J.A. (2009). *Fundamentals of Forensic Science*. Academic Press.

119 Chen, L.C., Ueda, T., Sagisaka, M. et al. (2007b). Visible laser desorption/ionization mass spectrometry using gold Nanorods. *The Journal of Physical Chemistry C* 111 (6): 2409–2415. https://doi.org/10.1021/jp065540i.

120 Brongersma, M.L., Halas, N.J., and Nordlander, P. (2015). Plasmon-induced hot carrier science and technology. *Nature Nanotechnology* 10 (1): 25–34. https://doi.org/10.1038/nnano.2014.311.

121 Li, Y., Cao, X., Zhan, L. et al. (2018). Hot electron transfer promotes ion production in plasmonic metal nanostructure assisted laser desorption ionization mass spectrometry. *Chemical Communications* 54 (77): 10905–10908. https://doi.org/10.1039/C8CC05793A.

122 Wei, X., Liu, Z., Jin, X. et al. (2017). Plasmonic nanoshells enhanced laser desorption/ionization mass spectrometry for detection of serum metabolites. *Analytica Chimica Acta* 950: 147–155. https://doi.org/10.1016/j.aca.2016.11.017. http://www.sciencedirect.com/science/article/pii/S0003267016313022.

123 Xu, X., Caswell, K.K., Tucker, E. et al. (2007). Size and shape separation of gold nanoparticles with preparative gel electrophoresis. *Journal of Chromatography A* 1167 (1): 35–41. https://doi.org/10.1016/j.chroma.2007.07.056. http://www.sciencedirect.com/science/article/pii/S0021967307012617.

124 Akchurin, G., Khlebtsov, B., Akchurin, G. et al. (2007). Gold nanoshell photomodification under a single-nanosecond laser pulse accompanied by color-shifting and bubble formation phenomena. *Nanotechnology* 19 (1): 15701. https://doi.org/10.1088/0957-4484/19/01/015701.

125 Lachaine, R., Boutopoulos, C., Lajoie, P.-Y. et al. (2016). *Nano Letters* 16 (5): 3187–3194. https://doi.org/10.1021/acs.nanolett.6b00562.

126 Malinsky, M.D., Kelly, K.L., Schatz, G.C., and Van Duyne, R.P. (2001). Chain length dependence and sensing capabilities of the localized surface Plasmon resonance of silver nanoparticles chemically modified with Alkanethiol self-assembled monolayers. *Journal of the American Chemical Society* 123 (7): 1471–1482. https://doi.org/10.1021/ja003312a.

127 Szunerits, S., Ghodbane, S., Niedziółka-Jönsson, J. et al. (2010). Development and characterization of a diamond-based localized surface Plasmon resonance interface. *The Journal of Physical Chemistry C* 114 (8): 3346–3353. https://doi.org/10.1021/jp908916y.

128 Galopin, E., Touahir, L., Niedziółka-Jönsson, J. et al. (2010). Amorphous silicon–carbon alloys for efficient localized surface plasmon resonance sensing. *Biosensors and Bioelectronics* 25 (5): 1199–1203. https://doi.org/10.1016/j.bios.2009.10.012. http://www.sciencedirect.com/science/article/pii/S0956566309005570.

129 Gao, S., Koshizaki, N., Koyama, E. et al. (2009). Innovative platform for transmission localized surface Plas- Mon transducers and its application in detecting heavy metal Pd(II). *Analytical Chemistry* 81 (18): 7703–7712. https://doi.org/10.1021/ac901137z.

130 Jeong, Y., Lee, C., Ito, E. et al. (2006). Time-dependent phase transition of self-assembled monolayers formed by Thioacetyl-terminated Tolanes on Au(111). *Japanese Journal of Applied Physics* 45 (7): 5906–5910. https://doi.org/10.1143/jjap.45.5906.

131 Zhang, Z.S., Wilson, O.M., Efremov, M.Y. et al. (2004). Heat capacity measurements of two-dimensional self-assembled hexadecanethiol monolayers on polycrystalline gold. *Applied Physics Letters* 84 (25): 5198–5200. https://doi.org/10.1063/1.1764938.

132 Lewis, M., Tarlov, M., and Carron, K. (1995). Study of the Photooxidation process of self-assembled Alkanethiol monolayers. *Journal of the American Chemical Society* 117 (37): 9574–9575. https://doi.org/10.1021/ja00142a030.

133 Zhang, X., Zhao, J., Whitney, A.V. et al. (2006). Ultrastable substrates for surface-enhanced Raman spectroscopy: Al2O3 Overlayers fabricated by atomic layer deposition yield improved anthrax biomarker detection. *Journal of the American Chemical Society* 128 (31): 10304–10309. https://doi.org/10.1021/ja0638760.

134 Unser, S., Bruzas, I., He, J., and Sagle, L. (2015). Localized surface plasmon resonance biosensing: current challenges and approaches. *Sensors* 15 (7): 15684–15716.

135 Huang, H., Liu, F., Huang, S. et al. (2012). Sensitive and simultaneous de- tection of different disease markers using multiplexed gold nanorods. *Analytica Chimica Acta* 755: 108–114. https://doi.org/10.1016/j.aca.2012.10.020. http://www.sciencedirect.com/science/article/pii/S0003267012014912.

136 Rosman, C., Prasad, J., Neiser, A. et al. (2013). Multiplexed Plasmon sensor for rapid label-free Analyte detection. *Nano Letters* 13 (7): 3243–3247. https://doi.org/10.1021/nl401354f.

137 Endo, T., Kerman, K., Nagatani, N. et al. (2006). Multiple label- free detection of antigen-antibody reaction using localized surface Plasmon resonance-based Core-Shell structured nanoparticle layer Nanochip. *Analytical Chemistry* 78 (18): 6465–6475. https://doi.org/10.1021/ac0608321.

138 Ben-Yoav, H., Dykstra, P.H., Bentley, W.E., and Ghodssi, R. (2012). A microfluidic-based electrochemical biochip for label-free diffusion-restricted DNA hybridization analysis. *Biosensors and Bioelectronics* 38 (1): 114–120. https://doi.org/10.1016/j.bios.2012.05.009. http://www.sciencedirect.com/science/article/pii/S0956566312002898.

139 Luo, Y., Yu, F., and Zare, R.N. (2008). Microfluidic device for immunoassays based on surface plasmon resonance imaging. *Lab on a Chip* 8 (5): 694–700. https://doi.org/10.1039/B800606G.

140 SadAbadi, H., Badilescu, S., Packirisamy, M., and Wüthrich, R. (2013). Integration of gold nanoparticles in PDMS microfluidics for lab-on-a-chip plasmonic biosensing of growth hormones. *Biosensors and Bioelectronics* 44: 77–84. https://doi.org/10.1016/j.bios.2013.01.016. http://www.sciencedirect.com/science/article/pii/S0956566313000183.

141 Chen, P., Chung, M.T., McHugh, W. et al. (2015). Multiplex serum cytokine immunoassay using Nanoplasmonic biosensor microarrays. *ACS Nano* 9 (4): 4173–4181. https://doi.org/10.1021/acsnano.5b00396.

142 N. Nanotechnology Home. LLC. Nanotechnology Products, Applications & Instruments. http://www.nanowerk.com/nanotechnology/nanomaterial/products{ }a.php.

143 Chen, S., Svedendahl, M., Käll, M. et al. (2009). Ultrahigh sensitivity made simple: nanoplasmonic label-free biosensing with an extremely low limit-of-detection for bacterial and cancer diagnostics. *Nanotechnology* 20 (43): 434015. https://doi.org/10.1088/0957-4484/20/43/434015.

144 Tong, L., Wei, H., Zhang, S., and Xu, H. (2014). Recent advances in Plasmonic sensors. *Sensors* 14 (5): 7959–7973. https://doi.org/10.3390/s140507959.

145 Reinhard, B.M., Siu, M., Agarwal, H. et al. (2005). Calibration of dynamic molecular rulers based on Plasmon coupling between gold nanoparticles. *Nano Letters* 5 (11): 2246–2252. https://doi.org/10.1021/nl051592s.

146 Sönnichsen, C., Reinhard, B.M., Liphardt, J., and Alivisatos, A.P. (2005). A molecular ruler based on plasmon coupling of single gold and silver nanoparticles. *Nature Biotechnology* 23 (6): 741–745. https://doi.org/10.1038/nbt1100.

147 Dutta, S., Saikia, K., and Nath, P. (2016). Smartphone based LSPR sensing platform for bio-conjugation detection and quantification. *RSC Advances* 6 (26): 21871–21880. https://doi.org/10.1039/C6RA01113F.

148 Lai, T.-S., Chang, T.-C., and Wang, S.-C. (2017). Gold nanoparticle- based colorimetric methods to determine protein contents in artificial urine using membrane micro-concentrators and mobile phone camera. *Sensors and Actuators B: Chemical* 239: 9–16. https://doi.org/10.1016/j.snb.2016.07.158. http://www.sciencedirect.com/science/article/pii/S0925400516312023.

149 Rampazzi, S., Danese, G., Leporati, F., and Marabelli, F. (2016). A localized surface Plasmon resonance-based portable instrument for quick on-site biomolecular detection. *IEEE Transactions on Instrumentation and Measurement* 65 (2): 317–327. https://doi.org/10.1109/TIM.2015.2465691.

150 Bingham, J.M., Anker, J.N., Kreno, L.E., and Van Duyne, R.P. (2010). Gas sensing with high-resolution localized surface Plasmon resonance spectroscopy. *Journal of the American Chemical Society* 132 (49): 17358–17359. https://doi.org/10.1021/ja1074272.

151 Li, Y., Jing, C., Zhang, L., and Long, Y.-T. (2012). Resonance scattering particles as biological nanosensors in vitro and in vivo. *Chemical Society Reviews* 41 (2): 632–642. https://doi.org/10.1039/C1CS15143F.

152 Lopez, G.A., Estevez, M.-C., Soler, M., and Lechuga, L.M. (2017). Recent advances in nanoplasmonic biosensors: applications and lab-on-a-chip integration. *Nanophotonics* 6 (1): 71–81.

153 Niu, L., Zhang, N., Liu, H. et al. (2015). Integrating plasmonic diagnostics and microfluidics. *Biomicrofluidics* 9 (5): 52611. https://doi.org/10.1063/1.4929579.

3

Highly Sensitive and Selective Plasmonic Sensing Platforms

Yeşeren Saylan and Adil Denizli

Department of Chemistry, Hacettepe University, Ankara, Turkey

3.1 Introduction

The growing need for monitoring highly selective and sensitive molecules requires the development of precise, fast, and easy-to-use innovative analytical devices for the medical and environmental applications [1]. Quantitative measurement of samples is usually carried out by conventional methods such as spectroscopic and chromatographic techniques for separation, extraction, and detection [2, 3]. These methods are sensitive but they need complicated and expensive equipment and also require expert staff to operate them with multistep sample preparation. It is hard to achieve real-time sensing of molecules by these labor-intensive and time-consuming methods [4–6]. Novel platforms with superior sensing capabilities and increased spatial localization are needed urgently. The development of highly sensitive and selective and also powerful, automated, and cost-effective sensing platforms for the fast and real-time analysis of molecules has been striving by researchers [7–10]. They combine chemistry, biology, physics, electronics, and nanotechnology to keep developments in risk management due to their distinctive characteristics sensing capabilities [11–15]. Sensors are analytical devices that integrate a sensing element with a physical transducer such as optical [16], electrochemical [17], piezoelectric [18], thermal [19], and magnetic [20]. The interaction occurs between a target molecule and a recognition molecule that translates into a signal. Despite their use in various fields is still in the early stages, sensors have excellent applications in the fields of the environment [21], food [22], drug [23], medical [24], and diagnosis [25]. Unfortunately, dispersed molecules that are diffusing freely into the solution are far from having sensitive surfaces in practical applications. In this regard, plasmonic sensors are favorable platforms that have different features such as sensitivity, specificity, inexpensive, label-free, and quantitative detection [26–30].

This chapter aims to summarize the most recent advances in the highly sensitive and selective plasmonic sensors. Following the explanation of reasons for highly sensitive and selective plasmonic sensing platforms (Section 3.2), a basic introduction to the principle

Plasmonic Sensors and their Applications, First Edition. Edited by Adil Denizli.

of plasmonic sensors (Section 3.3) is given briefly and a comprehensive overview of reported plasmonic sensors performances and characteristics are discussed for medical (Section 3.4.1) and environmental (Section 3.4.2) applications. Then, a discussion about conclusion remarks is added to the end of the chapter (Section 3.5).

3.2 What Is Highly Sensitive (Ultrasensitive)?

"Ultrasensitive" term is progressively utilized more as a descriptor concerning sensors [31]. A Web of Science search shows that the first publication in which the use of the "ultrasensitive" term is in 1971 [32]. Then, the first publication entitles "ultrasensitive" is in 1996 [33]. After that, the utilization of this term as an adjective has raised rapidly. All meanings come down to the "sensitive" meaning that explains several things to a chemist toward a biologist, and so, confusion appears in interdisciplinary fields. The International Union of Pure and Applied Chemistry definition of sensitivity is a slope of the calibration curve for chemists, and hence the ability of an analytical method to distinguish between small differences in concentration. But it is a bit different for biochemists that adjust how strong a stimulant has to be before a system acts to it the smaller the stimulant needed to elicit a reaction [34]. It is an agreement between the chemical and biological meanings that confirms the link between the meaning and sensitivity as it refers to the background signal.

3.3 Plasmonic Sensing Platforms

A sensor has three parts: (I) receptor, (II) transducer, and (III) detector with an output. In principle, the target is in close contact with the transducer of the sensor that translates the specific recognition of the target into an observable and/or quantifiable signal [35, 36]. They have several important abilities such as high sensitivity and specificity, high performance, fast response, relatively compact size, easy operation, portability, and real-time observation [37]. The efforts in the sensors field have enlarged quickly, and it has already indicated a wide range of applications in medical [38], drug [39], environmental [40], food [41], and security [42]. Today, scientists purpose to increase the sensitivity and selectivity of the existed techniques and methods on the sensor fabrication quality, improving advanced surface chemistry, raising the affinity between ligands and targets, and using nanomaterials for signal amplification studies [43].

Plasmonic sensors center on the change in the optic property measurements of the transducer surfaces when the target and recognition element react [44]. These sensors can be subclasses as direct and indirect sensors. In the direct plasmonic sensors, signal generation bases on the complex formation on the transducer surface. The indirect plasmonic sensors are generally fabricated with several labels (fluorophores or chromophores) to recognize the binding reactions and increase the signals [45, 46]. There are various plasmonic sensors such as evanescent wave fibers, optrode-based fibers, time-resolved fluorescence, resonant mirror, interferometric, and surface plasmon resonance in the literature and market. Their sensing ability is highly versatile and they also sense a lot of types of molecules [47–50].

3.4 Recent Applications

3.4.1 Medical Applications

The applications of plasmonic sensing platforms are growing rapidly to diagnose diseases such as inflammatory diseases, cardiovascular and neurodegenerative diseases, viral infections, cancer, diabetes mellitus, urinary tract infection, and so on due to their simplicity, specificity, and sensitivity. Potential of monitoring in plasmonic sensors varied in a broad range of target biomolecules from proteins, enzymes, nucleic acids, antibodies to viruses, pathogens, and cells [51–55].

Such an example, Inci et al. presented a plasmonic sensing platform that integrates with a microfluidic chip to detect hemoglobin (Figure 3.1). They widely performed theoretical simulations and kinetic calculations to test the performance of the platform and then provided simple configuration for user-interface, label-free detection, short assay-time, facile sampling, and inexpensive chips. They also mentioned that this plasmonic sensing platform can accelerate the spreading of portable systems for medical applications [56].

Lee et al. produced binary-nanoparticle-based carbon nanotubes through a two-step method and carried out as a plasmonic sensing platform (Figure 3.2). They used gold/iron-oxide magnetic nanoparticle-based carbon nanotubes for influenza and norovirus DNA monitoring. They measured DNA hybridization between a virus and probe DNA to sense a conductivity change of the gold/iron-oxide magnetic nanoparticle-based carbon nanotubes using 1 pM–10 nM concentrations and figured out the limits of detection for influenza virus and norovirus as 8.4 and 8.8 pM. They also confirmed the selectivity of using other mismatched DNA sequences [57].

He et al. described a plasmonic sensing platform for folic acid detection utilizing graphene-based chips. They exploited the properties of graphene to fabricate a highly sensitive and selective plasmonic sensing platform in serum. They also tested a post-adsorption of human and bovine serum albumin mixtures onto the plasmonic sensing platform resulted in a highly anti-fouling interface while keeping the abilities for biomarkers. They reported that the platform let femtomolar sensing that promising and well-adapted for clinical analysis [58].

Lee et al. showed a highly selective and sensitive plasmonic sensing platform for carbohydrate discrimination employing nanoantenna-based chips which work in the terahertz frequency range (0.5–2.5 THz). They reported that the platform sensing several carbohydrate types. They also verified that the performance of the nanoantenna-based plasmonic sensing platform by both transmittance terahertz spectra and images and applied recognition of several carbohydrates to perform in market beverages [59].

Yang et al. presented a hydrogen sulfide sensing platform utilizing plasmonic nanoprobes. They employed a silver etching in the silver/gold core-shell nanoprisms that participate with surface plasmon resonance signal shift. They first coated layers of gold onto silver nanoprisms and then introduced hydrogen sulfide to convert the silver core to silver sulfide from lateral walls. Furthermore, they located the plasmonic peak in the near-infrared region that performs these plasmonic nanoprobes more consulting for hydrogen sulfide detection in real samples [60].

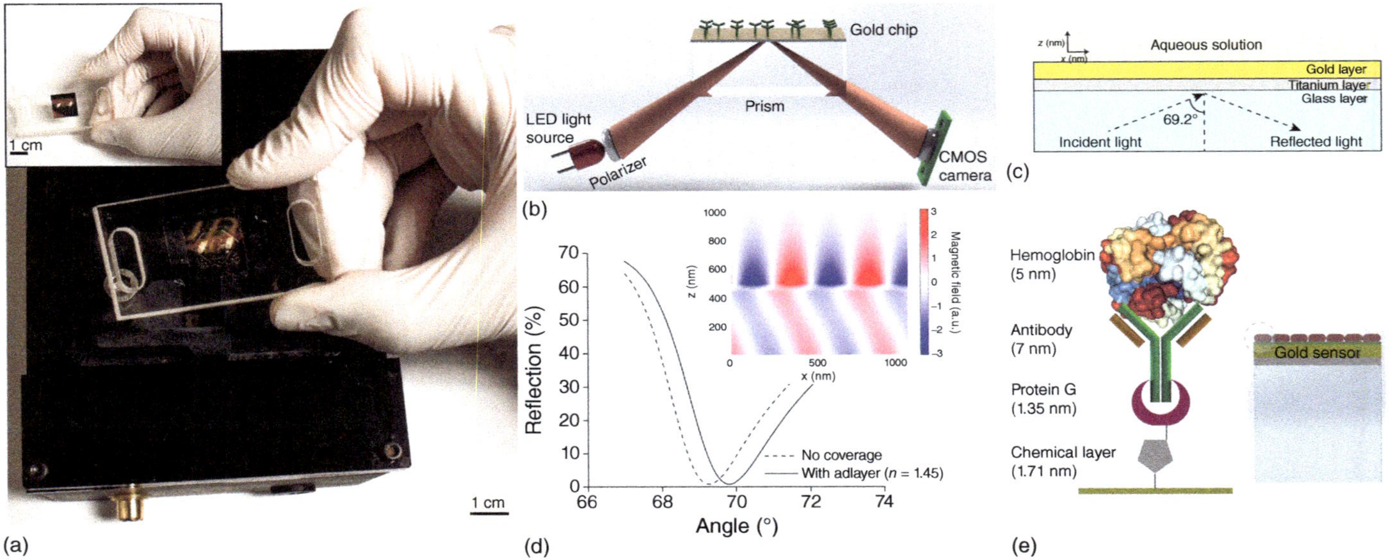

Figure 3.1 The microfluidic chip consists of poly(methyl methacrylate) and a gold chip with double-sided adhesives (a). The light-emitting diode is used to enlighten a lens that focuses the light onto a prism and the reflected light is captured on the sensor surface, and the captured image is transported to a computer (b). Gold chips are fabricated on a glass wafer by depositing titanium and gold layers (c). The binding of a homogenous adlayer generates the resonance angle change, and magnetic field distribution around the plasmonic surface is envisioned when the sensor surface is enlightened by a plane-wave from the bottom (d). The sensor surface is modified with (i) chemicals; (ii) protein G; and (iii) antibody to detect hemoglobin (e). Source: Reprinted with permission from Inci et al. [56]. © 2020, Elsevier.

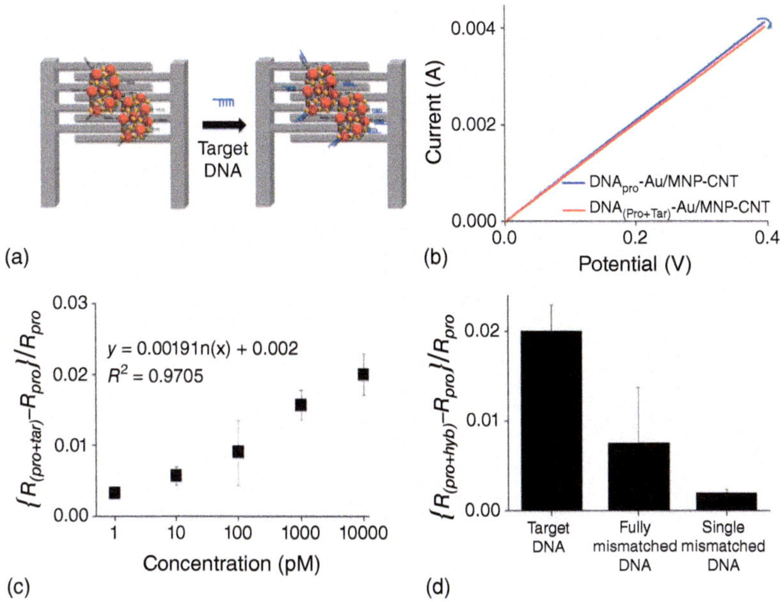

Figure 3.2 Scheme of DNA detection with hybrid materials (a). Curve change by hybridization (b). Resistance difference for the influenza virus DNA (c). Selectivity test with other types of DNA (d). Source: Reprinted with permission from Lee et al. [57]. © 2018, Elsevier.

Dong et al. designed a microRNA sensing platform employing DNA-bio-bar-code amplification. They modified this sensing platform a locked nucleic acid DNA probe to increase hybridization efficiency while preparing a signal reported molecular beacon with an endonuclease recognition site. They also reported that in the presence of microRNA, the magnetic and gold nanoprobes can hybridize with microRNA to form a sandwich structure. Following the optimization of experimental conditions, the exponential curve showed a linear range from 0.3 pM to 3 aM with a very low limit of detection (52.5 zM) [61].

Stebunov et al. described a plasmonic sensing platform depended on graphene-oxide linking layers for streptavidin detection (Figure 3.3). They claimed that this plasmonic sensing platform has higher sensitivity than the commercial sensors and also it is selective for nonspecific interaction and can be employed several times. They considered that the importance of this platform for highly sensitive plasmonic sensing applications [62].

Çetin et al. showed a gold nanohole platform for the tracking of antibodies easily (Figure 3.4). They mentioned that the nanohole platform demonstrated the plasmonic modes with well-preserved amplitudes for reliable spectral sensing when compared to conventional configurations. They also showed this platform is more sensitive to surface condition changes. They obtained a detection limit as low as 2×10^{-5} RIU for using wavelength shifts in a wide spectral window instead of sensing only the plasmonic resonance wavelength, and they also successfully demonstrated the real-time monitoring of interactions even at sub ng ml^{-1} levels [63].

Chen et al. stated a highly sensitive and selective plasmonic sensing platform for vascular endothelial growth factor detection using two DNA aptamers that employed as capture

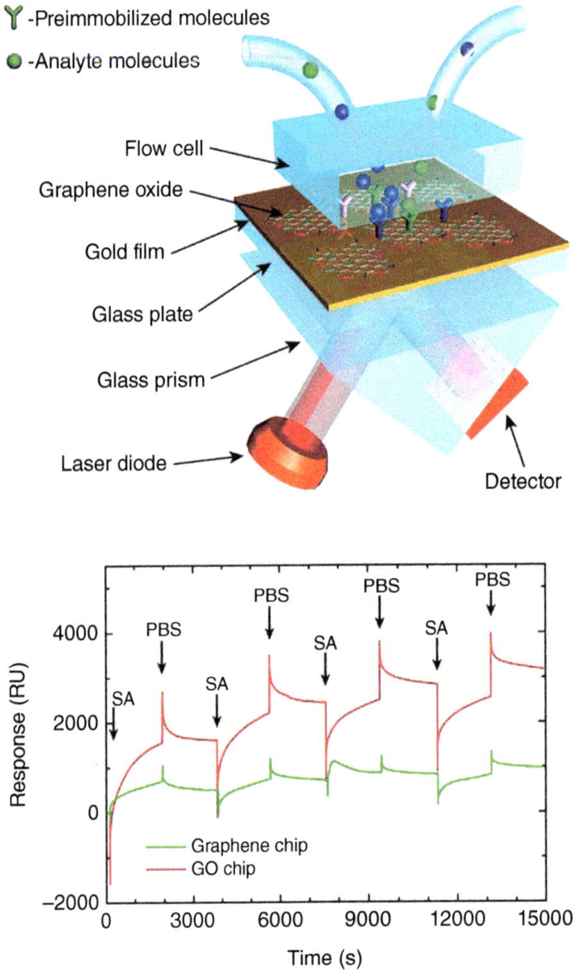

Figure 3.3 Scheme of the plasmonic sensing platform with the graphene-oxide-linking layer for streptavidin (SA) detection while using the amino-coupling that covered by monolayer graphene (green) and airbrushed graphene-oxide film (red). Source: Reprinted with permission from Stebunov et al. [62]. © 2015, American Chemical Society.

and detection probes (Figure 3.5). They loaded 3′-NH_2 immobilized aptamer and 3′-NH_2 modified primer DNA through amidation onto the monitoring layer for the amplification process to increase the signal using carboxyl-coated polystyrene microspheres. Moreover, they assigned a linear range from 100 pg ml^{-1} to 1 μg ml^{-1} with a low detection limit (100 pg ml^{-1}) [64].

Jun et al. fabricated a highly sensitive and selective electrolyte field-effect transistor-based sensing platform to determine platelet-derived growth factor applying aptamer-conjugated carboxylic polypyrrole-coated metal oxide-decorated carbon nanofibers (Figure 3.6). At first, they polymerized the carboxylic pyrrole monomer on metal oxide-decorated carbon nanofiber surfaces with no treatment that integrated with the aptamer

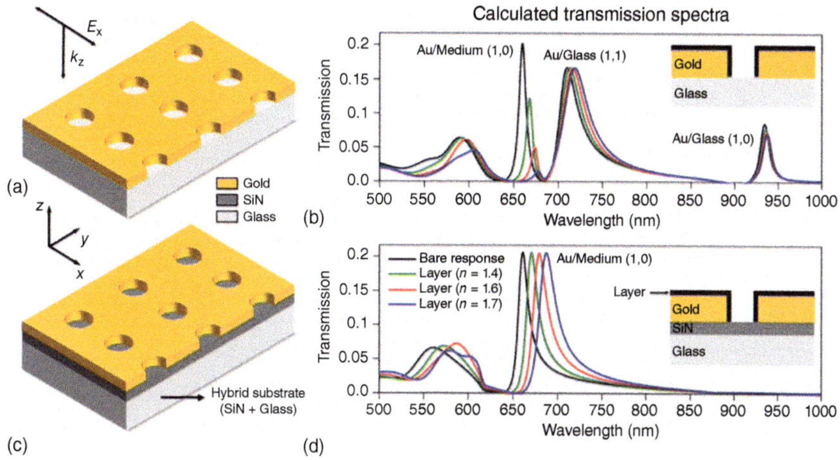

Figure 3.4 Scheme of the nanohole platform on glass (a) and hybrid (c) substrates and calculated transmission spectra of the nanohole platform on glass (b) and hybrid (d) substrates. Source: Reprinted with permission from Cetin et al. [63].© 2015, American Chemical Society.

and then immobilized on the interdigitated array substrate by covalent anchoring to obtain a transducer. They reported that this aptamer conjugated sensing platform is highly sensitive (5 fM) and selective for isoforms of platelet-derived growth factors [65].

Farhadi et al. introduced an interaction between bimetallic nanoparticles and cysteine for highly sensitive and selective cysteine sensing. They observed that cysteine induced the aggregation of bimetallic nanoparticles that resulted in a change in the yellowish-brown color of the colloid to green in the presence of salt. They also studied the sensitivity and selectivity of the platform toward the other amino acids and reported that this simple plasmonic sensing platform is very rapid and also highly sensitive and selective [66].

3.4.2 Environmental Applications

Environmental contamination has serious negative impacts on nature and humans. There is a quickly growing awareness of the world for the protection and sustainable use of natural resources for future human well-being. Several analytical techniques have been improved for the most used contaminants but in line with future requirements, real-time sensing methods or rapid screening are highly needed. Furthermore, novel, cost-effective, highly sensitive, and selective methods are developed for monitoring of natural sources and on-field sample screening to meet future analysis needs. There is expanding attention to fabricating plasmonic sensing platforms is reflected as the perfect basis for such field-usable devices for different contaminant detection [67–70].

Liu et al. established a highly sensitive and selective aptamer-based surface-enhanced Raman spectroscopic sensing platform for detecting 17-estradiol and carried out in the environmental sample solutions (Figure 3.7). They performed the kinetic studies in a broad linear range from 0.1 pM to 10 nM with a low detection limit (0.05 pM). Additionally, they showed that the high selectivity of the platform for 17-estradiol, where the intensity is greater than other interfering substances. They further investigated the affinity of the

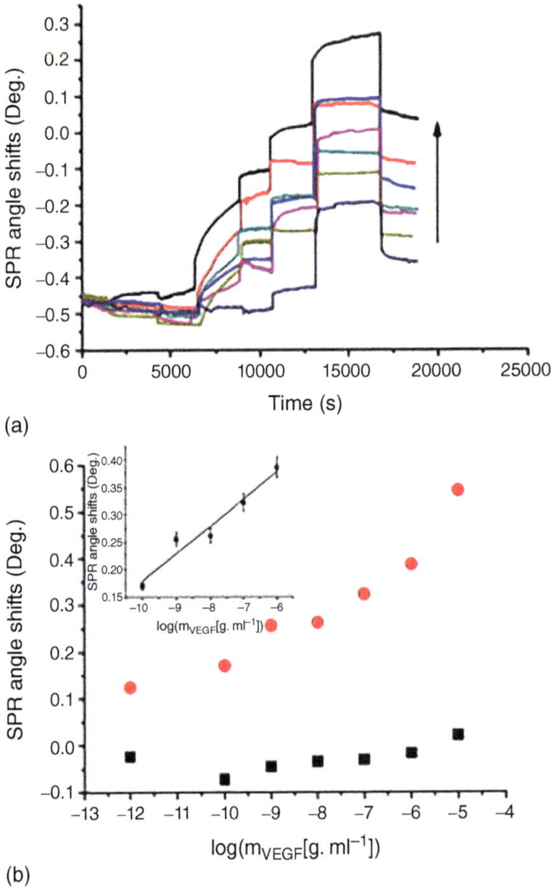

Figure 3.5 Concentration dependence of vascular endothelial growth factor detection (a) and the contrast of with or without the plasmonic signal amplification (b). Source: Reprinted with permission from Chen et al. [64]. © 2014, Elsevier.

17-estradiol-aptamer toward 17-estradiol employing UV–vis absorption analysis that 17-estradiol-aptamer has no binding affinity to other interferents. They also tested 17-estradiol detection performance in the real water that collected from the sewer and nearby river of the local obstetric hospitals which exhibited superior sensitivity [71].

Saylan et al. fabricated imprinted nanofilms and integrated them with surface plasmon resonance sensor surfaces for multiple pesticides (cyanazine, simazine, and atrazine) detection. They provided the measurements on the plasmonic sensor in a wide range from 0.10 to 6.64 nM, as well as denote a limit of detection values of 0.095, 0.031, and 0.091 nM for cyanazine, simazine, and atrazine, respectively, and also performed the cross-selectivity tests [72].

Zhang et al. constructed a turn-on fluorescent sensing platform using C-rich ssDNA-templated silver nanoclusters for highly sensitive and selective detection of lead ion (Figure 3.8). The fluorescence intensity of C-rich ssDNA-templated silver nanoclusters increases importantly in the presence of lead ion due to the interaction between the ion

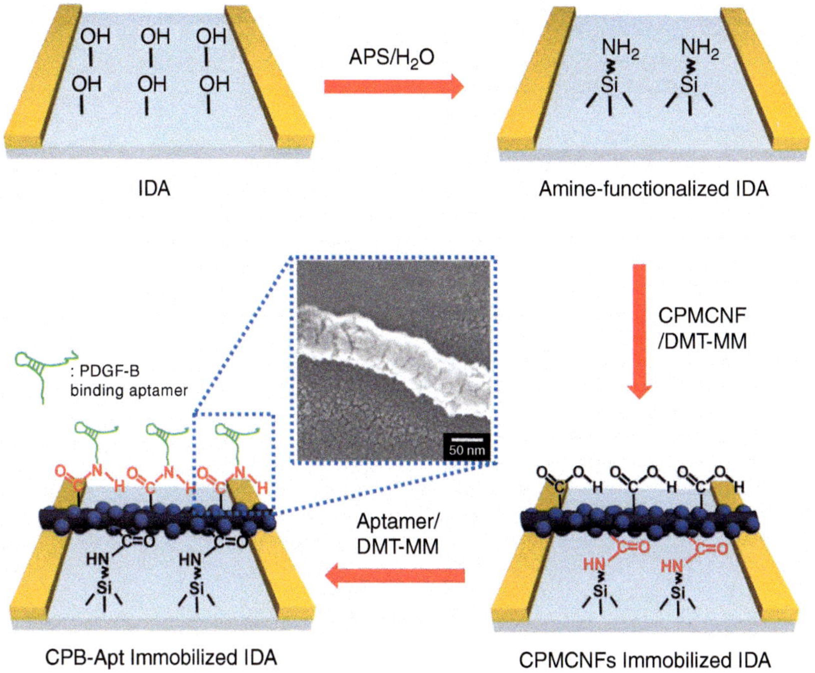

Figure 3.6 Scheme of the immobilization of aptamer transducers on the electrode substrate. Source: Reprinted with permission from Jun et al. [65]. © 2014, American Chemical Society.

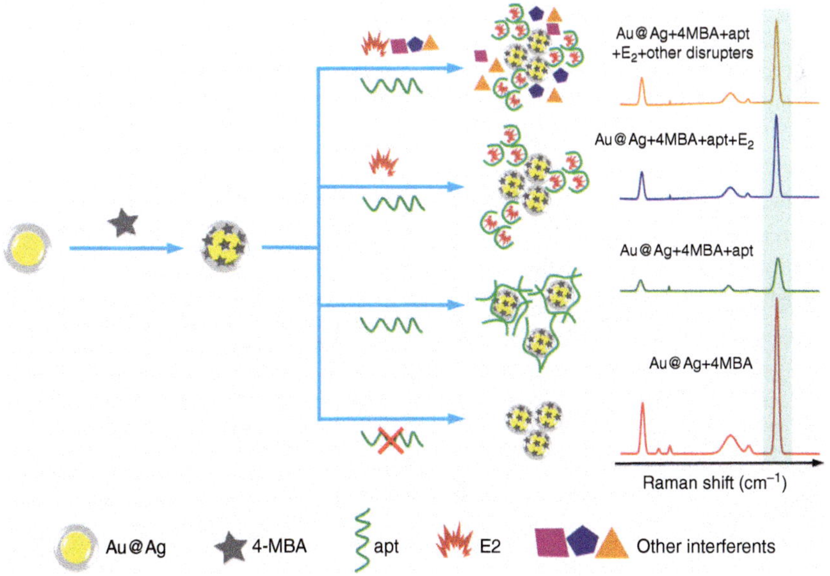

Figure 3.7 Scheme of 17-estradiol detection under different conditions. Source: Reprinted with permission from Liu et al. [71]. © 2018, Elsevier.

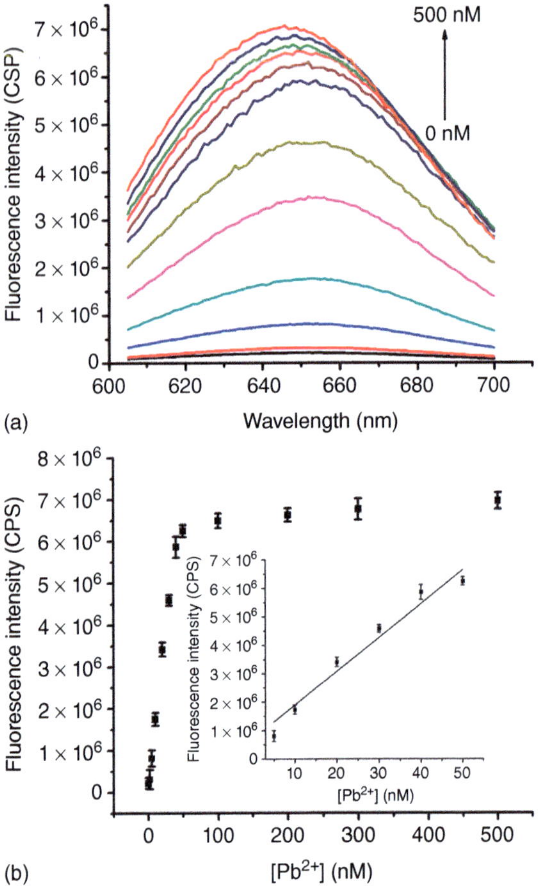

Figure 3.8 Changes of fluorescence emission spectra upon the addition of lead ion (a) and changes of maximum intensity as a function of ion concentration (b). Source: Reprinted with permission from Zhang et al. [73]. © 2018, Elsevier.

and its aptamer. They detected lead ion as low as 3.0 nM within a linear range (5–50 nM) using a fluorescent sensor and demonstrated the reliability of the sensor application employing in real water samples [73].

Ye et al. produced gold nanoparticles that have powerful resonance Rayleigh and surface-enhanced resonance Raman scattering effects and used Victoria blue B and rhodamine S. They performed kinetic studies and observed that the increased intensity responses linearly with the concentration of nanoparticles over 0.038–76, 19–285, and 3.8–456 ng ml^{-1}, respectively. They also developed a nanocatalysis surface plasmon resonance-scattering sensing platform for lead ion detection. They finally observed that increased resonance Rayleigh scattering intensities linearly with ion concentration over 16.7–666.7 nmol l^{-1} [74].

Zhuang et al. reported a highly sensitive and selective surface-enhanced Raman scattering-based sensing platform for zinc ion detection. They tested the kinetic performance of the plasmonic sensing platform in a range of 10^{-6}–10^{-16} M with a very low limit of detection 10^{-14} M. The platform maintained its high sensitivity with high selectivity and rapid response in diluted tap-water and opens up possibilities for the sensitive sensing of ion in environmental and also medical applications using surface-enhanced Raman scattering spectroscopy [75].

Chen et al. originated a colorimetric sensing platform for mercury ion monitoring that depended on the mercury-induced deprotection and 1-dodecanethiol-capped silver nanoprisms transition upon the presence of iodides. They obtained UV–vis absorption spectra for silver atoms consuming and moving from the surface that accompany the changes in the particle morphology. With increasing concentrations of mercury ion from 10 to 500 nM, the surface plasma resonance spectral band of silver nanoprisms emerged as a blue shift and exhibited a good linear relationship, and the limit of detection was 3.3 nM (Figure 3.9). They also carried out in real water experiments with high recovery (92%) [76].

Pienpinijtham et al. proposed a plasmonic sensing platform for iodide and thiocyanate determination by surface-enhanced Raman scattering of starch-reduced gold

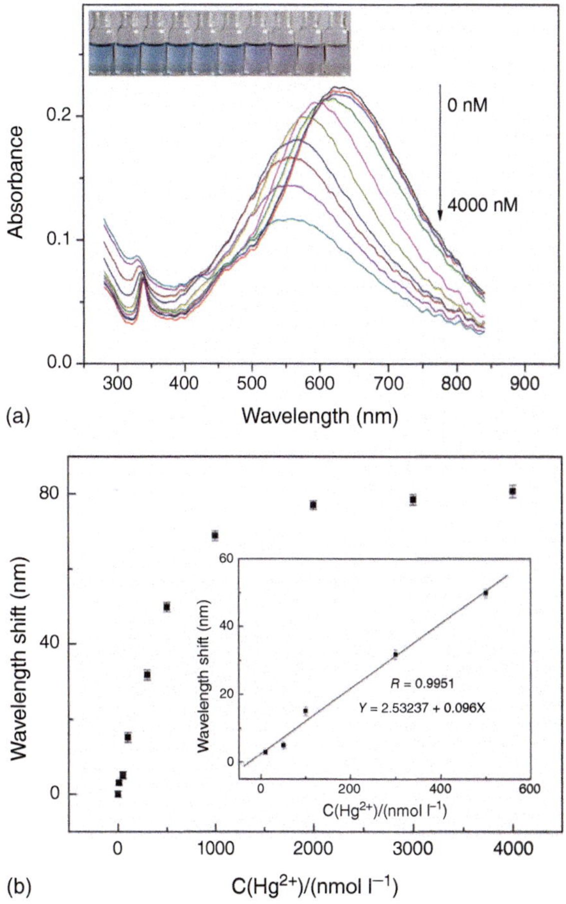

Figure 3.9 UV–vis absorption spectra of the silver nanoprisms following the different concentrations of mercury ion addition (a) and wavelength shifts with different concentrations of ion (b). Source: Reprinted with permission from Chen et al. [76]. © 2013, American Chemical Society.

nanoparticles (Figure 3.10). Following the optimization experiments, they observed that have an intrinsic Raman peak at $2125\,cm^{-1}$ due to the C≡C stretching and thiocyanate also strongly adsorbs on a surface, and a new peak appears at $2100\,cm^{-1}$, connected to the C≡N stretching in a surface-enhanced Raman scattering spectrum of nanoparticles. They used the two peaks to assign the iodide and thiocyanate concentrations separately. They calculated the detection limit as 0.01 μM with a range of 0.01–2.0 μM for iodide and as 0.05 μM with a range of 0.05–50 μM for thiocyanate. They reported that this plasmonic sensing platform is highly selective for iodide and thiocyanate ions without interference from other ions [77].

Yin et al. prepared a surface-enhanced Raman scattering sensing platform for cadmium ion detection by using the advantage of interparticle plasmonic coupling that produced cadmium ion-selective nanoparticle self-aggregation. They encoded the surface-enhanced Raman scattering-active nanoparticles with a Raman-active dye and grafted a layer of ion-chelating polymer brush coating on the nanoparticle. They optimized the surface-enhanced Raman scattering nanoparticles to keep spectrally silent when staying as single particles and added cadmium ion to leads the interparticle self-aggregation and directly turns on the signal with high enhancement [78].

Qu et al. reported a nanoplasmonic probe to detect trinitrotoluene molecules using the quenching of the Rayleigh scattering spectrum of the probe. In this regard, the detection mechanism is obtained between the Meisenheimer complex that formed between cysteine and trinitrotoluene and gold nanoparticles. They mentioned that high selectivity is achieved due to obtaining a donor–acceptor complex that overlaps extensively with the Rayleigh scattering peak of the gold nanoparticles and high sensitivity is realized due to the quantitative information of spectral resonant quenching in the Rayleigh scattering spectrum [79].

Dasary et al. also demonstrated a para-amino thiophenol modified gold nanoparticle-based dynamic light scattering sensing platform for trinitrotoluene monitoring. The formation of powerful donor–acceptor interaction between trinitrotoluene and para-amino thiophenol occurs and the gold nanoparticles carry aggregation in the presence of trinitrotoluene that differs the dynamic light scattering intensity. They discussed a detailed

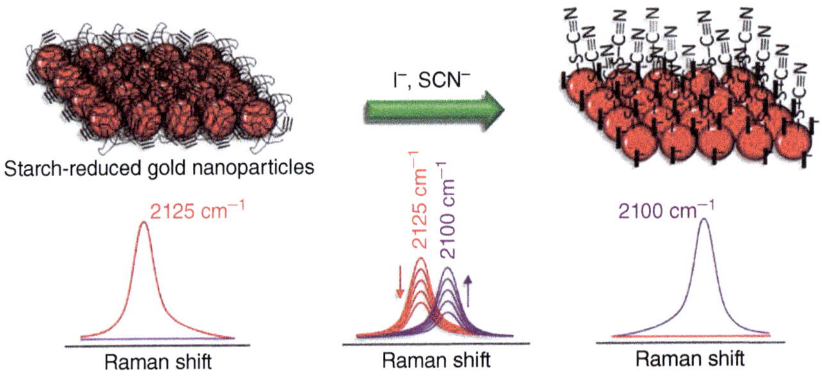

Figure 3.10 Scheme of Raman shift of the starch-reduced gold nanoparticles. Source: Reprinted with permission from Pienpinijtham et al. [77]. © 2011, American Chemical Society.

interaction for significant change and their experimental results showed that trinitrotoluene can be detected quickly and accurately without any dye tagging in 100 pM range with unique distinction toward other nitro compounds [80].

3.5 Conclusion Remarks

It is comprehensibly exemplified that plasmonic sensing platforms act a crucial role in medical and environmental applications compared to conventional methods. Plasmonic sensing platforms play as a first detection filter due to having various charming properties including;

- They propose powerful, stable, and cost-effective methods to seriously detect, sense, and monitor the targets.
- They include a significant potential for miniaturization owing to their minute size.
- They can be combined with microfluidics to assign a lab-on-a-chip system.

However, plasmonic sensing platforms act a necessary role in medical and environmental applications, future remarks and investments should be reflected to prohibit the restrictions of this system. Suggestions that can be imagined in the next studies to increase the role of sensing platforms are the construction of sensors suitable to work in different environments, reduction of the cost of signal detection, and optimization of the sensitivity, detection limit, and rapidity of reporting.

References

1 Thaler, M. and Luppa, P.B. (2019). *Analytical and Bioanalytical Chemistry* 411: 7623–7635.
2 Yuste, A.R. and Carrasco, S. (2019). *Polymers* 11: 1173.
3 Pirzada, M. and Altintas, Z. (2019). *Sensors* 19: 5311.
4 He, Y., Yang, T., Mo, H. et al. (2019). *Instrumentation Science & Technology* 47 (6): 581–596.
5 Katsarou, K., Bardani, E., Kallemi, P., and Kalantidis, K. (2019). *BioEssays* 41: 1900049.
6 Bellan, L.M., Wu, D., and Langer, R.S. (2011). *WIREs Nanomedicine and Nanobiotechnology* 3: 229–246.
7 Regasa, M.B., Soreta, T.R., Femi, O.E., and Ramamurthy, P.C. (2020). *ACS Omega* 5: 4090–4099.
8 He, C., Ledezma, U.H., Gurnani, P. et al. (2020). *Analyst* 145: 4504–4511.
9 Kim, J.B., Lee, S.Y., Min, N.G. et al. (2020). *Advanced Materials* 32: 26, 2001384.
10 Kim, J., Kim, S., Ahn, J. et al. (2020). *Advanced Materials* 32: 26, 2001360.
11 Luo, L., Zhang, F., Chen, C., and Cai, C. (2019). *Analytical Chemistry* 91 (24): 15748–15756.
12 Nawaz, T., Ahmad, M., Yu, J. et al. (2020). *New Journal of Chemistry* 44: 6992.
13 Luo, Y., Liu, F., Li, E. et al. (2020). *Biosensors and Bioelectronics* 148: 111832.
14 Tran, T.T., Clark, K., Ma, W., and Mulchandani, A. (2020). *Biosensors and Bioelectronics* 147: 111766.
15 Su, C., Li, Z., Zhang, D. et al. (2020). *Biosensors and Bioelectronics* 148: 111819.

16 Dibekkaya, H., Saylan, Y., Yılmaz, F. et al. (2016). *Journal of Macromolecular Science, Part A* 53 (9): 585–594.
17 Rebelo, P., Pacheco, J.G., Cordeiro, M.N.D.S. et al. (2020). *Analytical Methods* 12: 1486.
18 Akgönüllü, S., Battal, D., Yalcin, M.S. et al. (2020). *Microchemical Journal* 153: 104454.
19 Wang, Z., Jinlong, L., An, Z. et al. (2019). *Process Biochemistry* 79: 57–64.
20 Huber, S., Minac, C., Staat, C. et al. (2019). *Biosensors and Bioelectronics* 126: 240–248.
21 Supraja, P., Tripathy, S., Vanjari, S.R.K. et al. (2019). *Biosensors and Bioelectronics* 141: 111441.
22 Lan, L., Yao, Y., Ping, J., and Ying, Y. (2017). *ACS Applied Materials and Interfaces* 9 (28): 23287–23301.
23 Zaidi, S.A. (2020). *Materials Science for Energy Technologies* 3: 72–77.
24 Mansuriya, B.D. and Altintas, Z. (2020). *Materials* 13: 96.
25 Cui, F., Zhou, Z., and Zhou, H.S. (2020). *Sensors* 20: 996.
26 Andryukov, B.G., Besednova, N.N., Romashko, R.V. et al. (2020). *Biosensors* 10: 11.
27 Sharifi, M., Attar, F., Saboury, A.A. et al. (2019). *Journal of Controlled Release* 311–312: 170–189.
28 Inci, F., Karaaslan, M.G., Kojouri, A.M. et al. (2020). *Applied Materials Today* 20: 100709.
29 Asghar, W., Yuksekkaya, M., Shafiee, H. et al. (2016). *Scientific Reports* 6 (1): 1–10.
30 Saylan, Y., Yılmaz, F., Derazshamshir, A. et al. (2017). *Journal of Molecular Recognition* 30 (9): e2631.
31 Gooding, J.J. (2019). *ACS Sensors* 4: 528–528.
32 Zimmerman, J.E. and Frederick, N.V. (1971). *Applied Physics Letters* 19: 16–19.
33 Ellis, D.L., Zakin, M.R., Bernstein, L.S., and Rubner, M.F. (1996). *Analytical Chemistry* 68: 817–822.
34 Wu, Y., Tilley, R.D., and Gooding, J.J. (2019). *Journal of the American Chemical Society* 141: 1162–1170.
35 Saylan, Y., Erdem, Ö., Inci, F., and Denizli, A. (2020). *Biomimetics* 5 (2): 20.
36 Saylan, Y., Yilmaz, F., Özgür, E. et al. (2017). *Sensors* 17 (4): 898.
37 Saylan, Y., Akgönüllü, S., Yavuz, H. et al. (2019). *Sensors* 19 (6): 1279.
38 Regan, B., Boyle, F., O'Kennedy, R., and Collins, D. (2019). *Sensors* 19: 3485.
39 Bodoki, A.E., Iacob, B.C., and Bodoki, E. (2019). *Polymers* 11: 2085.
40 Naseri, M., Mohammadniaei, M., Sun, Y., and Ashley, J. (2020). *Chemosensors* 8: 32.
41 Xiao, F., Li, H., Yan, X. et al. (2020). *Analytica Chimica Acta* 1103: 84–96.
42 Kirsch, J., Siltanen, C., Zhou, Q. et al. (2013). *Chemical Society Reviews* 42: 8733–8768.
43 Assavapanumat, S., Ketkaew, M., Kuhn, A., and Wattanakit, C. (2019). *Journal of the American Chemical Society* 141: 18870–18876.
44 Saylan, Y., Erdem, Ö., Cihangir, N., and Denizli, A. (2019). *Chemosensors* 7 (3): 33.
45 Saylan, Y., Erdem, Ö., Ünal, S., and Denizli, A. (2019). *Biosensors* 9 (2): 65.
46 Khan, I.M., Niazi, S., Khan, M.K.I. et al. (2019). *Trends in Analytical Chemistry* 119: 115637.
47 Akgönüllü, S., Yavuz, H., and Denizli, A. (2020). *Talanta* 219: 121219.
48 Özgür, E., Saylan, Y., Bereli, N. et al. (2020). *Journal of Biomaterials Science Polymer Edition* 31 (9): 1–12.
49 Safran, V., Göktürk, I., Derazshamshir, A. et al. (2019). *Microchemical Journal* 148: 141–150.

50 Çalışır, M., Bakhshpour, M., Yavuz, H., and Denizli, A. (2020). *Sensors and Actuators B: Chemical* 306: 127561.

51 Goode, J.A., Rushworth, J.V.H., and Millner, P.A. (2015). *Langmuir* 31: 6267–6276.

52 Sciacca, B., François, A., Hoffmann, P., and Monro, T.M. (2013). *Sensors and Actuators B: Chemical* 183: 454–458.

53 Liu, L. and Moore, M.D. (2020). *Food* 9: 318.

54 Ravina, A.D., Mohan, H., Prasad, M., and Pundir, C.S. (2020). *Bioscience Reports* 40: 2, BSR20193852.

55 Saylan, Y. and Denizli, A. (2018). *Sensors* 18 (9): 3016.

56 Inci, F., Saylan, Y., Kojouri, A.M. et al. (2020). *Applied Materials Today* 18: 100478.

57 Lee, J., Morita, M., Takemura, K., and Park, E.Y. (2018). *Biosensors and Bioelectronics* 102: 425–431.

58 He, L., Pagneux, Q., Larroulet, I. et al. (2017). *Biosensors and Bioelectronics* 89: 606–611.

59 Lee, D.K., Kang, J.H., Lee, J.S. et al. (2015). *Scientific Reports* 5: 15459.

60 Yang, X., Ren, Y., and Gao, Z. (2015). *Chemistry A European Journal* 21: 988–992.

61 Dong, H., Meng, X., Dai, W. et al. (2015). *Analytical Chemistry* 87: 4334–4340.

62 Stebunov, Y.V., Aftenieva, O.A., Arsenin, A.V., and Volkov, V.S. (2015). *ACS Applied Materials and Interfaces* 7: 21727–21734.

63 Cetin, A.E., Etezadi, D., Galarreta, B.C. et al. (2015). *ACS Photonics* 2: 1167–1174.

64 Chen, H., Hou, Y., Qi, F. et al. (2014). *Biosensors and Bioelectronics* 61: 83–87.

65 Jun, J., Lee, J.S., Shin, D.H., and Jang, J. (2014). *ACS Applied Materials and Interfaces* 6: 13859–13865.

66 Farhadi, K., Forough, M., Pourhossein, A., and Molaei, R. (2014). *Sensors and Actuators B: Chemical* 202: 993–1001.

67 Kalyani, N., Goel, S., and Jaiswal, S. (2020). *Nanosensors for Environmental Applications* (eds. S.K. Tuteja, D. Arora, N. Dilbaghi and E. Lichtfouse), 197–224. Cham: Springer.

68 Liu, C., Guo, B., and Xue, J. (2018). *Water Environment Research* 90 (10): 1323–1347.

69 Saad, Y., Selmi, M., Gazzah, M.H. et al. (2019). *Optik* 190: 1–9.

70 Sakib, M.A., Yousuf, S.M.E.H., Gupta, S.D., and Islam, M.Z. (2018). *Plasmonics* 13: 2205–2213.

71 Liu, S., Cheng, R., Chen, Y. et al. (2018). *Sensors and Actuators B: Chemical* 254: 1157–1164.

72 Saylan, Y., Akgönüllü, S., Çimen, D. et al. (2017). *Sensors and Actuators B: Chemical* 241: 446–454.

73 Zhang, B. and Wei, C. (2018). *Talanta* 182: 125–130.

74 Ye, L., Wen, G., Ouyang, H. et al. (2016). *Scientific Reports* 6: 24150.

75 Zhuang, H., Wang, Z., Zhang, X. et al. (2017). *Chemical Communications* 53: 1797–1800.

76 Chen, L., Fu, X., Lu, W., and Chen, L. (2013). *ACS Applied Materials and Interfaces* 5: 284–290.

77 Pienpinijtham, P., Han, X.X., Ekgasit, S., and Ozaki, Y. (2011). *Analytical Chemistry* 83: 3655–3662.

78 Yin, J., Wu, T., Song, J. et al. (2011). *Chemistry of Materials* 23: 4756–4764.

79 Qu, W.G., Deng, B., Zhong, S.L. et al. (2011). *Chemical Communications* 47: 1237–1239.

80 Dasary, S.S.R., Senapati, D., Singh, A.K. et al. (2010). *ACS Applied Materials and Interfaces* 2 (12): 3455–3460.

4

Plasmonic Sensors for Detection of Chemical and Biological Warfare Agents

Semra Akgönüllü, Yeşeren Saylan, Nilay Bereli, Deniz Türkmen, Handan Yavuz, and Adil Denizli

Hacettepe University, Department of Chemistry, Ankara, Turkey

4.1 Introduction

The threat of biological and chemical agents continues to be a growing concern around the world. Therefore, there is a need to develop appropriate technology with early identification intended for use by first responders for biological and chemical warfare agents detection [1]. Biological and chemical warfare agents have been utilized as weapons and international efforts banned them after the horrible employment in World War I. The efforts were continued until today [2–4]. The chemical warfare agents include synthetic toxic chemicals but they do not contain toxins which are generally classified as biological warfare agents. There is no fundamental difference between synthetic and natural toxins when the toxicological risk of biological and chemical agents is evaluated. The impact on target individuals exposed to such threatening agents is terrible. The population exposed to such agents will be affected without differentiation [5]. Various biological and chemical warfare agents such as viruses, fungi, bacteria, toxins, and explosives extensively affect alive [6].

Biological and chemical warfare agents have been the subject of important scientific research over the last 20 years because of the increased threat of terrorism and usage in war zones. Numerous detection methods have been improved to determine biological and chemical warfare agents. The advanced determination methods including ion mobility spectroscopy, infrared spectroscopy, gas chromatography–mass spectroscopy [7], capillary electrophoresis [8], and liquid chromatography-mass spectroscopy [9]. A useful device for accurate and rapid detection of biological and chemical warfare agents has to fulfill several criteria such as portability, simplicity, sample processing, selectivity, sensitivity, reproducibility, analysis time, low false alarms, cost, and safety. The main difficulty for commercially available devices needs to increase the specificity, and sensitivity and to decrease the false alarm response for biological and chemical warfare agent detection. Sensors have the potential to reform medical diagnosis, environmental monitoring, and industrial

Plasmonic Sensors and their Applications, First Edition. Edited by Adil Denizli.

applications owing to their unique advantages [10–16]. This part will discuss the detection of biological and chemical warfare agents with plasmonic-based sensors.

4.2 Sensors

A sensor is a device that measures the presence or quantity of matter and converts it into an interpretable signal. The sensor obtains a biochemical interaction that is utilized as the basis for detecting the presence or level of a known analyte. This analyte can be an inorganic compound such as a salt, a small biological molecule such as a vitamin or sugar, and/or a larger biological macromolecule component such as proteins or nucleic acids. The biochemical interaction between the sensor system and the analyte provides a biological binding to the sensing event, as the detection mechanism used by the sensor is functionally similar to the interaction between the analyte and its biological environment [17].

Sensors can be classified according to the recognition elements including enzyme-based [18], antibody-based [19], DNA hybridization-based [20], aptamer-based, and so on [21] or transforming mechanism including optical, electrochemical, piezoelectric, thermal, and so on [22]. Having significant advantages over traditional detection approaches, the sensor plays a critical role in sensitive and selective biological and chemical threat agent detection [6]. Recent advances in optical transducers that combined with biomolecular interactions have led to the growth of a wide variety of sensors with various applications. The most used optical-based sensor is surface plasmon resonance (SPR) [23]. Besides, optical-based sensors can be designed using nanomaterials that are expected to meet the requirements for unlabeled and high-throughput analysis in detection [24].

4.2.1 Plasmonic-based Sensors

Plasmonic-based sensors are the class of optical sensors. They are created of sensing platforms that contain metal and/or metal-dielectric structures to support surface plasmons and recognition elements that can bind an analyte [25]. When an analyte solution applies to the sensor, the capture of the analyte is occurred by a receptor on the surface and the sensing element causes a refractive index (RI) change (Figure 4.1a). The surface plasmon is very sensitive to changes in the RI in the sensor surface, changes in the local RI induced can be detected by evaluating changes in the light coupled to surface plasmon, in terms of resonant wavelength, intensity, and/or phase change (Figure 4.1b). The plasmonic-based sensor has been utilized for rapid, real-time, and label-free monitoring of biologically and chemically relevant analytes, which to detect small molecules at ultra-low concentrations [25]. Table 4.1 is shown in the application fields of optical sensors.

4.3 Biological Warfare Agents

Several pathogenic bacteria, viruses, and toxins can be counted as potential biological warfare agents (BWA) [31, 32]. The most likely agents on the list are variola major virus (smallpox) and *Bacillus anthracis* (anthrax). Extremely dangerous agents are botulinum toxin,

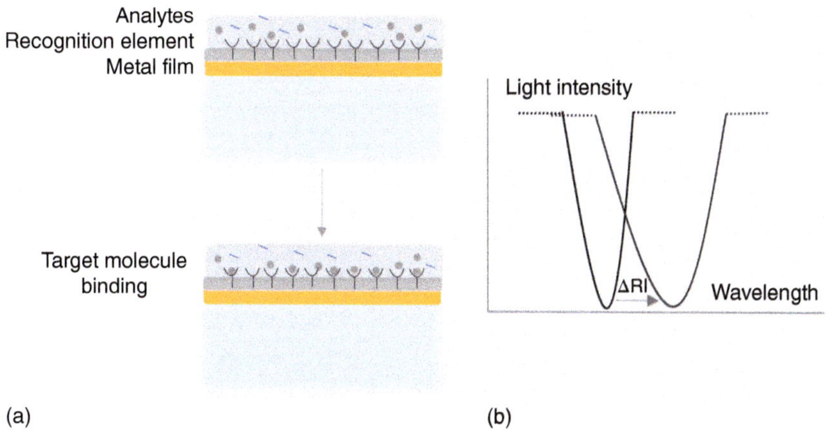

Figure 4.1 (a) Principle of a plasmonic-based sensor and (b) change in the spectrum due to the increase of RI in the sensor surface induced by the captured analyte. Source: © John Wiley & Sons.

Table 4.1 Different applications of optical sensors.

Application fields	Target	Sensors type	References
Food control	Mycotoxins	SPR	[26]
Medical diagnosis	Albumin	SPR	[27]
Environmental monitoring	Pesticide	Optical	[28]
Biomolecular interaction	Anti-lysozyme single-domain antibody	GC-SPFS	[29]
Homeland security	Volatile organic compounds	SERS	[30]

Yersinia pestis, *Francisella tularensis*, and *Salmonella typhimurium*. Others, such as Ebola, Venezuelan equine encephalitis, influenza viruses, and Marburg are lower on the list due to difficulties in their preparation, although infections for these viruses are severe and the mortality rate is high. BWA is much cheaper to produce when compared to chemical warfare agents (CWA), and the danger zone and expected loss of life in an attack is more effective. The infection dose (ID) is different for each agent. The risk ratio of each BWA is explained not only by ID but also by stability in the environment, its natural propagation path, and the probability of spore formation in the bacterial state. Usually, the ingestion of aerosol through the lungs can cause disease with a lower ID with given BWA. The state of the novel detection systems for sensing of BWA is based on the fit of perfect analytical methodology, and the application of advanced analytical tools [33–36]. The main biological warfare agents and give rise to diseases in humans, animals, or plants are depicted in (Table 4.2).

4.3.1 Detection of Biological Warfare Agents

Early detection of biological warfare agents can only be based on clinical diagnosis. The standard laboratory tests for viruses, bacteria, and their spores are culture-based methods.

Table 4.2 The main biological warfare agents and caused diseases.

Group	Agents	Diseases
Bacterial	*Bacillus anthracis*	Anthrax
	Yersinia pestis	Plague
	Francisella tularensis	Tularemia
	Brucella melitensis	Brucellosis
	Coxiella burnetii	Q fever
	Salmonella species	Food poisoning
	Escherichia coli O157:H7	Food poisoning
	Shigella	Food poisoning
	Burkholderia pseudomallei	Melioidosis
	Vibrio cholerae	Cholera
Viral	Alphaviruses and flaviviruses	Viral encephalitis
	Variolla major	Smallpox
	Filoviruses and arenaviruses	Viral hemorrhagic fevers
	Influenza virus	Influenza
	Human immunodeficiency virus (HIV)	AIDS
	SARS coronavirus (SARS-Cov)	SARS
	Ebolaviruses	Ebola
	Junin virus	Argentina hemorrhagic fever
	Hepatitis B virus	Hepatitis B
Toxins	*Staphylococcal enterotoxin B*	Food poisoning
	Ricinus communis (castor beans)	Ricin toxin poisoning
	Clostridium botulinum toxin	Botulism
	Clostridium perfringens	Perfringens toxins
	Mycotoxins	Mycotoxicoses
	Seaweed	Palytoxin

Source: Reprinted with permission from Pedrero et al. [35]. © 2012, John Wiley & Sons.

Toxins are determined by bioassay that bases on the toxin-specific antibodies. Microbiological methods are highly sensitive but are completed in one to four days. Therefore, real-time sensors are needed for environmental monitoring [37]. Specific and early detection of BWA is a critical aspect of defense in any attack [38–40]. Plasmonic sensors have the advantage of a label-free and real-time detection of analytes [41]. Portable plasmonic sensors may be used in manned [42] or un-manned [43] detection operations in a variety of environments [44, 45].

Bacteria are reproduced by replication and invasion of host tissue or release of a toxin. The ID is different according to the variety of bacteria [46]. *Brucellosis* is one of the most mutual bacterial zoonosis. This disease is caused by several species of the genus *Brucella*

[47]. Gupta et al. reported a study about an SPR sensor for *Brucella abortus* CSP-31 (*B. abortus* CSP-31) monitoring. They characterized *B. abortus* CSP-31 interaction with the antibody on a modified gold surface with SPR and electrochemical impedance spectroscopy (EIS). According to the EIS data, an impedance decrease is observed following the antigen–antibody interaction. The experimental parameters such as temperature and pH were optimized with an effect SPR angle change. The SPR sensor can detect 0.05 pM concentration of *B. abortus* CSP-31 in less than 10 minutes within a 2.0–16.0 pM of concentration [48]. *Brucella melitensis* is also one of the important BWA that cause brucellosis and can be transmitted from these animals to humans [36]. Sikarwar et al. developed an SPR sensor for the detection of *B. melitensis*. They fabricated a 4-mercaptobenzoic acid-modified gold SPR sensor for this biological warfare agent detection. Then, two different DNA probes are designed using different sequences of the DNA for the DNA target. Ten real samples in various concentrations are spiked with *B. melitensis* which showed the promising applicability of this methodology [49]. *B. anthracis* is a rod-shaped, gram-positive, spore-forming bacterium that gives rise to the disease anthrax in people and animals [50–53]. Adduci et al. reported a portable SPR sensor (SPIRIT 4.0, Seattle Sensor Systems) for the detection of *B. anthracis* and different species of *Bacillus* (Figure 4.2a). Spores of *B. anthracis* are detected using antibody-coated sensors via capturing at a concentration of 10^7 spores.ml^{-1}. The concentration of *B. anthracis* spores with a secondary antibody amplification calculated as 10^5 spores.ml^{-1}. They mentioned that emergency responders can implement protocols in limiting the number of exposed individuals and a timely fashion with a portable sensor to detect *B. anthracis* [54].

The anthrax toxins are classified as three distinct proteins as protective antigen (PA), lethal factor (LF), and edema factor (EF). An SPR sensor is utilized for the detection of protective antigen (PA) in different concentrations range by Ghoush et al. (Figure 4.2b). They used PA immobilized carboxymethyl dextran-modified gold sensor against *B. anthracis*. The equilibrium constant and maximum binding capacity of the analyte is calculated as 20 fM and 18.74 m°, respectively. The plasmonic-based sensor can detect as low as (1.0 pg.ml^{-1}) purified PA. The PA is an important early diagnosis marker for *B. anthracis* infections in blood samples [55].

The toxin that can be generated by a plant, animal, or microbe. Some toxins can also be propagated by chemical synthesis (low molecular weight toxins) and molecular

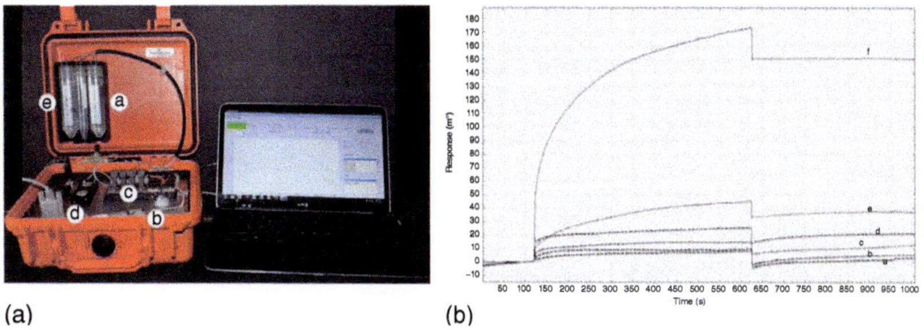

(a) (b)

Figure 4.2 (a) The portable SPR sensor and (b) sensorgram for antigen–antibody interaction for various concentrations of protective antigen. Source: Reprinted with permission from Aducci et al. [54] and Ghosh et al. [55].

biological techniques (protein toxins). The threat of these agents is a growing worry in the last years [56].

Botulinum neurotoxins are known as the deadliest toxins [57]. Tomar et al. developed an SPR sensor for the detection of Botulinum neurotoxins type A. In this study, they immobilized the fragment and synaptic vesicles on carboxymethyl dextran-modified gold surface and then obtained an immobilization of Botulinum neurotoxins type A antibody. The interaction with an immobilized antibody is characterized by SPR sensor and electrochemical impedance spectroscopy. The limit of detection value is calculated as 0.045 fM [58].

A mycotoxin is produced by one or more specific mold species. Diverse type of toxigenic fungi is talented in producing such different mycotoxins as the ochratoxins, aflatoxins, fumonisins, and trichothecenes [56]. Akgönüllü et al. developed an aflatoxin B1-imprinted polymeric nanofilm coated SPR sensor. They designed a gold nanoparticle-containing polymeric film coated sensor surface (Figure 4.3) and confirmed by scanning electron microscope analysis. The imprinted plasmonic-based sensor has responses in a broad linear range (0.0001–10.0 ng ml^{-1}). The limit of detection value is found to be 1.04 pg ml^{-1}. Furthermore, detection studies of aflatoxin B1 are carried out using various food samples, and then selectivity, reusability, and storage stability analysis are obtained by the SPR sensor [59].

Staphylococcus aureus enterotoxin B (SEB) with 28.4 kDa the most thermal stable protein toxin. SEB exposure can cause symptoms such as vomiting, nausea, diarrhea, anaphylactic shock, headache, sudden onset of fever, cough, and chills may appear [60]. Zhu et al. developed a hybrid gold-silver nanoparticles-based localized surface plasmon resonance (LSPR) sensor for SEB detection. SEB is directly detected at 0.1 ng.ml^{-1} level using the triangular hybrid nanoparticles. The hybrid gold-silver triangular nanoparticles are prepared with nanosphere lithography technique on a glass substrate (Figure 4.4a and b). They reported that the plasmonic-based sensor is an alternative application in the detection and identification of biological warfare agents [61].

SEB is also one of a family of serological types (SEA through SEJ) of emetic enterotoxins. A fiber optic SPR sensor is reported by Slavik et al. for SEB detection [62]. The SPR sensor is modified with a covalently cross-linked double-layer of antibodies against SEB (Figure 4.5a) and demonstrated detection capability ng.ml^{-1} in less than 10 minutes. They performed with bovine serum albumin (BSA) in PBS solutions containing SEB in concentrations of 0–100 ng.ml^{-1} (Figure 4.5b).

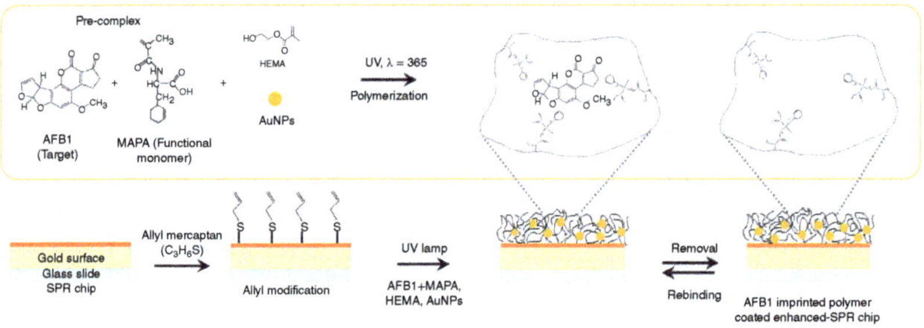

Figure 4.3 The preparation of gold nanoparticle-containing, and imprinted polymeric film coated sensor surface. Source: Reprinted with permission from Akgönüllü et al. [59]. © 2020, Elsevier.

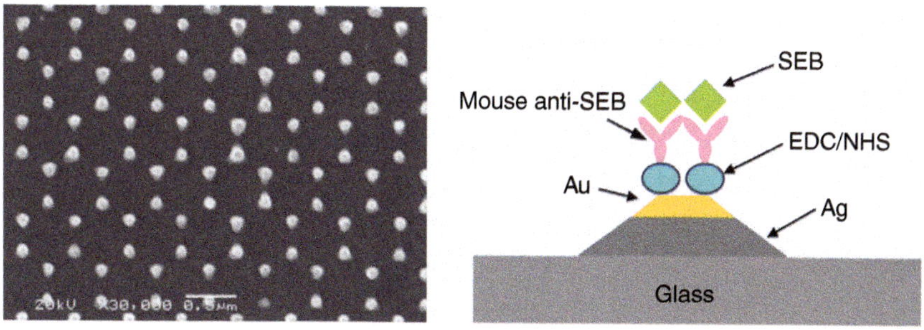

(a) (b)

Figure 4.4 (a) SEM image of the topography of the triangular hybrid Au–Ag nanoparticles fabricated by NSL and (b) schematic diagram of LSPR-sensor. Source: Reprinted with permission from Zhu et al. [61]. © 2009, Elsevier.

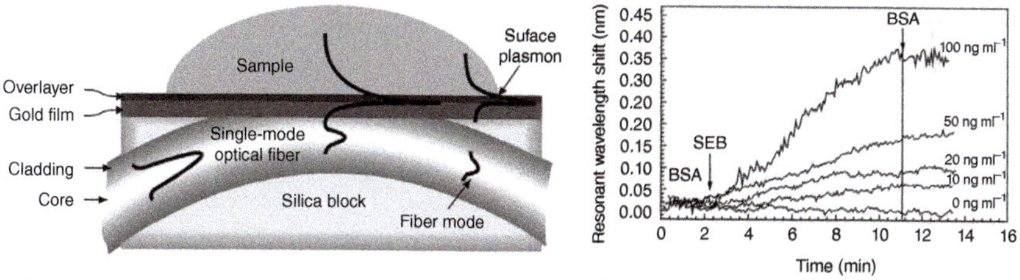

(a) (b)

Figure 4.5 (a) SPR sensing structure based on a side-polished single-mode optical fiber and (b) response of the SPR sensor to different concentrations of SEB present in BSA solution. Source: Reprinted with permission from Slavík et al. [62]. © 2002, Elsevier.

Ricin is one of the most toxic plant toxins. Ricin is produced from the seeds of castor oil (*Ricinus communis*). Its accessibility and relative ease of preparation make it a potential agent for criminal or biological terrorist attacks [63]. Tran et al. used a rapid, sensitive, and robust immunoassay-based commercial SPR instrument. Ten monoclonal antibodies are measured for their ability to recognize both horticultural and commercial ricins extracted from six different cultivars of *R. communis*. The detection of ricin concentrations is linear over a broad range and the limit of detection value is calculated as 0.5 ng.ml^{-1}. They also reported the sensor is highly reproducible and can detect ricin variants and environmental samples [64]. Stern et al. developed an SPR sensor for the detection of ricin and agglutinin. They used a cross-reactive antibody (R109) binding to both proteins and a ricin-specific antibody (R18). The SPR sensor is reusable and sensitive for different Ricinus communis cultivars and showed no false-positive responses. The sensing principle is visualized in Figure 4.6a–c [65]. Another work for ricin detection is developed by Feltis et al. They reported the detection of a low concentration of ricin at 200 ng.ml^{-1} within 10 minutes with an antibody-sandwich-based portable SPR sensor. They demonstrated both fast and trace level detection of ricin and its subsequent

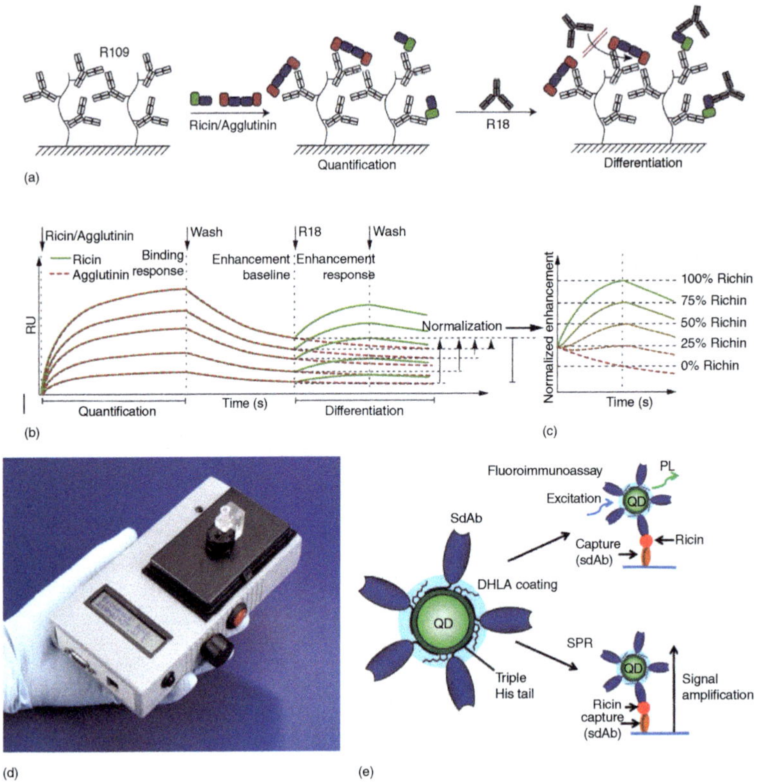

Figure 4.6 (a) Principle of immunological SPR sensor for simultaneous differentiation and quantification of ricin and agglutinin; (b) Schematic binding curves; (c) By normalization of enhancement binding responses to the enhancement baseline, the ability to differentiate ricin from agglutinin becomes concentration-independent. Source: Reprinted with permission from Stern et al. [65]. © 2016, Elsevier; (d) The portable SPR sensor, operating in hand-held mode [66]. Source: Reprinted with permission from Feltis et al. [66]. © 2008, Elsevier; and (e) QD–sdAb conjugate strategy. Source: Reprinted with permission from Anderson et al. [67]. © 2013, Elsevier.

validation with the commercial Biacore-1000 system. Figure 4.6d shows the sensor operating in standalone mode [66]. Fan et al. reported a plasmonic colorimetric sensing platform based on nanopin metasurfaces for the quantitative analysis of ricin. The antibody-functionalized plasmonic sensor has several properties including fast response, high sensitivity, and allowing detection in the wide range of 10–120 ng.ml^{-1} within less than 10 minutes. The limit of detection is found as 10 ng.ml^{-1} [68]. Anderson et al. prepared robust quantum dots conjugated with a single-domain antibody-based SPR sensor for ricin detection. Sandwich assay is conducted by flowing ricin over the capture reagent at different concentrations ranging from 0 to 100 nM that allow ricin detection in 3 minutes. The tracer single-domain antibody, single-domain antibody-quantum dots conjugate, and conventional antibody are flowed over the sensor to amplify the initial binding signal, respectively [67]. Figure 4.6e is shown a diagram of the single-domain antibody around DHLA-quantum dots.

4.4 Chemical Warfare Agents

Chemical warfare agents (CWAs) are utilized in World War I and II and during the Cold War and they are still produced even today although their use as weapons is prohibited [69, 70]. In the 1980s, the Iran-Iraq conflict is utilized sarin and sulfur mustard by Iraq [71]. In 1992, a treaty prohibiting the development, production, stockpiling, and use of chemical weapons by the Chemical Weapons Convention and mandating their destruction, is ratified and came into force in 1997 [72]. In 1995, Tokyo subway sarin gas attack [73], US postal anthrax letter attacks in 2001 [74], sarin and sulfur mustard used in the Syrian war [75], and in 2018, Novichok is a nerve agent for the assassination was employed in Great Britain [76].

The explosive compounds [77], halide ions [78], heavy metals ions [79], and radioactive compounds [80] are used as chemical warfare agents and have become a major problem for our society and environment today [81]. CWAs are classified into various categories based on how they affect the human body system [82]. The use of CWAs is still causing thousands of victims [83]. The CWAs are divided into seven classes as vesicant agents, nerve agents, suffocating/blood agents, cytotoxic proteins, incapacitating agents, pulmonary agents, and lachrymatory agent. The first three are the best well-known CWAs for their high toxicity (Table 4.3) [81].

4.4.1 Detection of Chemical Warfare Agents

Recently, the threat from CWAs has raised as the foremost security challenge due to their cheap and simple production, easy dispersal, complex detection, high-cost protection, and psychological, social, and economic impact. Early detection of CWAs while the intentional chemical event is necessary to start correctional emergency responses for management. Therefore, the development of sensing methods has huge importance for the detection of CWAs.

Table 4.3 Classes of main and common chemical weapons agents.

Group	Agent	Action mode
Nerve agents	G-group (Tabun, soman, and sarin) V-group (VX, Vx, CVx, and VR) A-group (Novichoks)	Inactivates the enzyme acetylcholinesterase (AChE), preventing the breakdown of the neurotransmitter acetylcholine (ACh) in the victim's synapses and causing both muscarinic and nicotinic effects
Vesicant/ blister	Sulfur mustard (HD H HT HL HQ)Nitrogen mustard (HN1 HN2 HN3)	Agents are acid-forming compounds that damage skin and respiratory system, resulting in burns and respiratory problems
Blood/ suffocating	Cyanogen chloride (CK) Hydrogen cyanide (AC)	Cyanide directly prevents cells from using oxygen. The cells then use anaerobic respiration, creating excess lactic acid and metabolic acidosis

Source: Reprinted with permission from Yue et al. [81]. © 2016, Elsevier.

Dimethyl methyl phosphonate (DMMP) a substitute molecule of the G-series nerve agents which are of concern because of its high toxicity, previous, and persistence deployment. Lafuente et al. reported a study about a surface-enhanced Raman scattering-based sensor for DMMP detection. They utilized a citrate-capped gold nanoparticle monolayer for the quantification of DMMP. The detection limit is calculated as 130 ppb. Moreover, the plasmonic-based sensor has excellent resistance to photodegradation and is quickly and easily regenerated by flushing air [84].

Neurotoxic organophosphates (OP) have been widely used for insect control in the environment. Besides, the threat of using OP-based CWAs is increasing in both ground war and terrorist attacks [85]. Mauriz et al. performed real-time sensing of the OP (chlorpyrifos) using a portable SPR in the real water samples. The analyte derivative the covalently immobilized onto the gold-coated surface for binding inhibition test. Also, sensor reusability is performed formation of alkanethiol self-assembled monolayers for 190 analysis including regeneration cycles. Three different gold-coated sensor surface chips were also designed for assay variability. The highly sensitive detection of chlorpyrifos was provided at nanogram per liter levels in pH 7.35 PBS buffer. The values of the limit of detection are ranged from 45 to 64 $ng.l^{-1}$ for the ground, river, and drinking water samples, and recovery values are calculated between 80 and 120%. They also validated this plasmonic-based sensor using conventional chromatographic methods and both techniques obtain by the high correlation coefficient [86]. SPR sensor study is reported by Daly et al. for organophosphorus vapors detection [87]. They synthesized cavitands at the upper rim of the cavity and appraised the interaction between the cavitands and the sarin nerve gas simulant. They used cavitand-based synthetic receptors to detect DMMP via complexation at the gas/solid interface. Eight-layer-thick films of the cavitand showed sensitivity to DMMP concentrations as low as 16 ppb.

Molecularly imprinted polymers (MIPs), referred to as "antibody mimics" or "artificial receptors," are utilized as the most promising alternative to the receptor. Sensitive, selective, feasibility, and cost-effective, with high mechanical strength and flexibility, MIPs are effective fixing materials [88, 89]. With the latest technology developing, MIPs are widely combined with various transducers such as plasmonic [90], luminescence [91], potentiometer [92], quartz crystal microbalance [93] to enable the customization of sensors for fast and selective identification of target analogs. Having thousands to millions of target shaped cavities, MIPs show affinity for receptors and are capable of specifically recognizing target molecules. Although MIP-based sensors present many unique challenges at the same time, detection sensitivity is greatly improved [94].

The growth of sensing platforms for nitroaromatic explosives, especially trinitrotoluene (TNT), and more hazardous explosives, hexahydro-1,3,5-trinitro-1,3,5-triazine (RDX) or pentaerythritol tetranitrate (PETN), is less improved and requires further efforts and improvement of the sensitivity [95]. TNT sensors have been reported with different transducers such as optical, electrochemical, or microgravimetric. Michael et al. reported that an ultrasensitive molecularly imprinted-based SPR sensor could provide a binding affinity for RDX with a limit of detection of 12 fM (Figure 4.7a–c). The recognition sites were created by using Kemp's acid as the imprinted molecules due to the low solubility of RDX. Then, this system was utilized to detect PETN, nitroglycerin, and ethylene glycol dinitrate [96]. Schematic presentation of imprinting Kemp's acid is shown in Figure 4.7d.

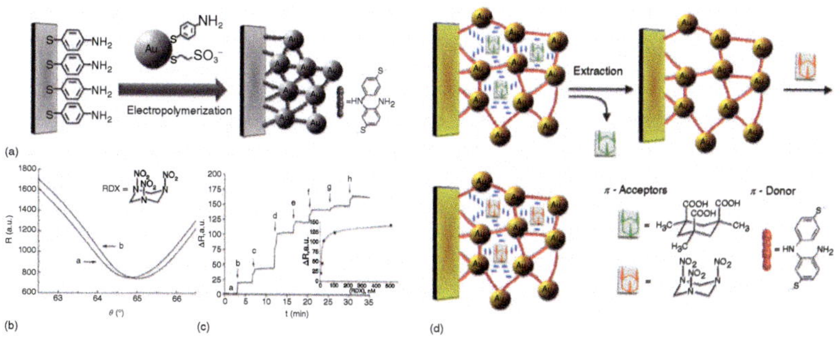

Figure 4.7 (a) Schematic presentation for the electropolymerization of a composite of bisaniline-crosslinked gold nanoparticles; (b) SPR curves corresponding to the bisaniline-crosslinked gold nanoparticles composite; (c) Sensorgram corresponding to the changes in the reflectance intensities; (d) Imprinting of Kemp's acid molecular recognition sites into the composite of bisaniline-crosslinked gold nanoparticles. Source: Reprinted with permission from Krause et al. [95]. © 2008, John Wiley and Sons.

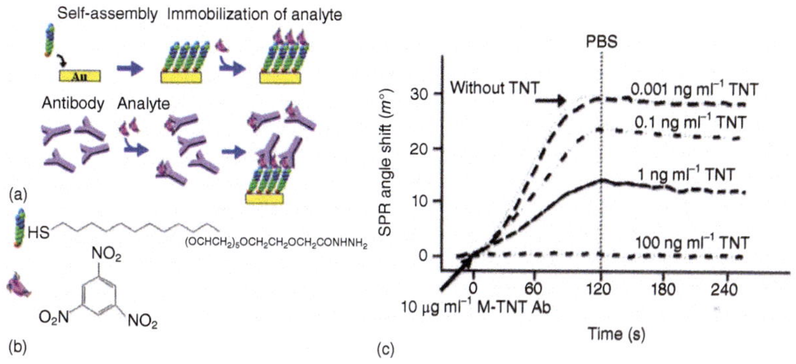

Figure 4.8 (a) Scheme of the construction of immuno-surface by self-assembly; (b) the principle of indirect competitive inhibition for TNT detection; (c) SPR response for the immunoreaction between M-TNT Ab and TNT immobilized PEG-NH2 SAM surface in the absence and presence of TNT. Source: Reprinted with permission from Kawaguchi et al. [97]. © 2007, Elsevier.

The interaction between a monoclonal anti-TNT antibody and trinitrophenyl-β-alanine immobilized thiolate monolayer surface (Figure 4.8a–c), a novel approach for the field robust portable imprinted-SPR TNT sensor was demonstrated by Kawaguchi. The response time could be 2 minutes with a limit of detection value as low as 0.008 ng.ml^{-1}. TNT was detected in the concentration range from 0.008 to 30.0 ng.ml^{-1}. They designed a reliable and stable SPR system by using poly(ethylene glycols) (PEG) based self-assembled layer for TNT detection [97].

4.5 Conclusion and Future Perspective

The detection of biological and chemical warfare agents is highly crucial due to their simple availability and their effects on lives. There has been a growing interest in sensor systems for the detection of biological and chemical warfare agents and research has focused on ways to produce portable devices that can provide rapid, accurate, and on-site detection. The importance of plasmonic-based sensors has been broadly recognized, and many reviews have explored their potential applications in clinical diagnosis, global health, personalized medicine, food safety, drug discovery, security of life, and forensic medicine. Although the plasmonic sensor system still has several difficulties, it is widely utilized in many fields. And the research for BWAs and CWAs determination by plasmonic sensors is certainly attractive and well adapted for simultaneous monitoring. Especially polymer films used for several decades for more sensitive determination, applications made by metal nanoparticles focused on combination with exciting technologies. Thus, it can be easily concluded that the superiority of most sensors in this combination is their high stability, sensitivity, and selectivity. Sensitivity and selectivity are also promising. By the way, these devices are economical and portable. It is mechanically strong, replacing the use of traditional methods. However, specialist training and high cost in the use of the devices and data analysis are still required in most cases. Real-time and on-the-spot sensing is a critical point in theory. For this reason, optically based sensors are undertaken as the most suitable sensors to be applied by revealing a measurable response. Plasmonic-based sensors for BWAs and CWAs determination are valuable for further studies.

References

1 Saito, M., Uchida, N., Furutani, S. et al. (2018). *Microsystems & Nanoengineering* 4: 1–11.
2 Thiermann, H., Kehe, K., Riehm, J., and Zöller, L. (2014). *Regulatory Toxicology and Pharmacology*: 855–868.
3 Cenciarelli, O., Rea, S., Carestia, M. et al. (2013). *Defence S&T Technical Bullettin* 6: 111–129.
4 Rowland, C.E., Brown, C.W., Delehanty, J.B., and Medintz, I.L. (2016). *Materials Today* 19: 464–477.
5 Sadik, O.A., Land, W.H., and Wang, J. (2003). *Electroanalysis* 15: 1149–1159.
6 Pal, V., Sharma, M.K., Sharma, S.K., and Goel, A.K. (2016). *Defence Science Journal* 66: 445–457.
7 Valdez, C.A., Leif, R.N., Hok, S., and Hart, B.R. (2018). *Reviews in Analytical Chemistry* 37: 1–27.
8 Li, P., Hu, B., and Li, X. (2012). *Journal of Chromatography A* 1247: 49–56.
9 Niemikoski, H., Söderström, M., and Vanninen, P. (2017). *Analytical Chemistry* 89: 11129–11134.
10 Petkovic, K., Swallow, A., Stewart, R. et al. (2019). *Micromachines* 10: 617.
11 Zhou, J., Qi, Q., Wang, C. et al. (2019). *Biosensors and Bioelectronics* 142: 111449.
12 Rahtuvanoğlu, A., Akgönüllü, S., Karacan, S., and Denizli, A. (2020). *ChemistrySelect* 5: 5683–5692.
13 Heo, N.S., Oh, S.Y., Ryu, M.Y. et al. (2019). *Biotechnology and Bioprocess Engneering* 24: 318–325.
14 Lobry, M., Lahem, D., Loyez, M. et al. (2019). *Biosensors and Bioelectronics* 142: 111506.
15 Saylan, Y. and Denizli, A. (2018). *Sensors* 18: 3016.

16 Chen, N., Chang, M., Lu, X. et al. (2020). *Applied Sciences* 10: 3897.
17 Resnick, A. (2013). *Contemporary Physics* 54: 169–169.
18 Songa, E.A. and Okonkwo, J.O. (2016). *Talanta* 155: 289–304.
19 Sumithra, B., Jayanthi, V.S.P.K.S.A., Manne, H.C. et al. (2020). *Biotechnology Letters* 42 (12): 2501–2509.
20 Heli, H., Sattarahmady, N., Hatam, G.R. et al. (2016). *Talanta* 156–157: 172–179.
21 Koyun, S., Akgönüllü, S., Yavuz, H. et al. (2019). *Talanta* 194: 528–533.
22 Malekzad, H., Sahandi Zangabad, P., Mohammadi, H. et al. (2018). *TrAC, Trends in Analytical Chemistry* 100: 116–135.
23 Homola, J. (2008). *Chemical Reviews* 108: 462–493.
24 Sepúlveda, B., Angelomé, P.C., Lechuga, L.M., and Liz-Marzán, L.M. (2009). *Nano Today* 4: 244–251.
25 Špačková, B., Wrobel, P., Bocková, M. et al. (2018). *Analytical Chemistry* v: 738 A–746 A.
26 Wei, T., Ren, P., Huang, L. et al. (2019). *Food Chemistry* 300: 125176.
27 Kaushik, S., Tiwari, U.K., Deep, A., and Sinha, R.K. (2019). *Scientific Reports* 9: 6987.
28 Saleh, S.M., Alminderej, F.M., Ali, R., and Abdallah, O.I. (2020). *Spectrochimica Acta – Part A Molecular Biomolecular Spectroscopy* 229: 117971.
29 Kaya, T., Nagatoishi, S., Nagae, K. et al. (2019). *PLoS One* 14: 1–16.
30 Wong, C.L., Dinish, U.S., Buddharaju, K.D. et al. (2014). *Applied Physics A: Materials Science and Processing* 117: 687–692.
31 Jansen, H.J., Breeveld, F.J., Stijnis, C., and Grobusch, M.P. (2014). *Clinical Microbiology and Infection* 20: 488–496.
32 Cenciarelli, O., Gabbarini, V., Pietropaoli, S. et al. (2015). *Virus Research* 210: 318–326.
33 Pohanka, M., Skládal, P., and Kroèa, M. (2007). *Defence Science Journal* 57: 185–193.
34 Upadhyayula, V.K.K. (2012). *Analytica Chimica Acta* 715: 1–18.
35 Pedrero, M., Campuzano, S., and Pingarrón, J.M. (2012). *Electroanalysis* 24: 470–482.
36 Thavaselvam, D. and Vijayaraghavan, R. (2010). *Journal of Pharmacy & Bioallied Sciences* 2: 179.
37 Petrenko, V.A. and Vodyanoy, V.J. (2003). *Journal of Microbiological Methods* 53 (2): 253–262.
38 Eubanks, L.M., Dickerson, T.J., and Janda, K.D. (2007). *Chemical Society Reviews* 36: 458–470.
39 Clifford Lane, H., La Montagne, J., and Fauci, A.S. (2001). *Nature Medicine* 7: 1271–1273.
40 Byrne, B., Stack, E., Gilmartin, N., and O'Kennedy, R. (2009). *Sensors* 9: 4407–4445.
41 Chinowsky, T.M., Soelberg, S.D., Baker, P. et al. (2007). *Biosensors and Bioelectronics* 22: 2268–2275.
42 Naimushin, A.N., Spinelli, C.B., Soelberg, S.D. et al. (2005). *Sensors and Actuators B: Chemical* 104: 237–248.
43 Palframan, M.C., Gruszewski, H.A., Schmale, D.G., and Woolsey, C.A. (2014). *Journal of Unmanned Vehicle Systems* 2: 103–118.
44 Usachev, E.V., Usacheva, O.V., and Agranovski, I.E. (2014). *Jornal of Applied Microbiology* 117: 1655–1662.
45 Rao, S.S., Mohan, K.V.K., and Atreya, C.D. (2010). *Journal of Microbiology Methods* 82: 1–10.
46 V. St. Georgiev (2009). *National Institute of Allergy and Infectious Diseases, NIH.*

47 Peruski, A.H. and Peruski, L.F. (2003). *Clinical and Diagnostic Laboratory Immunology* 10: 506–513.

48 Gupta, G., Energy, H., Boopathi, M. et al. (2011). *Analytical and Bioanalytical Electrochemistry* 3 (1): 26.

49 Sikarwar, B., Singh, V.V., Sharma, P.K. et al. (2017). *Biosensors and Bioelectronics* 87: 964–969.

50 Blackburn, J.K. (2010). Integrating geographic information systems and ecological niche modeling into disease ecology: a case study of bacillus anthracis in the United States and Mexico. In: *Emerging and Endemic Pathogens*, NATO Science for Peace and Security Series A: Chemistry and Biology (eds. K. O'Connell, E. Skowronski, A. Sulakvelidze and L. Bakanidze), 59–88. Dordrecht: Springer.

51 Kracalik, I., Lukhnova, L., Aikimbayev, A. et al. (2011). *Spatial and Spatiotemporal Epidemiology* 2: 11–21.

52 Read, T.D., Peterson, S.N., Tourasse, N. et al. (2003). *Nature* 423: 81–86.

53 Spencer, R.C. (2003). *Journal of Clinical Pathology* 56 (3): 182–187.

54 Adducci, B.A., Gruszewski, H.A., Khatibi, P.A., and Schmale, D.G. (2016). *Biosensors and Biolectronics* 78: 160–166.

55 Ghosh, N., Gupta, N., Gupta, G. et al. (2013). *Diagnostic Microbiology and Infectious Disease* 77: 14–19.

56 Venkataramana, M., Chandranayaka, S., Prakash, H.S., and Niranjana, S.R. (2013). *Toxinology*: 1–22.

57 Johnson, E.A. (2014). *Clostridium: Clostridium Botulinum*, 2e. Elsevier.

58 Tomar, A. and Gupta, G. (2016). *Journal of Bioterrorism and Biodefence* 7: 142.

59 Akgönüllü, S., Yavuz, H., and Denizli, A. (2020). *Talanta*: 121219.

60 Naimushin, A.N., Soelberg, S.D., Nguyen, D.K. et al. (2002). *Biosensors and Bioelectronics* 17: 573–584.

61 Zhu, S., Du, C.L., and Fu, Y. (2009). *Optical Materials* 31: 1608–1613.

62 Slavík, R., Homola, J., and Brynda, E. (2002). *Biosensors and Bioelectronics* 17: 591–595.

63 Stern, D., Pauly, D., Zydek, M. et al. (2015). *Biosensors and Bioelectronics* 78: 111–117.

64 Tran, H., Leong, C., Loke, W.K. et al. (2008). *Toxicon* 52: 582–588.

65 Stern, D., Pauly, D., Zydek, M. et al. (2016). *Biosensors and Bioelectronics* 78: 111–117.

66 Feltis, B.N., Sexton, B.A., Glenn, F.L. et al. (2008). *Biosensors and Bioelectronics* 23: 1131–1136.

67 Anderson, G.P., Glaven, R.H., Algar, W.R. et al. (2013). *Analytica Chimica Acta* 786: 132–138.

68 Fan, J.R., Zhu, J., Wu, W.G., and Huang, Y. (2017). *Small* 13: 1–7.

69 Gupta, R.C. (2015). *Handbook of Toxicology of Chemical Warfare Agents*, 2e.

70 Y. Seto (2020). *On-site detection of chemical warfare agents*, INC.

71 Haines, D.D. and Fox, S.C. (2014). *Forensic Science Review* 26 (2): 97–114.

72 OPCW (2020). *Organ. Prohib. Chem. Weapons.*

73 Nakamine, S., Kobayashi, M., Fujita, H. et al. (2018). *Journal of Social and Clinical Psychology* 37: 794–811.

74 North, C.S., Pfefferbaum, B., Vythilingam, M. et al. (2009). *Biosecurity and Bioterrorism* 7: 379–388.

75 John, H., van der Schans, M.J., Koller, M. et al. (2018). *Forensic Toxicology* 36: 61–71.

76 Franca, T.C.C., Kitagawa, D.A.S., Cavalcante, S.F.d.A. et al. (2019). *International Journal of Molecular Sciences* 20: 1222.

77 Patel, C.K.N. (2008). *European Physical Journal Special Topics* 153: 1–18.

78 Jafari Horestani, A.R., Jafari, M.T., Jazan, E., and Mossaddegh, M. (2018). *International Journal of Mass Spectrometry* 433: 19–24.

79 Quang, D.T. and Kim, J.S. (2010). *Chemical Reviews* 110: 6280.

80 Buesseler, K.O., Jayne, S.R., Fisher, N.S. et al. (2012). *Proceedings of the National Academy of Sciences of the United States of America* 109: 5984–5988.

81 Yue, G., Su, S., Li, N. et al. (2016). *Coordination Chemical Reviews* 311: 75–84.

82 Seto, Y. (2009). *Handbook of Toxicology Chemical Warfare Agents*: 813–825.

83 Aleksenko, S.S. (2012). *Journal of Analytical Chemistry* 67: 82–97.

84 Lafuente, M., Pellejero, I., Sebastián, V. et al. (2018). *Sensors and Actuators B: Chemical* 267: 457–466.

85 Simonian, A.L., Good, T.A., Wang, S.S., and Wild, J.R. (2005). *Analytica Chimica Acta* 534: 69–77.

86 Mauriz, E., Calle, A., Lechuga, L.M. et al. (2006). *Analytica Chimica Acta* 561: 40–47.

87 Daly, S.M., Grassi, M., Shenoy, D.K. et al. (2007). *Journal of Materials Chemistry* 17: 1809–1818.

88 Saylan, Y., Akgönüllü, S., Yavuz, H. et al. (2019). *Sensors* 19: 1279.

89 Lu, W., Xue, M., Xu, Z. et al. (2014). *Current Organic Chemistry* 19: 62–71.

90 Esentürk, M.K., Akgönüllü, S., Yılmaz, F., and Denizli, A. (2019). *Journal of Biomaterials Science Polymer Edition* 30: 646–661.

91 Ahmadpour, H. and Hosseini, S.M.M. (2019). *Talanta* 194: 534–541.

92 Konishi, A., Takegami, S., and Kitade, T. (2020). *Current Analytical Chemistry* 16 (6): 788–794.

93 Akgönüllü, S., Battal, D., Yalcin, M.S. et al. (2020). *Microchemical Journal* 153: 104454.

94 Wackerlig, J. and Lieberzeit, P.A. (2015). *Sensors and Actuators B: Chemical* 207: 144–157.

95 Krause, A.R., Van Neste, C., Senesac, L. et al. (2008). *Journal of Applied Physics* 103: 094906.

96 Riskin, M., Tel-Vered, R., and Willner, I. (2010). *Advanced Materials* 22: 1387–1391.

97 Kawaguchi, T., Shankaran, D.R., Kim, S.J. et al. (2007). *Talanta* 72: 554–560.

5

A Plasmonic Sensing Platform Based on Molecularly Imprinted Polymers for Medical Applications

Neslihan Idil[1], Monireh Bakhshpour[2], Sevgi Aslıyüce[2], Adil Denizli[2], and Bo Mattiasson[3,4]*

[1]*Department of Biology, Biotechnology Division, Hacettepe University, Ankara, Turkey*
[2]*Department of Chemistry, Biochemistry Division, Hacettepe University, Ankara, Turkey*
[3]*Division of Biotechnology, Lund University, Lund, Sweden*
[4]*Indienz AB, Annebergs Gård, Billeberga, Sweden*

5.1 Introduction

Molecular imprinting technology is based on complex formation between a template molecule and one or more functional monomers. In the presence of a sufficient amount of a crosslinker, a three-dimensional (3-D) polymeric structure is synthesized. Following the template molecule removal, specific cavities complementary in terms of shape, size, and chemical functions are exposed. In general, the molecular recognition domains of intermolecular interactions between molecules and functional groups, such as hydrogen bonds, dipole–dipole, and ionic interactions, are presented in the polymer matrix [1, 2]. In this case, the recognition sites within the synthesized polymer only recognize target molecules and have the ability to selectively bind the interesting molecule. High selectivity and affinity of the polymer for the target molecule is one of the main advantages of molecular imprinting technology. Molecularly imprinted polymers (MIPs) show higher physical resistance against high temperature, pressure, extreme pH conditions, metal ions, and organic solvents making them more robust than biomolecules used earlier, e.g., antibodies. In addition, it is less costly to synthesize MIPs which could be stored at room temperature for several years, keeping their much higher recognition ability and retention capacity [3]. In the last decades, MIPs have gained great importance in many areas especially sensor technology [4]. Combination of molecular imprinting technology with plasmonic sensing platforms provides sensitive and selective detection of the analytes of interest [5]. Sensors have recently been developed as ideal analysis devices to make practical real-time detections in the clinical field taking the advantage of portability. Surface plasmon resonance (SPR) is an analytical technique based on registering optical changes when the refractive index is changed as a result of binding of a target molecule to the surface of the plasmon resonance sensor. The parameters are affected by changes in the refractive index of the

Plasmonic Sensors and their Applications, First Edition. Edited by Adil Denizli.

dielectric close to the interface caused by interactions between immobilized molecules and target molecules. In recent years, plasmonic sensors play an important role in detecting analytes in different fields such as food processing and quality control [6], environmental monitoring [7], microbiology [8], narcotics [9], biological weapons [10], and clinical diagnostics [11]. Plasmonic sensors can be applied clinically to different analyte classes. Therefore, the plasmonic analytic method is applied for the detection of essential proteins, toxic molecules or pollutants, and for the applications in the clinical area; antibodies, drugs, microorganisms, hormones, and cancer markers as well as many biomolecular markers could be given as examples [12]. In this chapter, applications of molecular imprinted plasmonic sensors were discussed by the properties of high sensitivity and selectivity, simple use, adaptability for automation and short response time for the detection of biomolecules focusing on improvements in medicine.

5.2 Molecular Imprinting Technology

Molecular imprinting technology has been presented as an extremely rewarding route for the construction of tailor-made recognition sites in polymeric materials [13, 14]. In the polymerization step, target molecule, polymerizable monomer(s), crosslinker(s), initiator, and solvent were mixed. In this regard, whenever the polymerization has got started, folding of the polymeric chains takes place around the template molecule by the spatial association of functional groups. It is noteworthy to highlight that application of correct and active functional monomers having ability to interact with the target molecule plays a key role for creating efficient recognition regions. Furthermore, functional monomers behave as anchors for the appropriate arrangement [4, 15]. Consequently, template-shaped regions were formed for the generation of specific cavities with pre-established high selectivity and affinity. In this context, following the removal of template, it leaves behind recognition sites which are generated as a 3D-mirrors of the shape, size, and chemical functionality of the template in/on the imprinted polymer network. In this manner, a specific interaction occurs between the target molecule and the cavity in the polymer surface [16, 17].

The specific binding sites show different properties depending on the interactions created during polymerization. Generally, the molecular imprinting technique is classified according to the nature of the interactions between the monomer and the template molecule. Today, there are two basic strategies used in molecular imprinting technology depending on the interaction between the template molecule and the monomer before polymerization. The first approach is based on non-covalent interactions [18], and the second one relies on covalent interactions between the template molecule and the functional monomer [19]. In non-covalent imprinting, there are no limitations found in covalent imprinting. Non-covalent imprinting is created in a suitable solvent based on a variety of interactions between molecule–monomer complexes, such as hydrogen bonds, ionic, Van der Waals, metal coordinated, and hydrophobic interactions. Afterwards, removal of the template was performed, the functionalized polymer matrix can rebind the target molecule through the same non-covalent interactions. Today, non-covalent imprinting technique takes place as the most common polymerization approach for molecular imprinting technology [20]. In covalent imprinting, template molecules are generally attached to monomers by covalent bonds. After polymerization, the covalent bonds are broken and the template molecule is removed from the polymer. These bonds are re-established in order to rebind the target molecule. Because covalent bonds are more stable, by this reason binding and removing steps perform much slower. The bonds between template molecules and functional monomers have to form rapidly by reversible interactions enabling to produce new bonds in the rebinding step. Therefore, spectrum of suitable template molecules which are appropriate for covalent imprinting is restricted [21].

MIPs have been used in order to obtain highly selective binding sites by arranging the three-dimensional structures of functional groups. Bioreceptors such as antibodies are popular ligands for interacting with their specific antigens due to their high selectivity. However, these highly selective biomolecules can be easily affected by environmental conditions since their protein structures tend to denature under, e.g. acidic conditions, high temperatures, etc. Molecular imprinting results in copying the recognition mechanisms of biological systems such as antibody/antigen and enzyme/substrate to polymeric structures. The ultimate principal of this methodology is dependent upon developing highly selective, easy preparative, cost-effective, reusable polymers which are stable in extreme conditions [22]. When these unique properties were taken into consideration, molecular imprinting has been introduced as a promising approach for the generation of highly crosslinked polymeric matrices.

5.3 Plasmonic Sensing

When polarized light is sent into a gold-plated prism, some of the light is absorbed, some of it is reflected. Free electrons referred as surface plasmons on the outer surface of the metal layer oscillate and this phenomenon is called surface plasmon resonance. When the intensity of the reflected light is observed by changing the angle of incidence, it is seen that the reflected light intensity decreases. Resonance angle is the angle at which the maximum loss in reflected light intensity occurs. At a specific angle of incidence, due to matching of electron frequencies on the metal surface, it will interact with the incident light. A decrease in the intensity of the reflected beam occurs, because of the absorbed energy in this resonance state [23]. In case of the accumulation on the surface or changes of the environment properties in contact with the metal surface, the resonance angle changes lead to changes in resonance. Achieving controlled accumulation on the surface provides an important advantage for the development of quantitative sensors. Resonance angle is extremely sensitive to refractive index changes of the medium in the very close to the metal surface. The difference of angle changes which occurs by binding of molecules which produces changes in refractive index.

SPR is the collective oscillation of conduct light at the interface between the negative and positive transmittance material excited by the incident light. SPR provides the basis of several standard means for the measurement of the material adsorption to the surface of metals (typically gold or silver, partially copper, titanium, or chrome) and metal nanoparticles. The surface plasmon polarization is an electromagnetic surface wave without radiation propagating in a direction parallel to the negative permeability/dielectric material interface. The oscillations are very sensitive to small changes, to give an example, the adsorption of molecules to the conductive surface, because the wave is at the boundary of the conductor and the external area [24].

Local surface plasmon resonance (LSPR) exploits the collective electron charge oscillations induced by light in metallic nanoparticles (i.e. silver, gold) exhibiting close field amplitude at resonance wavelength. This field is localized to the nanoparticle and the far field scattering by the particle is also enhanced by its resonance. When metallic nanomaterials interact with light, photons are absorbed and then they scatter in different directions. Both absorption and reflection are greatly increased when LSPR is stimulated [25].

When LSPR was used, one could benefit from the advantage of generating portable, sensitive, label-free, inexpensive, real-time, and fast response strategies for the detection of biomolecules by nanoscale spatial resolution. Nano-sized sensing offers a tremendous chance to miniaturize sensors by easy fabrication to a rather small magnitude inaccessible with the other strategies including SPR. Low-cost microfluidic devices could also be combined with LSPR and therefore, they enable precise diagnostic evaluation. Eventually, point-of-care diagnostic tools could be improved with the favor and competence of LSPR [26].

Over the last decades, molecular imprinting technology has been combined with sensors for creating molecularly imprinted sensors [27]. Especially the recent developments and investigations in the field of MIP-based sensing systems have gained great attention. Substances analyzed are mainly the catalytic proteins as well as glycosylated proteins. Furthermore, besides micro-organisms also drugs and compounds affecting growth (acids and growth factors) have been analyzed [28]. To give an example, in our research group, molecular imprinting was applied for creating an artificial creatinine receptor on the gold surface of SPR chip. In Figure 5.1, a schematic representation of sensor fabrication is given. It is seen that the surface of the SPR chip was modified with allyl mercaptan. *N*-methacryloyl-(L)-histidine methyl ester (MAH), ethylene glycol dimethacrylate (EGDMA), 2,2′-azobis(2-methylpropionitrile) (AIBN) were used as functional monomer, crosslinker and initiator, respectively. In the presence of the template molecule (creatinine), UV polymerization was performed and a polymeric film was produced. It can be clearly seen that creatinine imprinted SPR sensor was obtained successfully with specific recognition cavities [29].

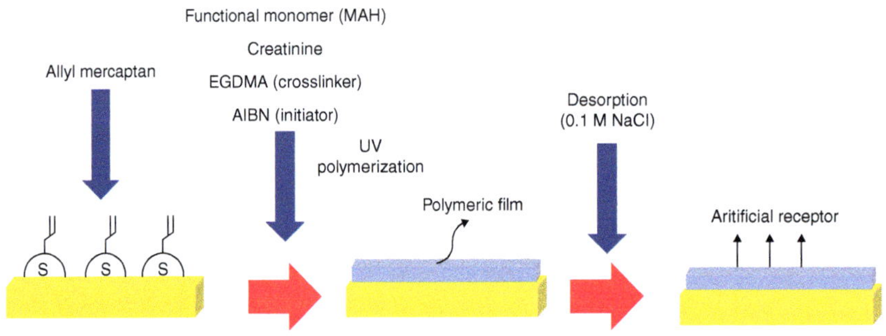

Figure 5.1 Schematic representation of sensor fabrication. Source: Reproduced with permission from Topçu et al. [29]. © 2019, Elsevier.

5.4 Medical Applications

The sensor technology based on SPR technique has recently been introduced as a potential tool for the characterization and detection of the interactions of biomolecules. SPR-based sensing platforms have filled a huge gap, and the combination of technologies is very favorable for the development of highly sensitive and specific biosensors constructed with easy and quick response methodologies providing the generation of laboratory tests operated with fewer biological recognition biomolecules. In the field of medical applications [28], drug analysis and development [30], nucleic acid examination [31], immunoglobulins detection [32], hormone levels determination [33], microorganism detection and, quantification [34], and toxin determination [35] could be performed.

5.4.1 Drug Detection Via MIP-based SPR Sensor

SPR sensor is used for a label-free determination technique based on the surface plasmon of metal-dielectric waveguides [16, 36, 37].

In a previous study, Luo et al. [30] developed a nanoscale molecularly imprinted polymeric film-based SPR sensor for highly selective and sensitive detection of the ciprofloxacin. They used ciprofloxacin (CIP) as a small molecule model to evaluate the performance of the sensor system. This molecularly imprinted polymeric film-based SPR sensor was obtained by an in situ photo-initiated polymerization method. Applied method has advantages such as good uniformity, controllable thickness, and short polymerization time. The thickness and wettability of molecularly imprinted polymeric film on the SPR sensor were monitored by stylus profiler and contact angle measurement. The authors showed a well sensitive, stable, and selective sensor system for the detection of CIP. They selected azithromycin, dopamine, and penicillin as competitor molecules and showed high imprinting selective factors for CIP against these molecules. The detection studies were performed in the concentration range between 10^{-11} and 10^{-7} $\mathrm{mol\,l^{-1}}$ of CIP. Also, the stability and repeatability of the presented sensor system were proven.

In another study, Roche et al. [38] used an SPR sensor for detection of a common cough suppressant, dextromethorphan molecule. Dextromethorphan, defined as a pharmacological important marker drug, enables the activity of the CYP2D6 class of p450 monoxygenases. They used β-cyclodextrin for the production of thin layer imprinted polymers to design a sensitive and selective detection method. They reported 0.035 μM as the limit of the detection (LOD) value when the concentration range of the template was between 0.035 μM and 6.00 mM. Also, they used commercial fluorescence-based methodologies and liquid chromatography-mass spectroscopy for the evaluation of obtained results.

To give an example for detection of drug active ingredient, Altınbaş et al. [39] designed a nanoparticle-based SPR sensor via molecular imprinting method for detection of diclofenac. They reported approximately 132 ± 3.2 nm size of nanoparticles with 0.1 polydispersity index. The immobilization of nanoparticles on the surface of the SPR chip was done with a combination of *N*-ethyl-*N*′(3-[dimethylamino]-propyl)carbodiimide (EDC), and *N*-hydroxy succinimide (NHS). They reported 1.24–80 $\mathrm{ng\,ml^{-1}}$ concentration range value for detection of diclofenac.

In the other study, they designed a MIP nanoparticle-based SPR sensor for drug detection and recognition. They reported 169.4 ± 3.5 nm size and with 0.3 polydispersity index of the nanoparticles. They detected metoprolol in the range of 1.9–1.0 μg ml^{-1} [40].

Jiang and coworkers prepared a low-cost and sensitive SPR sensor system based on molecularly imprinted polymeric film for the detection of histamine. They used methacrylic acid (MAA), EGDMA monomers in dimethyl sulfoxide (DMSO), and histamine as a template. They coated the monomer solution on the surface of the SPR chip for the preparation of molecularly imprinted film. They investigated the effect of pH conditions and the concentration of histamine in the range of 25–1000 μg l^{-1} on the detection of histamine [41].

Antibiotics as micropollutants affect the environment and destroy ecosystems. The suitable methods are needed for the determination of antibiotic concentration. Ayankojo et al. used molecular imprinting technology for sensitive and real-time detection of amoxicillin via SPR sensor system. Firstly, they prepared amoxicillin imprinted film using organic functional monomer (methacrylamide), inorganic precursor (tetraethoxysilane), and coupling agent (vinyltrimethoxysilane). They combined this molecularly imprinted film with an SPR sensor to selectively determine concentration of amoxicillin in the sample. They prepared a non-imprinted film for showing the binding capacity, sensitivity, and selectivity of a molecularly imprinted film. The LOD value was reported as 73 pM for amoxicillin [42].

In another study, Kara et al. [43] developed a SPR sensor for the detection of chloramphenicol via molecularly imprinted-based nanoparticles. The mini-emulsion polymerization method was used for the preparation of chloramphenicol imprinted nanoparticles. After that, the nanoparticles were coated onto the surface of the SPR chip. Contact angle measurements and atomic force microscopy were carried out for the characterization of SPR sensors. The effect of chloramphenicol concentration was examined in the 0.155–6.192 nM range, and the LOD was obtained 40 ng kg^{-1} when analyzing honey samples.

Sari et al. [44] designed a novel SPR sensor for highly selective, sensitive, and rapid detection of ciprofloxacin. They synthesized ciprofloxacin imprinted nanoparticles using MAA as a functional monomer. They used scanning electron microscopy (SEM), Fourier transform infrared spectroscopy (FTIR), contact angle, and zetasizer for the characterization of ciprofloxacin imprinted nanoparticles. After preparation of the nanoparticle-based SPR sensor, the detection experiments were performed. They reported a LOD value as 3.21 ppb in ultrapure water and 7.1 ppb in synthetic waste-water. The repeatability of the sensor showed 5.81% relative standard deviation (RSD). They demonstrated a simple, sensitive, low-cost, and selective molecularly imprinted-based sensor to detect ciprofloxacin. In another study, they designed a sensitive sensor system for detection of erythromycin using the same method via molecularly imprinted nanoparticles. They reported 0.29 ppm as LOD value in this study with 0.99 linearity range [45].

Recently, Faalnouri and coworkers developed a rapid molecularly imprinted-based SPR sensor for sensitive and selective determination of amoxicillin using the local and commercial milk samples. They combined polymeric nanoparticles and SPR sensors with film. These two different polymeric materials-based SPR sensors were used for rapid detection of antibiotics. The 0.1–200 ng ml^{-1} amoxicillin concentration range was evaluated in both systems. They reported 0.0009 and 0.0012 ng ml^{-1} LOD values for polymeric nanoparticles and film-based SPR sensor, respectively. Also, they showed the selectivity of sensing systems using cholesterol, ampicillin, and cephalexin as competitive agents [46].

In another study, Pernites et al. [47] prepared a highly selective and robust film-based sensor for the detection of theophylline. Theophylline is a phosphodiesterase inhibiting drug used in treatment for respiratory diseases. They used an electropolymerized molecularly imprinted film sensor. Figure 5.2A shows the schematic preparation of a molecularly imprinted film sensor. The formation of the electropolymerized molecularly imprinted film was observed by in situ electrochemical-SPR spectroscopy. Therefore, the changes in optical and electrochemical properties of the film were shown in real time. They used atomic force microscopy, static contact angle, quartz crystal microbalance, and X-ray photoelectron spectroscopy for the characterization of electropolymerized molecularly imprinted-based films. The LOD was obtained to be 3.36 μM^{-1}. The selectivity of the sensor

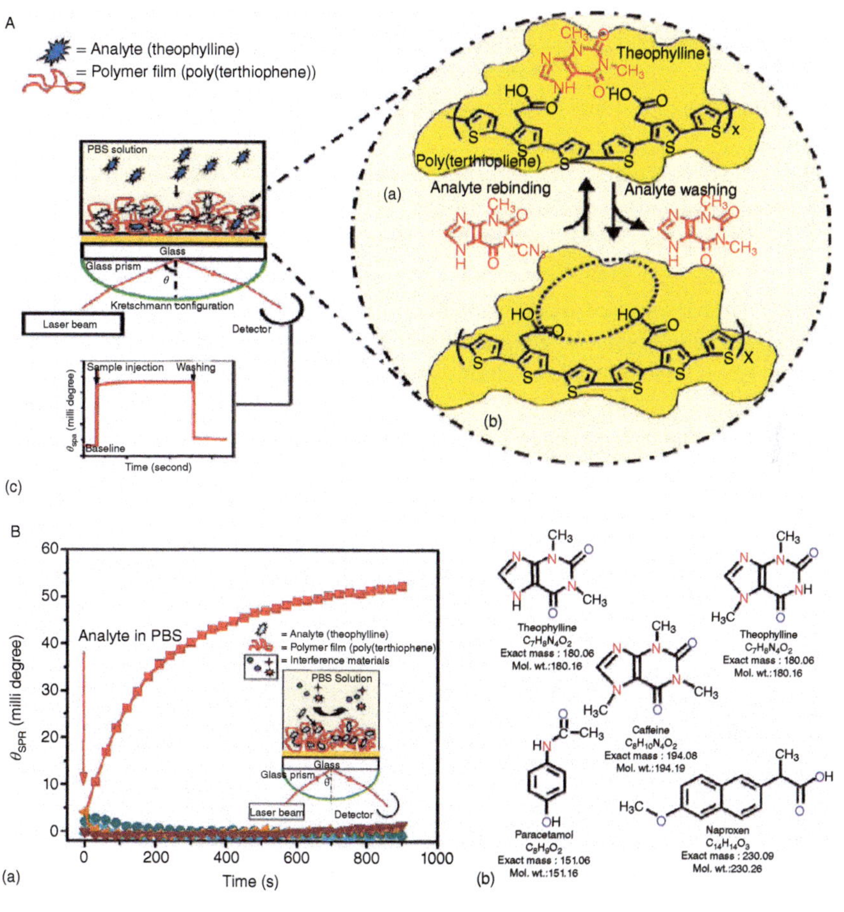

Figure 5.2 A. (a) Molecular imprinting of the template. (b) Formation of the cavity after washing the template. (c) SPR setup for sensing of the template. B. (a) Selectivity study of the molecularly imprinted sensor film, (b) Chemical structure of the compounds used for selectivity study. Source: Reproduced with permission from Pernites et al. [47]. © 2010, American Chemical Society.

system was obtained using 500 μM caffeine, 500 μM theobromine, 50 μM paracetamol, and 50 μM naproxen (Figure 5.2B).

5.4.2 Hormone Detection Via MIP-based SPR Sensor

Zangh et al. [48] developed a water-compatible macroporous molecularly imprinted film for highly sensitive, and selective detection of testosterone in urine. The film was obtained using 2-hydroxyethyl methacrylate (HEMA), MAA, and EGDMA in the presence of polystyrene nanoparticles which are coupled with testosterone. The authors reported the LOD as $10^{15}\,g\,ml^{-1}$ for testosterone. Figure 5.3 shows the scheme of SPR sensor setup.

Cenci et al. [49] used the molecular imprinting technique to create tailor-made synthetic recognition cavities via nanoparticles for detection of the hormone Hepcidin-25. They functionalized molecularly imprinted nanoparticles with biotin and then the resultant nanoparticles were immobilized on the surface of the SPR chip. The thickness of the SPR chip was found to be 20–50 nm. The real-time and sensitive detection of Hepcidin-25 was obtained in 3 minutes. They reported a highly selective detection of the target in the concentration range of 7.2–720 pM. They successfully demonstrated 5 pM LOD value and the applicability of the prepared sensing system was clarified using real serum samples spiked with Hepcidin-20.

In the other study, Türkoğlu and coworkers developed a sensitive 17β-estradiol (E2) imprinted film on the surface of SPR chip for selective detection of the template. They used *N*-methacryloyl-(L)- leucine methyl esters as a functional monomer in the presence of HEMA monomer. The non-imprinted film-based SPR chip was obtained without using 17β-estradiol. The kinetic properties of the SPR sensor were investigated in the concentration range of 17β-estradiol between 20 and 10 000 $ng\,ml^{-1}$. They showed the selectivity of systems via analysis of stigmasterol and cholesterol as competitive molecules. The analysis of a molecularly imprinted film-based SPR sensor showed that the molecularly imprinted film-based SPR sensor displayed greater selectivity to the target than other competitors and non-imprinted film-based SPR sensor [33].

In another work, Jiao et al. [50] fabricated a SPR sensor for the detection of 17β-estradiol using molecular imprinting technique. They used a UV polymerization method to synthesize the film on the surface of the modified SPR chip. They characterized the SPR chip via

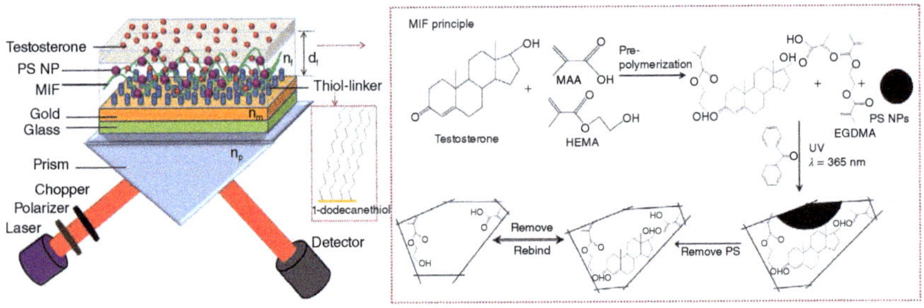

Figure 5.3 Scheme of SPR sensor setup, the PS–MIF (polystyrene nanoparticles-macroporous molecularly imprinted-based film) functionalized sensor chip, and the procedure for the formation of macroporous MIF. Source: Reproduced with permission from Zhang et al. [48]. © 2014, Elsevier.

a SEM, FTIR, and water contact angle measurements. They reported high selectivity for 17β-estradiol in the presence of the other molecules via a molecularly imprinted film-based SPR sensor. The sensing experiments were performed in the 2.5×10^{-16} – 2.5×10^{-8} mol l^{-1} concentration range in pH 7.4 and reported 9.14×10^{-18} mol l^{-1} LOD value showing high performance. Also, human urine and tap water were used as real samples for showing the selectivity and sensitivity of the proposed sensor system.

Nawaz et al. [51] fabricated a novel SPR sensor using an itaconic acid as a bifunctional group monomer for preparing a molecularly imprinted-based material. They used UV photo-polymerization in synergy with a chain transfer reagent-based 2-methyl-2 [(dodecylsulfanylthiocarbonyl)sulfanyl]propanoic acid and EGDMA crosslinker monomer. The prepared SPR sensor was utilized for the detection of progesterone. After the characterization of the SPR sensor, the detection experiments were carried out within the concentration range between 10^{-18} and 10^{-8} mol l^{-1}. This developed SPR sensor system has a great potential to provide real-time and label-free detection with high sensitivity and selectivity.

5.4.3 Microorganism and Virus Detection Via MIP-based SPR Sensor

In our research group, significant contributions have been made to the design of molecularly imprinted-based SPR sensors for the detection of micro-organisms. In a study, they used MAH as a functional monomer for the synthesis of molecularly imprinted-based film. They prepared a MAH-Cu(II) complex that resembled a natural ligand that has the ability to bind specifically to the bacterial cell surface. The surface imprinting approach was done with using Au nanoparticles to prepare *Escherichia coli* imprinted polymer on the modified SPR chip surface. Different bacterial strains were used for showing the selectivity of the sensor system. They observed 1.0 CFU ml^{-1} the LOD value in the SPR sensor [52]. To increase the sensitivity of molecularly imprinted-based SPR sensors, Özgür et al. introduced Ag nanoparticles into the polymer mixture during the creation of *E. coli* artificial receptors on the surface of the SPR chip. They reported 0.57 CFU ml^{-1} LOD value in aqueous solution. Clearly, the incorporation of Au and Ag nanoparticles significantly enhanced the SPR and thus raised the sensitivity of the SPR sensor [53].

In another study, Altintas et al. [34] used an automatic solid-phase synthesis method to prepare *E. coli* bacteriophage (MS2) imprinted-based sensor. Bacteriophage was imprinted in/onto the silica microbeads; therefore, molecular imprinting was integrated into a novel automated solid-phase method. The size of synthesized nanoMIPs was reported between 200 and 230 nm determined by using dynamic light scattering with a Zetasizer Nano. They reported LOD as 5×10^{6} plaque forming units ml^{-1}. New technology for the purification of water sources was proposed by removing viruses. The schematic preparation of sensor for the detection of viruses is given in Figure 5.4. Firstly, self-assembled monolayer was produced on the sensor chip. Then, amine coupling strategy was applied to bind biomolecules strongly to the surface of the sensor. The activation of carboxyl groups on the sensor surface was obtained by using an EDC/NHS enabling the formation of reactive succinimide esters. The immobilization of bacteriophage-imprinted nanoparticles was carried out on the activated surface of the sensor. Accordingly, the bacteriophages could rebind to the recognition cavities which are formed by successful imprinting.

5.4.4 Antibody Detection Via MIP-based SPR Sensor

SPR sensor systems play very important roles in the analysis of antibodies in human serum for the diagnosis of many diseases, especially autoimmune diseases. There have been many studies in the literature focused on the determination of antibodies with a molecularly imprinted SPR sensor and these studies are briefly given below.

Uzun et al. [54] prepared a Hepatitis B surface antibody imprinted SPR sensor for the detection of Hepatitis B antibody. The *N*-methacryloyl-L-tyrosine methyl ester monomer and hepatitis B antibody are complexed at a ratio of 2:1 (mole). They synthesized a film on the surface of the chip using the prepared complex and HEMA as a monomer. They applied different adsorption isotherm models and found that it is closest to the Langmuir model. Accordingly, they found the LOD to be 208.22 $mIU\,ml^{-1}$.

Ertürk et al. [32] prepared Fab fragments imprinted nanofilms on the SPR chip surface to develop SPR-based sensor for the detection of immunoglobulin G (IgG) molecules and Fab fragments. Fab fragments are antigen-binding arms of immunoglobulins. While performing the imprinting process, firstly complexing the template molecule with the MAH monomer was performed and then UV polymerization on the gold chip surface was carried out. This study could be given as an example for epitope imprinting. Sensing experiments were done using aqueous solutions of both Fab fragment and IgG molecules. According to the selectivity against Fab fragments, selectivity coefficients were determined to be 2.93 and 1.90 for

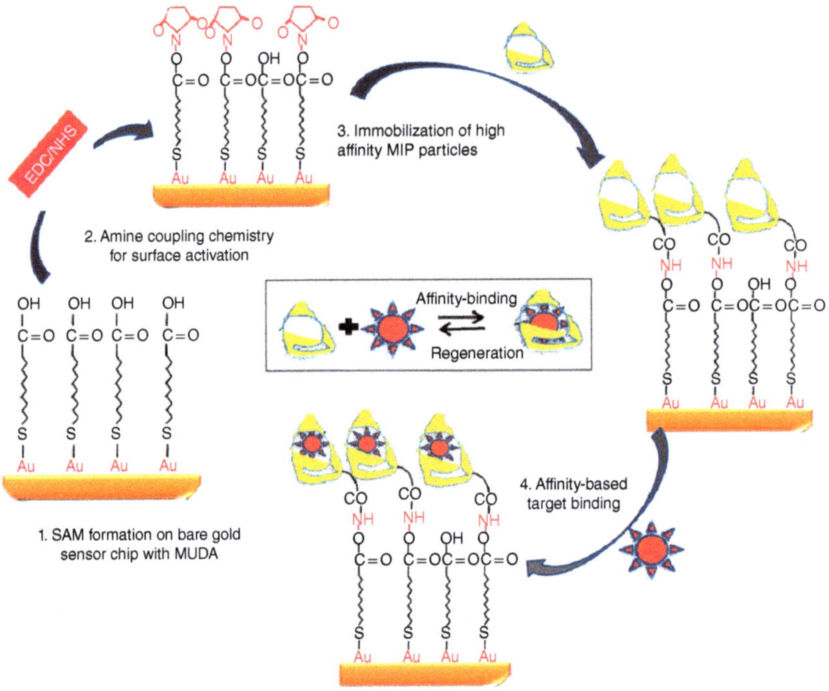

Figure 5.4 The schematic preparation of sensor for detection of viruses. Source: Reproduced with permission from Altintas et al. [34]. © 2015, American Chemical Society.

bovine serum albumin (BSA) and Fc fragment of IgG, respectively. According to the selectivity against IgG molecules, selectivity coefficients were calculated to be 21.00 and 13.63 for BSA and Fc fragments of IgG, respectively.

Dibekkaya et al. [55] prepared anti-cyclic citrullinated peptide (anti-CCP) imprinted SPR sensors for the diagnosis of Rheumatoid Arthritis. They prepared a pre-complex using acrylamide (AAm) as a monomer interacting with anti-CCP. Using this pre-complex, they carried out the polymerization on the surface of the gold chip modified with allyl mercaptan. They performed the experiments by preparing anti-CCP solutions of different concentrations in a pH 7.0 phosphate buffer (10 mM). According to the obtained data, they found the LOD as 0.177 RU ml^{-1}. The selectivity experiments were done using immunoglobulin M (IgM) and BSA; the selectivity ratios were found to be 3.0 and 8.0 respectively.

5.4.5 Nucleic Acid Detection Via MIP-based SPR Sensor

Nucleic acids have been widely applied for the detection of DNA and RNA molecules. They could be used in diagnostics for different analyzes such as genetic diseases, various mutations, bacterial biomarkers [56–58]. Despite the high availability in the medical area, plasmonic sensing applications of molecularly imprinted nucleic acids are very limited.

Diltemiz et al. [31] imprinted the guanine/guanosine DNA template by binding platinum (II) to the functional monomer of MAH. In this study, the detection was successfully carried out by combining the ligand exchange with the molecular imprinting method. A thin film was produced on the gold chip surface during the polymerization process. They determined that the films did not show effective affinity for adenosine and adenine. They showed that films have high affinity for ssDNA and dsDNA. They found the affinity constants as 65.827 M^{-1}, 61.548 M^{-1}, 37.206 M^{-1}, and 14.368 M^{-1} for guanine, guanosine, ssDNA, and dsDNA, respectively.

5.4.6 Biomarker Detection Via MIP-based SPR Sensor

A protein biomarker is an indicator for clinically identifying a particular disease. In order to be able to perform the treatment correctly at an early stage, it is of great importance to detect them quickly and even if they are found in low amounts. Due to the increasing number of protein biomarkers, there is a growing need for reliable methods for the sensitive and selective determination of various body fluids. For this purpose, the combination of SPR and molecular imprinting technique makes real-time and selective analysis available for daily routine clinical applications. Below, the analyses of protein biomarkers based on molecularly imprinted SPR sensor applications for the detection of various diseases such as cancer, renal, cardiac, blood diseases are summarized in Table 5.1.

5.5 Conclusion

Sensors have found a unique place in biomedical examinations, health care settings, and pharmaceutical checkover for several purposes. In this chapter, the usage of molecularly imprinted-based sensing was prescribed. Looking from the viewpoint of the biomedical

Table 5.1 Applications of molecularly imprinted-based SPR sensors for the detection of various diseases.

Marker	Monomer/ Comonomer	Imprinting method	LOD	LOQ	Selectivity (k′)	Disease	References
Myoglobin	HEMA/MATrp	Microcontact imprinting	26.8 ng ml^{-1}	87.6 ng ml^{-1}	Lysozyme, 3.19 Cytochrome C, 3.81 BSA, 5.59	Cardiac	[59]
Uric acid	HEMA/ MAC + Fe^{3+}	Nanoparticle-based imprinting	0.498 mg l^{-1}	1.66 mg l^{-1}	Ascorbic Acid, 6.288 Theophylline, 20.44 Urea, 243.02	Parcinson	[60]
Procalcitonin	HEMA	Microcontact imprinting	2.97 ng ml^{-1}	9.90 ng ml^{-1}	Myoglobin, 5.56 HSA, 1.81 Cytochrome C, 4.85	Sepsis	[61]
Prostate specific antigen	MAA	Microcontact imprinting	91 pg ml^{-1}	—	HSA, 5.92 Lysozyme, 6.63	Cancer	[62]
Lysozyme	EGDMA/MAH	Nanoparticle imprinting	84 pM	9.2 pM	Cytochrome C, 6.42	Rheumatoid arthritis	[63]
Lysozyme	EGDMA/MAPA	Nanoparticle imprinting	0.66 nM	—	Hemoglobin, 4.2 Myoglobin, 3.2	Rheumatoid arthritis	[64]

Microalbumin	EGDMA/MALM	Nanoparticle imprinting	0.7 pM	1.9 pM	Hemoglobin, 5.33 Transferrin, 9.56	Renal disease	[65]
Hemoglobin	Acrylamide	Surface imprinting	—	—	Transferrin, 4.36 BSA, 11.72 Myoglobin, 17.88	Thalassemia, hemoglobinopathies	[66]
Creatinine	MAH + Cu^{2+}	Nanofilm-based imprinting	57 μM	190 μM	Creatine, 5.61 NHS, 3.72	Renal disease	[29]
Myoglobin	HEMA/MATrp	Nanoparticle-based imprinting	4.72 ng ml^{-1}	15.74 ng ml^{-1}	BSA, 2.6 Cytochrome C, 1.8	Cardiac	[67]
Hemoglobin	Acrylamide	Nanofilm-based imprinting	0.00035 mg ml^{-1}	—	Lysozyme, 4.08 Transferrin, 7.04 BSA, 8.71 Myoglobin, 6.93	Thalassemia, hemoglobinopathies	[68]

LOD: Limit of detection, LOQ: Limit of quantification, HEMA: 2-hydroxyethyl methacrylate, EGDMA: Ethylene glycol dimethacrylate, MAA: Methacrylic acid, MAH: *N*-methacryloyl-(L)-histidine methyl ester, MAC: *N*-methacryloyl-(L)-cysteine, MATrp: *N*-methacryloyl-(L)-tryptophan, MAPA: *N*-methacryloyl-(L)-phenylalanine methyl ester, MALM: *N*-methacryloyl-(L)-leucine methyl ester BSA: Bovine serum albumin, HAS: Human serum albumin.

field, early detection at a low number of specific molecules in complex media carries a great importance. MIPs have the ability to function excellently mimic their natural counterparts. In comparison to conventional methods, molecularly imprinted-based sensors have been taken one step forward due to providing specific, selective, inexpensive, stable, portable, real-time, and fast response characteristics. It is noteworthy to indicate that LOD obtained by molecularly imprinted-based sensors approximately close to routine techniques. Moreover, molecularly imprinted-based sensor systems are open to functionalize for the improvement of LOD, selectivity, and sensitivity using appropriate functional monomers/ligands. In the future perspectives, molecularly imprinted-based sensing platforms are projected to enhance permanently in biomedicine with broad application prospects. Taken together, labs on chips provide point-of-care diagnostics could be obtained by this promising approach and these systems may have applications in patient-specific diagnostics or general usage in routines of high-throughput centers.

References

1 Yu, C. and Mosbach, K. (1997). *J. Organomet. Chem.* 62 (12): 4057–4064.
2 Ekberg, B. and Mosbach, K. (1989). *Trends Biotechnol.* 7 (4): 92–96.
3 Mosbach, K. and Haupt, K. (1998). *J. Mol. Recognit.* 11 (1–6): 62–68.
4 Chen, L., Wang, X., Lu, W. et al. (2016). *Chem. Soc. Rev.* 45 (8): 2137–2211.
5 Cennamo, N., D'Agostino, G., Arcadio, F. et al. (2020). *IEEE Trans. Instrum. Meas.* 69 (4): 1464–1473.
6 Akgonullu, S., Cimen, D., Bakhshpour, M. et al. (2020). *Commercial Biosensors and Their Applications: Clinical, Food, and Beyond*, 89. Elsevier.
7 Safran, V., Göktürk, I., Derazshamshir, A. et al. (2019). *Microchem. J.* 148: 141–150.
8 Yoo, H., Shin, J., Sim, J. et al. (2020). *Biosens. Bioelectron.* 168: 112561.
9 Özgür, E., Saylan, Y., Bereli, N. et al. (2020). *J. Biomater. Sci. Polym. Ed.* 31 (9): 1211–1222.
10 Trzaskowski, M. and Ciach, T. (2017). SPR system for on-site detection of biological warfare. *Anal. Chem.* 13 (2): 144–149.
11 Çalışır, M., Bakhshpour, M., Yavuz, H., and Denizli, A. (2020). *Sensors Actuators B Chem.* 306: 127561.
12 Masson, J.F. (2017). *ACS Sensors* 2 (1): 16–30.
13 Roque, A.C.A., Silva, C.S.O., and Taipa, M.Â. (2007). *J. Chromatogr. A* 1160 (1–2): 44–55.
14 Arora, S., Saxena, V., and Ayyar, B.V. (2017). *Methods* 116: 84–94.
15 Bakhshpour, M., Yavuz, H., and Denizli, A. (2018). *Artif. Cells Nanomed. Biotechnol.* 46: 946–954.
16 Idil, N., Hedström, M., Denizli, A., and Mattiasson, B. (2017). *Biosens. Bioelectron.* 87: 807–815.
17 Aslıyüce, S., Mattiasson, B., and Denizli, A. (2019). *Biomed. Chromatogr.* 33 (10): e4605.
18 Mosbach, K. (1994). *Trends Biochem. Sci.* 19 (1): 9–14.
19 Wulff, G. (1995). *Angew. Chem. Int. Ed. Eng.* 34 (17): 1812–1832.
20 Chen, L., Xu, S., and Li, J. (2011). *Chem. Soc. Rev.* 40 (5): 2922–2942.
21 Ikegami, T., Mukawa, T., Nariai, H., and Takeuchi, T. (2004). *Anal. Chim. Acta* 504 (1): 131–135.

22 Amatatongchai, M., Thimoonnee, S., Jarujamrus, P. et al. (2002). *Microchem. J.* 158: 105298.

23 Nguyen, H., Park, J., Kang, S., and Kim, M. (2015). *Sensors* 15 (5): 10481–10510.

24 Amatatongchai, M., Thimoonnee, S., Jarujamrus, P. et al. (2020). *Results Phys.* 16: 102874.

25 Caucheteur, C., Guo, T., and Albert, J. (2015). *Anal. Bioanal. Chem.* 407 (14): 3883–3897.

26 Singh, P. (2017). *Reviews in Plasmonic 2016*, 211–238. Cham: Springer.

27 Bhogal, S., Kaur, K., Malik, A.K. et al. (2020). *TrAC Trends Anal. Chem.*: 116043.

28 Idil, N., Bakhshpour, M., Perçin, I. et al. (eds.) (2020). *Nanosensor Technologies for Environmental Monitoring*, Nanotechnology in the Life Sciences. Cham: Springer.

29 Topçu, A.A., Özgür, E., Yılmaz, F. et al. (2019). *Mater. Sci. Eng. B Solid-State Mater. Adv. Technol.* 244: 6–11.

30 Luo, Q., Yu, N., Shi, C. et al. (2016). *Talanta* 161: 797–803.

31 Emir Diltemiz, S., Denizli, A., Ersöz, A., and Say, R. (2008). *Sensors Actuators B Chem.* 133 (2): 484–488.

32 Ertürk, G., Uzun, L., Tümer, M.A. et al. (2011). *Biosens. Bioelectron.* 28 (1): 97–104.

33 Türkoğlu, E.A., Bakhshpour, M., and Denizli, A. (2019). *Biointerface Res. Appl. Chem.* 7 (1): 1989–1994.

34 Altintas, Z., Gittens, M., Guerreiro, A. et al. (2015). *Anal. Chem.* 87 (13): 6801–6807.

35 Hodnik, V. and Anderluh, G. (2009). *Sensors* 9: 1339–1354.

36 Bakhshpour, M. and Denizli, A. (2020). *Microchem. J.* 159: 105572.

37 Perçin, I., Idil, N., Bakhshpour, M. et al. (2017). *Sensors* 17: 1375.

38 Roche, P.J.R., Ng, S.M., Narayanaswamy, R. et al. (2009). *Sensors Actuators B Chem.* 139: 22–29.

39 Altintas, Z., Guerreiro, A., Piletsky, S.A., and Tothill, I.E. (2015). *Sensors Actuators B Chem.* 213: 305–313.

40 Altintas, Z., France, B., Ortiz, J.O., and Tothill, I.E. (2016). *Sensors Actuators B Chem.* 224: 726–737.

41 Jiang, S., Peng, Y., Ning, B. et al. (2015). *Sensors Actuators B Chem.* 221: 15–21.

42 Ayankojo, A.G., Reut, J., Öpik, A. et al. (2018). *Biosens. Bioelectron.* 118: 102–107.

43 Kara, M., Uzun, L., Kolayli, S., and Denizli, A. (2013). *J. Appl. Polym. Sci.* 129: 2273–2279.

44 Sari, E., Üzek, R., Duman, M., and Denizli, A. (2018). *J. Biomater. Sci. Polym. Ed.* 29: 1302–1318.

45 Sari, E., Üzek, R., Duman, M., and Denizli, A. (2016). *Talanta* 120: 607–614.

46 Faalnouri, S., Çimen, D., Bereli, N., and Denizli, A. (2020). *ChemistrySelect.* 5 (15): 4761–4769.

47 Pernites, R.B., Ponnapati, R.R., and Advincula, R.C. (2010). *Macromolecules* 43 (23): 9724–9735.

48 Zhang, Q., Jing, L., Wang, Y. et al. (2014). *Anal. Biochem.* 463: 7–14.

49 Cenci, L., Andreetto, E., Vestri, A. et al. (2015). *J. Nanobiotechnology.* 13: 51.

50 Jiao, S.Q., Chen, X.L., Yu, J.Y. et al. (2019). *IOP Conf. Ser. Earth Environ. Sci.* 295 (3): 03217.

51 Nawaz, T., Ahmad, M., Yu, J. et al. (2020). *New J. Chem.* 44: 6992–7000.

52 Gür, S.D., Bakhshpour, M., and Denizli, A. (2019). *Mater. Sci. Eng. C* 104: 109869.

53 Özgür, E., Topçu, A.A., Yılmaz, E., and Denizli, A. (2020). *Talanta* 212: 120778.

54 Uzun, L., Say, R., Ünal, S., and Denizli, A. (2009). *Biosens. Bioelectron.* 24 (9): 2878–2884.

55 Dibekkaya, H., Saylan, Y., Yılmaz, F. et al. (2016). *Sci. Part A* 53 (9): 585–594.

56 Thum, T., Gross, C., Fiedler, J. et al. (2008). *Nature* 456 (7224): 980–984.
57 D'Agata, R. and Spoto, G. (2013). *Anal. Bioanal. Chem.* 405 (2–3): 573–584.
58 Caygill, R.L., Blair, G.E., and Millner, P.A. (2010). *Anal. Chim. Acta* 681 (1–2): 8–15.
59 Osman, B., Uzun, L., Beşirli, N., and Denizli, A. (2013). *Mater. Sci. Eng. C* 33 (7): 3609–3614.
60 Göçenoğlu Sarıkaya, A., Osman, B., Çam, T., and Denizli, A. (2017). *Sensors Actuators B Chem.* 251: 763–772.
61 Sener, G., Ozgur, E., Rad, A.Y. et al. (2013). *Analyst* 138 (21): 6422–6428.
62 Ertürk, G., Özen, H., Tümer, M.A. et al. (2016). *Sensors Actuators B Chem.* 224: 823–832.
63 Sener, G., Uzun, L., Say, R., and Denizli, A. (2011). *Sensors Actuators B Chem.* 160 (1): 791–799.
64 Saylan, Y., Yılmaz, F., Derazshamshir, A. et al. (2017). *J. Mol. Recognit.* 30 (9): e2631.
65 Esentürk, M.K., Akgönüllü, S., Yılmaz, F., and Denizli, A. (2019). *J. Biomater. Sci. Polym. Ed.* 30 (8): 646–661.
66 Saylan, Y. and Denizli, A. *Proc. IEEE Sensors.* 2017: 6–8.
67 Atay, N.O., Osman, B., Akgol, S. et al. (2018). *Curr. Appl. Polym. Sci.* 2 (2): 102–111.
68 Saylan, Y. and Denizli, A. (2018). *Sensors* 18 (9): 3016.

6

Magnetoplasmonic Nanosensors

Recep Üzek[1], Esma Sari[2], and Arben Merkoçi[3,4]

[1] *Department of Chemistry, Faculty of Science, Hacettepe University, Ankara, Turkey*
[2] *Vocational School of Health Services, Medical Laboratory Techniques, Yüksek 'Ihtisas University, Ankara, Turkey*
[3] *Catalan Institute of Nanoscience and Nanotechnology (ICN2), CSIC and BIST, Barcelona, Spain*
[4] *Catalan Institution for Research and Advanced Studies (ICREA), Barcelona, Spain*

6.1 Introduction

Nanotechnology, one of the multidisciplinary researches and development fields, has become very popular in the production of novel materials in recent years. Nanoparticles (NPs) in the 1–100 nm range are seen as pioneers of innovations particularly in the materials science, because of their adjustable physicochemical characteristics. Over the last decades, magnetic nanoparticles (MNPs), especially magnetic iron and iron oxide NPs (magnetite (Fe_3O_4), maghemite (γ-Fe_2O_3), etc. have become much attractive not only in the biomedical applications such as separation, purification, drug delivery, imaging, disease diagnosis, and treatments, etc. but also in the sensing applications owing to their unique properties such as low-cost production, high surface area, magnetic moment, chemical stability, ease of synthesis and modification, biocompatibility [1]. Moreover, MNPs are designed as multifunctional nanocomposites (MFNCs) in almost all of these applications [2].

The MFNCs present heterogeneous structures in the forms of core-shell, heterodimer, doped, or embed [3–5]. There are generally two different approaches which are applied to synthesize the MFNCs in the form of core–shell structure: (i) coating of the magnetic nanoparticle with the different types of NPs (Au NPs, QDs, etc.) or the shell including silica, polymer or molecules; (ii) coating of different types of NPs (Au NPs, QDs, etc.) with magnetic film or magnetic NPs. To form heterodimeric NPs (dumbbell-like bifunctional NPs), two different types of NPs, separately synthesized, are bound interfacially [6]. The other multifunctional nanocomposite structure is obtained by loading of MNPs in the synthesis process of NPs such as silica, polymer, etc. to embed MNPs into the structure of NPs or by encapsulating MNPs into porous NPs such as carbon nanotubes, mesoporous silica, etc. [7]. The chemical composition, structure, size, and shapes of MFNCs are designed according to the desired applications and features. For example, the MFNCs composed MNPs

Plasmonic Sensors and their Applications, First Edition. Edited by Adil Denizli.

with PNPs or QDs are used for the multimodal imaging in biomedical applications, while the MFNCs formed by coating of MNPs with the polymer for the drug encapsulation are used for the drug delivery [8–10]. In sensor applications, MFNCs are generally synthesized with the MNPs and the NPs commonly used in the sensor applications, such as carbon-based NPs, plasmonic NPs (PNPs), and quantum dots (QDs), etc. Moreover, the sensitivity and stability of sensors are enhanced by using MFNCs prepared with the MNPs and the NPs. Magnetoplasmonic nanocomposites (MPNCs) synthesized by using MNPs and PNPs are the most widely used MFNCs in the sensor applications.

PNPs, exhibit the surface plasmon created by the oscillations of the conduction-band electrons on the surface of noble metals exited by electromagnetic radiations. These can be synthesized in different shapes, sizes, compositions, or structures with the tunable optical properties from the visible to near-infrared regions [11]. PNPs are widely used in many applications especially sensors, solar cells, biomedical and biological imaging, due to their superior optical and chemical properties. In recent years, PNPs, exhibiting effective heating at λ_{max}, have been also efficiently used in the cancer treatments including photothermal therapy and heat-assisted antitumor drug release [12, 13]. Gold (Au) and silver (Ag) NPs have been predominantly studied as PNPs for the diagnostic and biomedical applications owing to their wide-range spectral optical properties, biocompatibility, chemical stability, easy modification, and synthesis.

Since the last decades, MPNCs, the novel type of MFNCs, have exhibited the great potential in the development of innovative approaches for the biomedical applications such as biosensor, cancer treatment, optical imaging, magnetic resonance imaging (MRI) [14–16]. MPNCs are specially designed for the magnetically and optically dual imaging, simultaneously MRI and photothermal therapy or thermally drug release in the cancer treatments [17]. In sensor applications, the dual-mode sensing including magnetic and optic as well as optical detection after the magnetic separation are used [14, 18]. The optic and magnetic properties of MPNCs can be tuneable by changing of morphology, composition, and size of MPNCs to suit the desired final applications [19, 20]. The morphologies can be mainly achieved in the form of the heterodimer, multicomponent doped hybrid, spherical, or non-spherical core–shell or core–satellite. In this chapter, we discuss the various synthesis procedures and the main sensing applications of MPNCs.

6.2 Synthesis

There are three main approaches applied for the synthesis of MPNCs with interest for biosensing applications: (i) combining of the NPs after the synthesis of individual magnetic and plasmonic NPs to form the heterodimeric structures; (ii) coating of the surface of the core consisting of MNPs with PNPs in the form of a shell or antenna (the core–shell or core–satellite structures); (iii) doping of various individual PNPs or MNPs into the carrier materials (i.e., polymer or silica) to form the multicomponent hybrid structures. These structures are illustrated in Figure 6.1. In this section, the synthesis methods for these structures in the biosensing applications involving chemical reduction [21–23], thermal decomposition [24, 25], solvothermal method [26–29], microemulsion method [30, 31], electrochemical deposition [32], gamma-ray radiation [33–35], etc. are described.

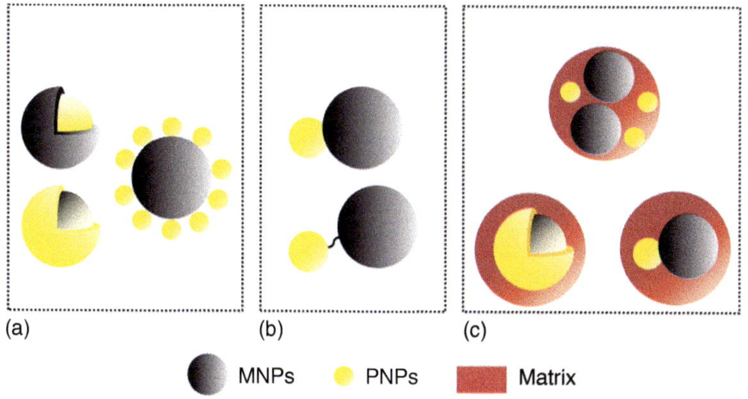

Figure 6.1 The structural form of MPNCs: core-shell or core-satellite structures (a); heterodimer (b); multicomponent (c). Source: © John Wiley & Sons.

6.2.1 Core–Shell or Core–Satellite

While a single or multi-magnetic nanoparticle cores are covered completely with gold shell in the form of a core–shell structure, the core–satellite structures are formed by coating of the magnetic cores with gold NPs. This type of coating increases specific gold areas on magnetic NPs besides providing the functionalizable areas on the uncovered magnetic core for further applications. In core–shell structures, the full coating of the magnetic cores reduces magnetic amplitude while increasing stability and biocompatibility. Moreover, the optical properties of MPNCs in the core–shell structures can be tunable by changing the gold shell thickness. As an example, Billen et al. [36] prepared MPNCs in the form of a core–shell structure. Firstly, the MNPs were synthesized by using the thermal decomposition method, and then the surface of the MNPs core was coated with the Au shell by chemical reduction of gold salt to obtain the MPNCs. The gold shell thickness was adjusted by changing the amount of gold salt added during the reduction process. Then, the thiolated PEG was used for the surface modification of MPNCs to enhance the solubility and biocompatibility for further experiments. Therefore, the structure of MPNCs should be designed considering its advantages for the desired applications.

In both morphologies, the magnetic cores are generally coated with the plasmonic layer by using chemical reduction [37, 38], electrostatic interactions [39], solvothermal method [40], etc., after their synthesis using methods such as thermal decomposition [37], solvothermal process [41], coprecipitation [42], micelle synthesis [43], hydrothermal synthesis [44], etc.

Liang et al. [45] designed the core–shell MPNCs for surface plasmon resonance (SPR) bioassay. Firstly, Fe_3O_4 NPs were synthesized by the coprecipitation method and then the surface was modified with (3-aminopropyl)triethoxysilane (APTES). The gold shell on the MNPs was formed to obtain MPNCs by the reduction of the gold precursor. Finally, the surface of MPNCs was functionalized with α-fetoprotein antibody to specifically bind to α-fetoprotein for SPR bioassay. In a similar study where the MPNCs were synthesized by the same method, the MPNCs were functionalized with glucose oxidase and trypsin after hexadecanethiol modification of MPNCs [46]. In another core–shell approach, the precursors including Fe $(NO_3)_3.9H_2O$ and different amounts of $AgNO_3$ were used to synthesize the MPNCs based on Ag NPs and Fe_3O_4 NPs via the solvothermal method [47]. The core–shell structure was formed by the encapsulation of Ag NPs by Fe_3O_4 shell. It was reported that the core–shell structure was subjected to deviations in high $AgNO_3$ amounts and these deviations increased with the increase in $AgNO_3$ concentrations.

Chen et al. synthesized MPNCs by thermal decomposition in the form of a core–satellite structure [48]. Firstly, the spherical and cubic MNPs were synthesized by using thermal decomposition and hot-injection polyol method, respectively and then the MNPs were coated with Polysiloxane-Diblock Copolymer. After the Au NPs were synthesized in the size of 2–4 nm by using the aqueous reduction method, the Au NPs were immobilized by the electrostatic interactions on the surface of Polysiloxane-Diblock Copolymer coated MNPs to obtain the core–satellite MPNCs. Finally, the bioconjugation performance of core–satellite MPNCs was tested by using peptides, streptavidin, oligonucleotides, antibodies. The results indicate that the efficiency of conjugation is higher than EDC/NHS coupling method. Moreover, the core–shell satellite MPNCs based on the polymer-coated cubic MNPs present a strong magnetization and a high biomolecule which is also associated with the amount of Au NPs immobilized on the surface. In a similar approach, polyethyleneimine was used for the modification of MNPs to bind the Au NPs for the synthesis of MPNCs. In another study, the core–satellite MPNCs were synthesized by the deposition of Au NPs on the surface of Fe_3O_4 MNPs [49]. Sun et al. [50] synthesized the core–satellite MPNCs by using the Au NPs and the silica-coated Fe_3O_4 NPs. Firstly, the Au NPs were synthesized via the aqueous reduction method, and the Fe_3O_4 MNPs were coated with the silica shell after the synthesis of the Fe_3O_4 MNPs by the solvothermal method. Then, the core–shell silica-coated Fe_3O_4 MNPs were modified with poly(diallyldimethylammonium chloride) (PDDA) for the immobilization of Au NPs by layer-by-layer (LBL) method to achieve the core–satellite structure. This method provides the adjustable NP gap on the surface of core–satellite MPNCs by changing the amount of Au NPs in the synthesis [50, 51]. Abbas et al. [52] designed a one-pot synthesis method for the core–satellite MPNCs. The Fe_3O_4 MNPs were synthesized in the cubic form by using the sonochemical method. Then, the surface of Fe_3O_4 MNPs was functionalized with PVP as a template for the Au NPs nucleation via the reduction method. Similarly, the MPNCs were synthesized by using the cubic Fe_3O_4 MNPs coated with PEI and PEI-DTC [53]. In another approach, the Fe_3O_4 MNPs were synthesized by the modified polyol method [54]. Then, the SiO_2 layer onto the MNPs was formed by using TEOS via the Stöber method. For the conjugation of Au seeds onto the core–shell Fe_3O_4@SiO_2 MNPs, the amine functionalization of the surface was carried out by using APTES. Then, the Au seeds were synthesized and mixed with the amine modified core–shell Fe_3O_4@SiO_2 MNPs to obtain the MPNCs. The surface of MPNCs was modified with benzene-1,4-dithiol (BDT) for the growing of second Au or Ag layer onto the Au seeds. The illustration of synthetic procedures and TEM images of MPNCs are given in Figure 6.2. As seen in the TEM images, this synthesis method indicates that the uniform MPNCs were synthesized in the form of core–shell.

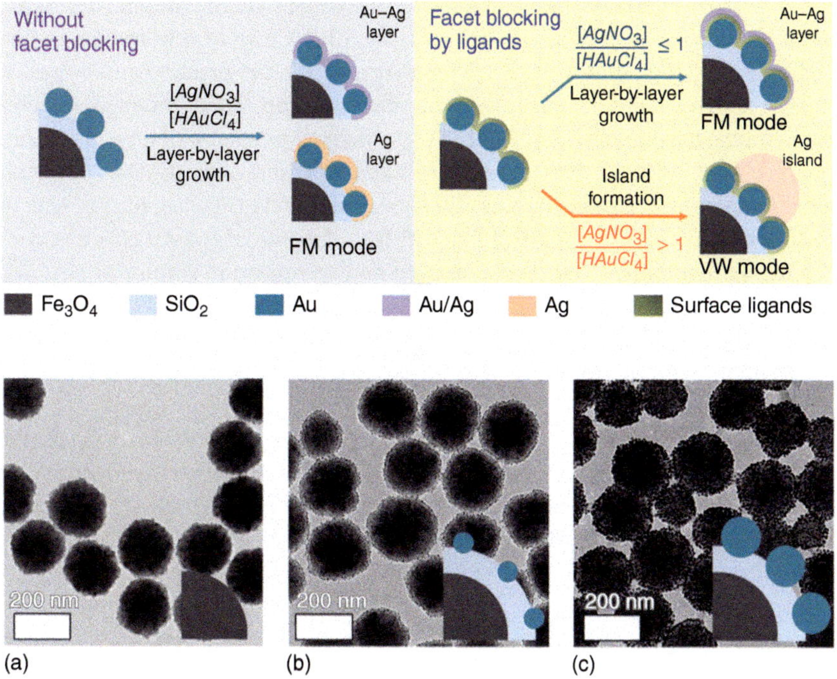

Figure 6.2 The schematic illustration of the synthesis mechanism for the MPNCs by using a modified polyol method. TEM images of NPs: (a) Fe_3O_4 MNPs. (b) Au seed NPs-the core-shell Fe_3O_4@SiO_2 MNPs (c) the core-satellite Fe_3O_4@SiO_2@Au NPs MPNCs after Au seed-mediated growth. Source: Reprinted from Kim et al. [54]. © 2020, with permission from John Wiley & Sons.

6.2.2 Heterodimers

The independently synthesized NPs, including single PNP and MNP, are interfacially bonded to form the dumbbell structure known as heterodimeric structures. In another approach, one type of the NPs from PNPs or MNPs is directly synthesized and then the other one can be decomposed on this selected NPs to form the heterodimer MPNCs. The MPNCs in the form of heterodimeric structures are generally synthesized by using the solvothermal method [55, 56], flame spray pyrolysis [6], thermal decomposition [57–62], in situ deposition [63], hydrothermal method [64], etc.

Yu et al. [65] developed the Dumbbell-like MPNCs by using the decomposition of $Fe(CO)_5$ on the surface of the Au NPs. In a similar approach, the MPNCs were synthesized by the decomposition of $Fe(acac)_3$ on the surface of gold NPs [66]. In both approaches, the Au NPs were synthesized firstly and then the MPNCs were prepared in the form of a heterodimeric structure. In another approach, the Ag NPs were synthesized and used instead of the Au NPs and then the decomposition of $Fe(acac)_3$ was carried out on the surface of Ag NPs [67]. The results obtained from the synthesized MPNCs with Ag NPs indicate the enhancement of the thermal stabilization and the magnetic anisotropy according to the MNPs alone.

Kostevsek et al. [60] synthesized the dumbbell shape MPNCs via the thermal decomposition method in the one-step synthesis process. The MPNCs were synthesized by the reduction of $Au(acec)_3$ and $Fe(CO)_5$ with 1,2-hexadecandiol in the presence of oleyamine (OLA) and oleic acid (OA) capping agents in the same solution. Then, the modified chitosan was used for the functionalization of MPNCs to increase biocompatibility. In another similar approach [62], $Fe(CO)_5$ was decomposed and oxidized onto the Au NPs with 1-octadecene in the presence of oleyamine (OLA) and oleic acid (OA) capping agents in the same solution to obtain the dumbbell like MPNCs. The size of the Au NPs was controlled by changing the temperature of the reaction and the injection volume of $HAuCl_4$. Moreover, the size of Fe_3O_4 MPNCs was adjusted by changing the ratio between $Fe(CO)_5$ and Au, and the size was increased by increasing the amount of $Fe(CO)_5$. As seen in Figure 6.3, the functionalized dumbbell like MPNCs were quite uniform structure with the size of 8–20-nm (Au-Fe_3O_4).

In another approach, Nosrati et al. [68] fabricated the heterodimer MPNCs by the reduction of Au NPs on the surface of Fe_3O_4 MNPs. Firstly, the Fe_3O_4 MNPs were synthesized via the coprecipitation method and then $HAuCl_4$ was reduced to the Au NPs on the surface of Fe_3O_4 MNPs. Finally, the heterodimer MPNCs was functionalized with BSA to enhance the stability in the physiological medium.

6.2.3 Multicomponent Doped Hybrids

The multicomponent hybrid MPNCs are formed by embedding or encapsulation of PNPs and MNPs into matrixes such as such as carbon-based structures (i.e., graphene, graphene oxides, carbon nanotubes, etc.) [17, 69, 70], metal-organo-frameworks (MOF) [71] polymer [72], silica [73, 74], called as the multicomponent hybrid MPNCs. The modification or functionalization of these matrixes with the PNPs and MNPs also creates the multicomponent hybrid MPNCs.

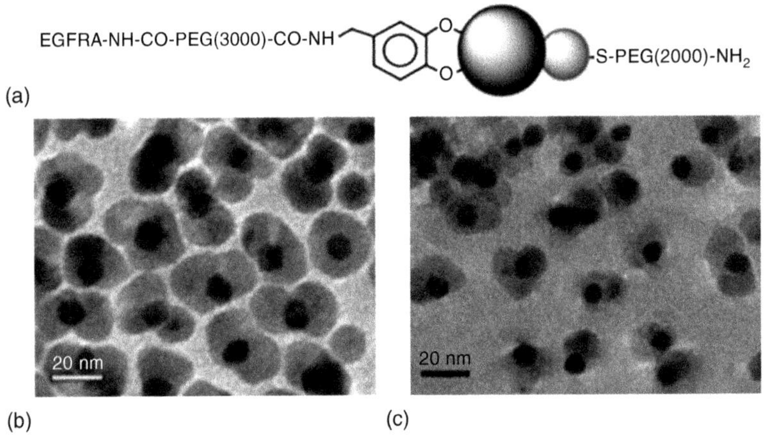

Figure 6.3 The surface functionalization of the dumbbell like MPNCs (a); TEM images of the dumbbell like MPNCs before (b) and after (c) surface functionalization. Source: Reprinted from Xu et al. [62]. © 2008, with permission from John Wiley & Sons.

Shi et al. [17] fabricated the multicomponent MPNCs by the modification of graphene oxide (GO) with the Fe_3O_4 MNPs and Au NPs. Firstly, the magnetic reduced graphene oxide (mRGO) was synthesized by simultaneously forming Fe_3O_4 MNPs and reducing graphene oxide via the solvothermal method. Then, the Au seeds were prepared and grown on mRGO. Finally, the multicomponent MPNCs were functionalized with the lipoic acid-modified PEG (LA–PEG) and folic acid conjugated LA–PEG (LA–PEG–FA) for the enhancement of stability and the decreasing the toxicity. In another approach, the core–shell MPNCs were immobilized on GO by EDC/NHS coupling [75].

In another study, Chen et al. [72] synthesized the multicomponent MPNCs by the encapsulation of PNPs and MNPs in the polymer. Firstly, the carboxylate functionalized Fe_3O_4 (c-Fe_3O_4) MNPs were synthesized by the solvothermal process. Then, $HAuCl_4$, c-Fe_3O_4, and Poly(styrene-alt-maleic acid) (PSMA) $HAuCl_4$ were mixed and the encapsulation was carried out by the adding of HCl for the crosslinking of the polymer. Finally, doxorubicin hydrochloride (DOX) was loaded onto the MPNCs for chemo-photothermal therapy.

Poorakbar et al. [74] fabricated the mesoporous silica-based MPNCs. Firstly, the magnetic Fe_3O_4 NPs were synthesized by the coprecipitation method and then the surface was coated with Au and Ag shell by the thermal decomposition, respectively. Then, the surface of the core–shell MPNCs was covered with the mesoporous silica. Finally, the cellulase enzyme was immobilized onto their surfaces.

6.3 Biosensing Applications

The on-site and rapid analysis of many biomolecules such as proteins, nucleotides, enzymes, especially biomarkers, etc. is critical to prevent the spread and treatment of many diseases. Nowadays, biosensors are come into more prominence in the diagnosis of many diseases, such as the covid19 epidemic, owing to their unique features of simple, cheap, and fast compared to traditional methods. The synergic effects caused by the combining of the magnetic and plasmonic properties in a single nanoparticle (i.e. MPNC) enhance the stability and sensitivity of probes used in biosensing applications. The MPNCs, the novel type of multifunctional nanocomposites, fabricated by combining of the MNPs and PNPs, has attracted attention for the development of innovative approaches for the biomedical applications such as for biosensors, cancer treatment, optical imaging, magnetic resonance imaging due to their superior properties, tunable optical properties with a strong magnetic moment, high biocompatibility, and stability. These unique features provide that dual-mode sensing including magnetic and optical as well as optical detection after magnetic separation was used in the sensor application. In this section, their uses in the detection of major analytes such as proteins, nucleotides, and pathogens that form the basis of biosensor applications are examined.

6.3.1 Protein

The detection of biological entities such as proteins, enzymes, cholesterol, and other metabolic compounds are of great importance in clinical diagnosis, treatment, and monitoring of biological systems [76–78]. In recent years of sensing application research, it has been proposed that integrating the interdisciplinary incorporation of the various impacts of optical, electrical, or magnetic properties into a single technique could give rise to a new multitasking platform [79–81]. Due to the desired properties of magnetoplasmonic nanocomposites (MPNCs) such as excellent surface chemistry, special optical properties, and superparamagnetic properties, these compounds are widely used for protein detection [82–86].

The concept of MPNCs was demonstrated using an aptamer-based assay for electrochemical sensing of thrombin. Two separate NPs were used to form the duplex structure as a magnetoplasmonic sensing probe by attaching aptamer on MNPs and labeling gold NPs with complementary oligonucleotides. The addition of the aptamer, that recognized target, resulted in the binding of thrombin and cause the dissociation of the magnetic nanoparticle-aptamer/gold nanoparticle-oligonucleotide probe. Then, the electrochemical signal of the released gold nanoparticle labeled with complementary oligonucleotide was measured by the DPV responses. The readout peak current of DPV was observed proportionally with the amount of thrombin, and the quantitative assay of thrombin was successfully operated by using this suggested MPNCs-based sensing platform. The proposed sensing system exhibited good selectivity and rapid detection by combining magnetic separation with the excellent electrochemical properties of gold NPs. These promising results lead authors to conclude that the proposed sensing platform may well be combined with such systems to extend the application areas [84]. Besides this heterodimeric morphology, another method to ultrasensitive electrochemical detection of thrombin that is based upon synthesizing core/shell Fe_3O_4/Au MPNCs. Two separate aptamers were utilized to recognize the different parts of thrombin in order to form a sandwich structure. According to the results, the plasmonic resonance of the noble metal coated over a magnetic core was found to be specific, sensitive, and having a low detection limit for thrombin achieved by the electrochemical catalytic signals [85].

Furthermore, the different techniques such as Surface Plasmon Resonance (SPR), Localized Surface Plasmon Resonance (LSPR), Surface-Enhanced Raman Scattering (SERS) combined with MPNCs have been used to serve as a sensing system for protein detection. In order to overcome insufficient sensitivity for the determination of small molecules or low-concentration compounds in the matrix, the different referred methods should be coalesced with MPNCs [45, 86–88]. For example, Liang et al. addressed the detection of α-fetoprotein (AFP) depending on an amplification technique using the core–shell magnetoplasmonic nanocomposites (Fe_3O_4@Au MPNCs) for an SPR bioassay. In this study, Fe_3O_4@Au MNPs were used not only as of the magnetic cores for easy manipulation in an external magnetic field but also as the surface modifier with the Au layer that provides an interface for the assembly of antibody molecules. In the sandwich-type design, AFP is captured by a primary AFP antibody (Ab1) and sensitively detected through the addition of secondary antibody conjugates (Fe_3O_4@Au–Ab2). Compared with conventional SPR, Fe_3O_4@Au–Ab2 conjugates result in a significant SPR angle increase, which is mainly attributed to the high refractive index and high molecular weight of the Fe_3O_4@Au MNPs therefore these conjugates afforded through as an amplifier. SPR angle shift was linearly changed by binding of AFP in the range of 1.0–200.0 $ng\,ml^{-1}$ and the detection limit was achieved as low as 0.65 $ng\,ml^{-1}$. The results proved that the developed sensor could be applied successfully to sensitively and selectively detect not only AFP but also other biomolecules [45].

6.3.2 Pathogens

Pathogens that may be categorized as bacteria, viruses, fungi, parasites, or prions are microorganisms that cause diseases in humans. Global pandemic diseases, particularly caused by viruses such as Coronavirus Disease 2019 (COVID-19), Acquired Immunodeficiency Syndrome (AIDS), or simply influenza, have posed serious threats to human health and may even cause death [89, 90]. There have been many methods for the determination of viruses and bacteria since they cause a large increase in human pathogenesis and death throughout human history [91, 92]. Among numerous techniques, the magnetoplasmonic nanocomposites (MPNCs) have a great potential application due to their merits such as plasmonic activity and magnetic field guide opportunity for pathogen detection [82, 92–104]. As an example of MNPs tagging on the gold surface was an application related to impedimetric HIV-1 protease detection [91]. In the protocol, the HIV-1 protease peptide substrate was immobilized on magnetic nanoparticle via its N-terminus, and the gold surface of the SPR chip was covered with this functionalized magnetic nanoparticle through the sulfur atom of cysteine. In the presence of HIV-1 protease, the HIV-1 protease peptide substrate refers to cleavage of the probe, resulting in dissociation of the physical link between the magnetic beads and the sensor surface. After the magnetic separation by an externally applied magnetic field, a significant shift of the electrochemical signal is expected because of the release of the magnetic beads from the sensor surface. The proposed sensing platform is simple, rapid, and sensitive and does not require washing or blocking step as it is label-free detection [91]. Another approach to use the MPNCs in combination with graphenes (GRPs) has been proposed by Lee et al. [105]. A significant advantage in this study consists of the binary NPs that can be employed not only in the magneto-optical effect and plasmon resonance energy transfer but also in electrical conductivity. Reported sensor based on specifically gold (Au)/magnetic nanoparticle (MNP) – decorated graphenes (GRPs) nanohybrids revealed a low detection limit of 1.16 $pg\,ml^{-1}$, exceptional linear response of 0.01 pg to 1 ng with remarkable sensitivity for the norovirus-like particle (NoV-LP) [105].

Eissa and Zourob designed a rapid, simple, and cost-effective sensing platform for dual colorimetric and electrochemical detection of *Staphylococcus aureus* based on the proteolytic activity of *S. aureus* protease on a specific peptide substrate. The probe consists of a specific peptide substrate carrying a magnetic nanoparticle. This probe is attached to the gold screen printed electrodes. Upon cleavage of the probe peptide sequence via the addition of the *S. aureus* protease solution, the release of the MNPs revealing the golden color of the electrode. At the same time, a significant shift of the square wave voltammetric signal is expected as a result of the protease-induced release of the MNPs beads from the sensor [99]. In 2015, Yang and co-workers reported a giant magnetoimpedance (GMI)-based biosensor and improved sandwich assay with magnetic nanoparticle to detect *Escherichia coli (E. coli) O157:H7*. While a separate Au film substrate was covered with the monoclonal mouse anti-*E. coli* antibody to obtain immunoplatform, Dynabeads as magnetic labels were prepared by conjugating streptavidin-coupled Dynabeads with biotin-labeled polyclonal mouse anti-*E. coli* antibodies. The GMI-based biosensor shows a rapid and sensitive detection of *E. coli O157:H7* by the classical sandwich assay for the first time [100].

As an example of pathogen analysis, Wang et al. developed an aptasensor based on the plasmon properties, magnetic separation, and photokilling of multiple pathogens with combined functionalities of hybrid NPs as-synthesized probes for the detection of multiple pathogens [106]. After the synthesis of NPs and nanorods, the site-selective assembly of Fe_3O_4 was aligned on the ends or ends and sides of gold nanorods with different aspect ratios (ARs), and after the alignment process, multifunctional nanorods incorporating optical and magnetic materials that provide tunable plasmonic and magnetic properties. An external magnetic field was used to isolate bacteria bound to the necklace-like Fe_3O_4-Au_{rod} from the solution. To detect bacteria (*E. coli O157:H7 and S. typhimurium*), the optical detection of the Fe_3O_4-Au_{rod} necklacelike and Fe_3O_4-Au_{rod}-Fe_3O_4 nanodumbbells bioprobes was monitored depending on the plasmon absorbance and thermal ablation under the various circumstances. This multi-functional nanohybrid possessed magnetic and tuneable plasmonic properties. The schematic illustration of protocol for this assay is shown in Figure 6.4 [106].

6.3.3 DNA

Rapid, sensitive, and reliable DNA detection has gained attention because it is one of the main factors not only in the successful diagnosis of disease but also gene analysis, fast detection of biological warfare agents, and forensic applications [94, 101]. With the gradually improving of the detection technology, the available strategies for the DNA analysis mainly include polymerase chain reaction (PCR) or comparable target-amplification system. Despite the usefulness of the conventional methods, these methods are mainly hampered by the requirement of additional instruments and reagents [107]. Several approaches to overcome this challenge for DNA detection have been offered by researchers based on the use of NPs [104, 108–111].

In 2017, Faridi and co-workers reported magnetoplasmonic sensor based on the surface plasmon effect and different light absorbers of graphene, MoS_2, and Au ultrathin layers with magnetic effect via the presence of the NiFe (Py) layer to detect DNA. While Au/Py layer was covered with graphene, MoS_2, and Au to obtain SPR surface, NiFe (Py) layer was utilized to capture the analyte of interest easily from the medium. The researchers showed that the detection of DNA by three layers of graphene and two layers MoS_2 resulted in the nine and four times, respectively higher sensitivity when compared with conventional the conventional Au/Co/Au trilayer SPR sensor. The changes in the SPR signal caused by the coupling of light with 2-D materials in magnetoplasmonic devices. This novel magnetoplasmonic sensor exhibited the highest reported DNA sensitivity [111]. In another study, Fe_3O_4 MNPs-cysteine immobilized the carbon paste electrode (CPE) was developed by utilizing Multiwalled carbon nanotubes (MWCNTs), gold NPs (GNPs), and chitosan (Chi) as the biotinylated response probe for the DNA recognition as a model analyte. Proposed biosensors exhibited a considerable response to the target DNA with LOD of 1 nM and a sensitivity of 2.707×10^3 mA M^{-1} cm^{-2}. As compared with the other method, the resulting detection limit was not remarkable but this electrochemical sensor is relatively inexpensive, sensitive, and specific for the detection of target DNA [109].

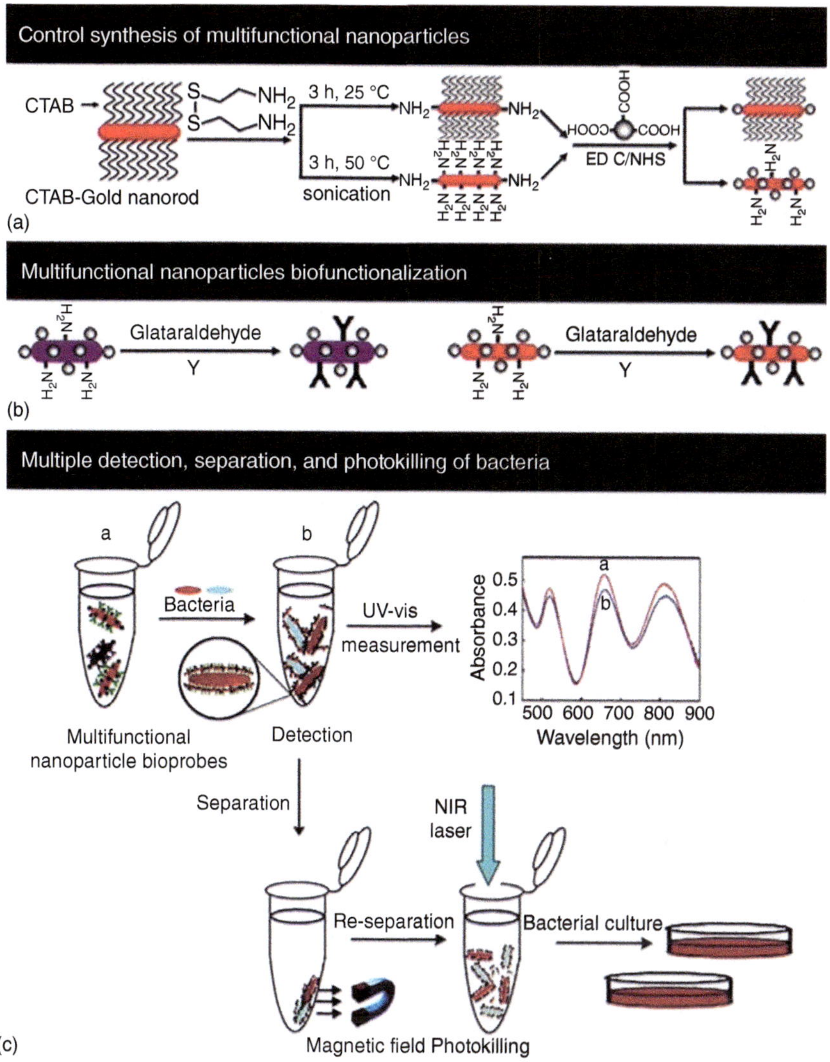

Figure 6.4 (a) Schematic showing the controlled assembly of Fe_3O_4 MNPs onto GNRs; (b) the biofunctionalization of MPNCs; and (c) the detection, separation, and thermal ablation of multiple bacterial targets. EDC: 1-ethyl-3-(3-dimethylaminopropyl) carbodiimide hydrochloride, NHS: *N*-hydroxysuccinimide. Source: Reprinted from Wang and Irudayaraj [106]. © 2010, with permission from John Wiley & Sons.

6.4 Conclusion

Recently, the efficacy of MPNCs, a novel form of MFNCs, has made these particles very useful as innovative approaches for biomedical applications. MPNCs manufactured by combining MNPs with PNPs combine advantages of magnetic and plasmonic NPs such as high biocompatibility and chemical stability, ease of synthesis and modification, wide-range spectral optical properties between others. The methods such as chemical reduction, thermal decomposition, solvothermal method, microemulsion method, etc. for the synthesis of these structures in the form of a heterodimer, multicomponent doped hybrid, spherical or nonspherical core–shell or core–satellite have been with intcrested for biosensing applications. The optical and magnetical properties of MNPs can be tuneable by the change of morphology, composition, and size of MPNCs to suit the desired final applications. Moreover, the surfaces can be easily modified with biomolecules, polymers, or fluorophores for further applications. However, certain restrictions still exist regarding their uses in biomedical applications. Various studies have focused on overcoming their limitation including the long-term toxicity and related metabolic activity. Therefore, the synthesis of novel MPNCs without these limitations will enable development of innovative multimodal sensing MPNCs for the simultaneous diagnosis and treatment of different diseases in the future.

Acknowledgments

A.M. thanks funding by the CERCA programme/Generalitat de Catalunya and the Severo Ochoa Centres of Excellence programme and by the Spanish Research Agency (AEI, Grant No. SEV-2017-0706) given to ICN2.

References

1 Lu, A.H., Salabas, E.E.L., and Schüth, F. (2007). Magnetic nanoparticles: synthesis, protection, functionalization, and application. *Angewandte Chemie International Edition* 46 (8): 1222–1244.

2 Gao, J., Gu, H., and Xu, B. (2009). Multifunctional magnetic nanoparticles: design, synthesis, and biomedical applications. *Accounts of Chemical Research* 42 (8): 1097–1107.

3 Lin, L.-S., Cong, Z.-X., Cao, J.-B., et al. (2014). Multifunctional Fe_3O_4@ polydopamine core–shell nanocomposites for intracellular mRNA detection and imaging-guided photothermal therapy. *ACS Nano* 8 (4): 3876–3883.

4 Nooralian, Z., Gashti, M.P., and Ebrahimi, I. (2016). Fabrication of a multifunctional graphene/polyvinylphosphonic acid/cotton nanocomposite via facile spray layer-by-layer assembly. *RSC Advances* 6 (28): 23288–23299.

5 Gibson, R.F. (2010). A review of recent research on mechanics of multifunctional composite materials and structures. *Composite Structures* 92 (12): 2793–2810.

6 Sotiriou, G.A., Hirt, A.M., Lozach, P.-Y., et al. (2011). Hybrid, silica-coated, Janus-like plasmonic-magnetic nanoparticles. *Chemistry of Materials* 23 (7): 1985–1992.

7 Lin, Y.-S. and Haynes, C.L. (2009). Synthesis and characterization of biocompatible and size-tunable multifunctional porous silica nanoparticles. *Chemistry of Materials* 21 (17): 3979–3986.

8 Park, H., Yang, J., Seo, S., et al. (2008). Multifunctional nanoparticles for photothermally controlled drug delivery and magnetic resonance imaging enhancement. *Small* 4 (2): 192–196.

9 Zhang, J. and Misra, R. (2007). Magnetic drug-targeting carrier encapsulated with thermosensitive smart polymer: core–shell nanoparticle carrier and drug release response. *Acta Biomaterialia* 3 (6): 838–850.

10 Zou, F., Zhou, H., Tan, T.V., et al. (2015). Dual-mode SERS-fluorescence immunoassay using graphene quantum dot labeling on one-dimensional aligned magnetoplasmonic nanoparticles. *ACS Applied Materials & Interfaces* 7 (22): 12168–12175.

11 Lu, X., Rycenga, M., Skrabalak, S.E., et al. (2009). Chemical synthesis of novel plasmonic nanoparticles. *Annual Review of Physical Chemistry* 60: 167–192.

12 Ali, M.R., Wu, Y., and El-Sayed, M.A. (2019). Gold-nanoparticle-assisted plasmonic photothermal therapy advances toward clinical application. *The Journal of Physical Chemistry C* 123 (25): 15375–15393.

13 Kuppe, C., Rusimova, K.R., Ohnoutek, L., et al. (2020). "Hot" in plasmonics: temperature-related concepts and applications of metal nanostructures. *Advanced Optical Materials* 8 (1): 1901166.

14 Trang, N.T. Thuy, T.T., Higashimine, K., et al. (2013). Magnetic–plasmonic FePt@ Ag core–shell nanoparticles and their magnetic and SERS properties. *Plasmonics* 8 (2): 1177–1184.

15 Hou, X., Wang, X., Liu, R., et al. (2017). Facile synthesis of multifunctional Fe_3O_4@SiO_2@ Au magneto-plasmonic nanoparticles for MR/CT dual imaging and photothermal therapy. *RSC Advances* 7 (31): 18844–18850.

16 Fan, Z., Shelton, M., Singh, A.K., et al. (2012). Multifunctional plasmonic shell–magnetic core nanoparticles for targeted diagnostics, isolation, and photothermal destruction of tumor cells. *ACS Nano* 6 (2): 1065–1073.

17 Shi, X., Gong, H., Li, Y., et al. (2013). Graphene-based magnetic plasmonic nanocomposite for dual bioimaging and photothermal therapy. *Biomaterials* 34 (20): 4786–4793.

18 Gühlke, M., Selve, S., and Kneipp, J. (2012). Magnetic separation and SERS observation of analyte molecules on bifunctional silver/iron oxide composite nanostructures. *Journal of Raman Spectroscopy* 43 (9): 1204–1207.

19 Xu, Z., Hou, Y., and Sun, S. (2007). Magnetic core/shell Fe_3O_4/Au and Fe_3O_4/Au/Ag nanoparticles with tunable plasmonic properties. *Journal of the American Chemical Society* 129 (28): 8698–8699.

20 Yang, D., Pang, X., He, Y., et al. (2015). Precisely size-tunable magnetic/plasmonic core/shell nanoparticles with controlled optical properties. *Angewandte Chemie* 127 (41): 12259–12264.

21 Huang, L., Ao, L., Hu, D., et al. (2016). Magneto-plasmonic nanocapsules for multimodal-imaging and magnetically guided combination cancer therapy. *Chemistry of Materials* 28 (16): 5896–5904.

22 Tomitaka, A., Arami, H., Raymond, A., et al. (2017). Development of magneto-plasmonic nanoparticles for multimodal image-guided therapy to the brain. *Nanoscale* 9 (2): 764–773.

23 Miola, M., Ferraris, S., Pirani, F., et al. (2017). Reductant-free synthesis of magnetoplasmonic iron oxide-gold nanoparticles. *Ceramics International* 43 (17): 15258–15265.

24 Cozzoli, P.D., Pellegrino, T., and Manna, L. (2006). Synthesis, properties and perspectives of hybrid nanocrystal structures. *Chemical Society Reviews* 35 (11): 1195–1208.

25 Choi, S.-H., Na, H.B., Park, Y.I., et al. (2008). Simple and generalized synthesis of oxide–metal heterostructured nanoparticles and their applications in multimodal biomedical probes. *Journal of the American Chemical Society* 130 (46): 15573–15580.

26 Zhang, H., Zhong, X., Xu, J.J., et al. (2008). Fe_3O_4/polypyrrole/Au nanocomposites with core/shell/shell structure: synthesis, characterization, and their electrochemical properties. *Langmuir* 24 (23): 13748–13752.

27 Wang, C., Qian, J., Wang, K., et al. (2016). Colorimetric aptasensing of ochratoxin A using Au@ Fe_3O_4 nanoparticles as signal indicator and magnetic separator. *Biosensors and Bioelectronics* 77: 1183–1191.
28 Zhan, F., Wang, R., Yin, J., et al. (2019). Facile solvothermal preparation of Fe_3O_4–Ag nanocomposite with excellent catalytic performance. *RSC Advances* 9 (2): 878–883.
29 Zhang, L., Wu, L., Chen, L., et al. (2016). Preparation of magnetic Fe_3O_4/TiO_2/Ag composite microspheres with enhanced photocatalytic activity. *Solid State Sciences* 52: 42–48.
30 Wu, C.H., Cook, J., Emelianov, S., et al. (2014). Multimodal magneto-plasmonic nanoclusters for biomedical applications. *Advanced Functional Materials* 24 (43): 6862–6871.
31 Gong, P., Li, H., He, X., et al. (2007). Preparation and antibacterial activity of Fe_3O_4@ Ag nanoparticles. *Nanotechnology* 18 (28): 285604.
32 Grumelli, D., Vericat, C., Benítez, G., et al. (2009). Electrochemical preparation and delivery of melanin–iron covered gold nanoparticles. *ChemPhysChem* 10 (2): 370–373.
33 Kinoshita, T., Seino, S., Mizukoshi, Y., et al. (2005). Magnetic separation of amino acids by gold/iron-oxide composite nanoparticles synthesized by gamma-ray irradiation. *Journal of Magnetism and Magnetic Materials* 293 (1): 106–110.
34 Seino, S., Kinoshita, T., Otome, Y., et al. (2003). Magnetic composite nanoparticle of Au/γ-Fe_2O_3 synthesized by gamma-ray irradiation. *Chemistry Letters* 32 (8): 690–691.
35 Seino, S., Kinoshita, T., Otome, Y., et al. (2005). Gamma-ray synthesis of magnetic nanocarrier composed of gold and magnetic iron oxide. *Journal of Magnetism and Magnetic Materials* 293 (1): 144–150.
36 Billen, A., de Cattelle, A., Jochum, J.K., et al. (2019). Novel synthesis of superparamagnetic plasmonic core-shell iron oxide-gold nanoparticles. *Physica B: Condensed Matter* 560: 85–90.
37 Plácido, A., Pereira, C., Guedes, A., et al. (2018). Electrochemical genoassays on gold-coated magnetic nanoparticles to quantify genetically modified organisms (GMOs) in food and feed as GMO percentage. *Biosensors and Bioelectronics* 110: 147–154.
38 Robinson, I., Tung, L.D., Maenosono, S., et al. (2010). Synthesis of core-shell gold coated magnetic nanoparticles and their interaction with thiolated DNA. *Nanoscale* 2 (12): 2624–2630.
39 Ye, Y., Mao, S., He, S., et al. (2020). Ultrasensitive electrochemical genosensor for detection of CaMV35S gene with Fe_3O_4-Au@ Ag nanoprobe. *Talanta* 206: 120205.
40 Zhang, Y., Ding, H., Liu, Y., et al. (2012). Facile one-step synthesis of plasmonic/magnetic core/shell nanostructures and their multifunctionality. *Journal of Materials Chemistry* 22 (21): 10779–10786.
41 Deng, H., Li, X., Peng, Q., et al. (2005). Monodisperse magnetic single-crystal ferrite microspheres. *Angewandte Chemie* 117 (18): 2842–2845.
42 Zhang, L., Xu, J., Mi, L., et al. (2012). Multifunctional magnetic–plasmonic nanoparticles for fast concentration and sensitive detection of bacteria using SERS. *Biosensors and Bioelectronics* 31 (1): 130–136.
43 Chen, J., Sorensen, C., Klabunde, K., et al. (1994). Magnetic properties of nanophase cobalt particles synthesized in inversed micelles. *Journal of Applied Physics* 76 (10): 6316–6318.
44 Tadic, M., Panjan, M., Damnjanovic, V., et al. (2014). Magnetic properties of hematite (α-Fe_2O_3) nanoparticles prepared by hydrothermal synthesis method. *Applied Surface Science* 320: 183–187.
45 Liang, R.-P., Yao, G.-H., Fan, L.X., et al. (2012). Magnetic Fe_3O_4@ Au composite-enhanced surface plasmon resonance for ultrasensitive detection of magnetic nanoparticle-enriched α-fetoprotein. *Analytica Chimica Acta* 737: 22–28.

46 Kalska-Szostko, B., Rogowska, M., Dubis, A., et al. (2012). Enzymes immobilization on Fe_3O_4–gold nanoparticles. *Applied Surface Science* 258 (7): 2783–2787.

47 Mazhani, M., Alula, M.T., and Murape, D. (2020). Development of a cysteine sensor based on the peroxidase-like activity of AgNPs@ Fe_3O_4 core-shell nanostructures. *Analytica Chimica Acta* 1107: 193–202.

48 Chen, H., Hu, H., Tao, C., et al. (2019). Self-assembled Au@ Fe core/satellite magnetic nanoparticles for versatile biomolecule functionalization. *ACS Applied Materials & Interfaces* 11 (27): 23858–23869.

49 Ballesteros, C.A., Cancino, J., Marangoni, V.S., et al. (2014). Nanostructured Fe_3O_4 satellite gold nanoparticles to improve biomolecular detection. *Sensors and Actuators B: Chemical* 198: 377–383.

50 Sun, Z., Du, J., Yan, L., et al. (2016). Multifunctional Fe_3O_4@ SiO_2–Au satellite structured SERS probe for charge selective detection of food dyes. *ACS Applied Materials & Interfaces* 8 (5): 3056–3062.

51 Ge, X., Tao, Y., Zhang, A., et al. (2013). Electrochemical detection of dual exposure biomarkers of organophosphorus agents based on reactivation of inhibited cholinesterase. *Analytical Chemistry* 85 (20): 9686–9691.

52 Abbas, M., RamuluTorati, S., and Kim, C. (2017). Multifunctional Fe_3O_4/Au core/satellite nanocubes: an efficient chemical synthesis, characterization and functionalization of streptavidin protein. *Dalton Transactions* 46 (7): 2303–2309.

53 Liu, Y., Zhang, Y., Kou, Q., et al. (2018). Fe_3O_4/Au binary nanocrystals: facile synthesis with diverse structure evolution and highly efficient catalytic reduction with cyclability characteristics in 4-nitrophenol. *Powder Technology* 338: 26–35.

54 Kim, M.S., Park, B.C., Kim, Y.J., et al. (2020). Design of magnetic-plasmonic nanoparticle assemblies via interface engineering of plasmonic shells for targeted cancer cell imaging and separation. *Small* 16 (20): 2001103.

55 Li, Y., Yang, S., Lu, X., et al. (2019). Synthesis and evaluation of the SERS effect of Fe_3O_4–Ag Janus composite materials for separable, highly sensitive substrates. *RSC Advances* 9 (6): 2877–2884.

56 Guardia, P., Nitti, S., Materia, M., et al. (2017). Gold–iron oxide dimers for magnetic hyperthermia: the key role of chloride ions in the synthesis to boost the heating efficiency. *Journal of Materials Chemistry B* 5 (24): 4587–4594.

57 Sheng, Y. and Xue, J. (2012). Synthesis and properties of Au–Fe_3O_4 heterostructured nanoparticles. *Journal of Colloid and Interface Science* 374 (1): 96–101.

58 Schick, I., Gehrig, D., Montigny, M., et al. (2015). Effect of charge transfer in magnetic-plasmonic Au@ MO_x (M=Mn, Fe) heterodimers on the kinetics of nanocrystal formation. *Chemistry of Materials* 27 (13): 4877–4884.

59 Yong, C., Chen, X., Xiang, Q., et al. (2018). Recyclable magnetite-silver heterodimer nanocomposites with durable antibacterial performance. *Bioactive Materials* 3 (1): 80–86.

60 Kostevsek, N., Locatelli, E., Garrovo, C., et al. (2016). The one-step synthesis and surface functionalization of dumbbell-like gold–iron oxide nanoparticles: a chitosan-based nanotheranostic system. *Chemical Communications* 52 (2): 378–381.

61 Velasco, V., Muñoz, L., Mazarío, E., et al. (2014). Chemically synthesized Au–Fe3O4 nanostructures with controlled optical and magnetic properties. *Journal of Physics D: Applied Physics* 48 (3): 035502.

62 Xu, C., Xie, J., Ho, D., et al. (2008). Au–Fe_3O_4 dumbbell nanoparticles as dual-functional probes. *Angewandte Chemie International Edition* 47 (1): 173–176.

63 Yin, H., Wang, C., Zhu, H., et al. (2008). Colloidal deposition synthesis of supported gold nanocatalysts based on Au–Fe_3O_4 dumbbell nanoparticles. *Chemical Communications* (36): 4357–4359.

64 Trang, V.T., Phan, V.N., Van Quy, N., et al. (2017). Two-step hydrothermal synthesis of bifunctional hematite–silver heterodimer nanoparticles for potential antibacterial and anticancer applications. *Journal of Electronic Materials* 46 (6): 3323–3332.

65 Yu, H., Chen, M., Rice, P.M., et al. (2005). Dumbbell-like bifunctional Au– Fe_3O_4 nanoparticles. *Nano Letters* 5 (2): 379–382.

67 Lopes, G., Vargas, J.M., Sharma, S.K., et al. (2010). Ag—Fe_3O_4 dimer colloidal nanoparticles: synthesis and enhancement of magnetic properties. *The Journal of Physical Chemistry C* 114 (22): 10148–10152.

68 Nosrati, H., Davaran, S., Kheiri Manjili, H., et al. (2019). Bovine serum albumin stabilized iron oxide and gold bimetallic heterodimers: synthesis, characterization and stereological study. *Applied Organometallic Chemistry* 33 (10): e5155.

69 Tessonnier, J.-P., Ersen, O., Weinberg, G., et al. (2009). Selective deposition of metal nanoparticles inside or outside multiwalled carbon nanotubes. *ACS Nano* 3 (8): 2081–2089.

70 Lee, J., Takemura, K., and Park, E.Y. (2018). Plasmonic/magnetic graphene-based magnetofluoro-immunosensing platform for virus detection. *Sensors and Actuators B: Chemical* 276: 254–261.

71 Liu, Y.L., Fu, W.L., Li, C.M., et al. (2015). Gold nanoparticles immobilized on metal-organic frameworks with enhanced catalytic performance for DNA detection. *Analytica Chimica Acta* 861: 55–61.

72 Chen, C.-W., Syu, W.-J., Huang, T.-C., et al. (2017). Encapsulation of Au/Fe_3O_4 nanoparticles into a polymer nanoarchitecture with combined near infrared-triggered chemo-photothermal therapy based on intracellular secondary protein understanding. *Journal of Materials Chemistry B* 5 (29): 5774–5782.

73 Tsoncheva, T., Ivanova, L., Lotz, A., et al. (2007). Gold and iron nanoparticles in mesoporous silicas: preparation and characterization. *Catalysis Communications* 8 (11): 1573–1577.

74 Poorakbar, E., Shafiee, A., Saboury, A.A., et al. (2018). Synthesis of magnetic gold mesoporous silica nanoparticles core shell for cellulase enzyme immobilization: improvement of enzymatic activity and thermal stability. *Process Biochemistry* 71: 92–100.

75 Demeritte, T., Viraka Nellore, B.P., Kanchanapally, R., et al. (2015). Hybrid graphene oxide based plasmonic-magnetic multifunctional nanoplatform for selective separation and label-free identification of Alzheimer's disease biomarkers. *ACS Applied Materials & Interfaces* 7 (24): 13693–13700.

76 Ren, X., Yang, L., Tang, F., et al. (2010). Enzyme biosensor based on NAD-sensitive quantum dots. *Biosensors and Bioelectronics* 26 (1): 271–274.

77 Jin, D., Seo, M.-H., Huy, B.T., et al. (2016). Quantitative determination of uric acid using CdTe nanoparticles as fluorescence probes. *Biosensors and Bioelectronics* 77: 359–365.

78 Fields, S.J.S. (2001). Proteomics in genomeland. *Science* 291 (5507): 1221–1224.

79 Liu, H.L., Sonn, C.H., Wu, J.H., et al. (2008). Synthesis of streptavidin-FITC-conjugated core–shell Fe_3O_4-Au nanocrystals and their application for the purification of CD4+ lymphocytes. *Biomaterials* 29 (29): 4003–4011.

80 Luo, J., Wang, L., Mott, D., et al. (2008). Core/shell nanoparticles as electrocatalysts for fuel cell reactions. *Advanced Materials* 20 (22): 4342–4347.

81 Üzek, R., Sari, E., and Merkoçi, A.J.M. (2019). Optical-based (bio) sensing systems using magnetic nanoparticles. *Magnetochemistry* 5 (4): 59.

82 Zhou, H., Zou, F., Koh, K., et al. (2014). Multifunctional magnetoplasmonic nanomaterials and their biomedical applications. *Journal of Biomedical Nanotechnology* 10 (10): 2921–2949.

83 Mu, X., Tong, Z., Huang, Q., et al. (2016). An electrochemiluminescence immunosensor based on gold-magnetic nanoparticles and phage displayed antibodies. *Sensors* 16 (3): 308.

84 Zheng, J., Cheng, G.-F., He, P.-G., et al. (2010). An aptamer-based assay for thrombin via structure switch based on gold nanoparticles and magnetic nanoparticles. *Talanta* 80 (5): 1868–1872.

85 Zhao, J., Liu, M., Zhang, Y., et al. (2013). Apoferritin protein nanoparticles dually labeled with aptamer and horseradish peroxidase as a sensing probe for thrombin detection. *Analytica Chimica Acta* 759: 53–60.

86 Chen, H., Qi, F., Zhou, H., et al. (2015). Fe_3O_4@Au nanoparticles as a means of signal enhancement in surface plasmon resonance spectroscopy for thrombin detection. *Sensors and Actuators B: Chemical* 212: 505–511.

87 Liang, R.-P., Yao, G.-H., Fan, L.-X., et al. (2015). Enzyme-guided plasmonic biosensor based on dual-functional nanohybrid for sensitive detection of thrombin. *J Analytica chimica acta* 70: 404–410.

88 Neng, J., Harpster, M.H., Wilson, W.C., et al. (2013). Surface-enhanced Raman scattering (SERS) detection of multiple viral antigens using magnetic capture of SERS-active nanoparticles. *Biosensors and Bioelectronics* 41: 316–321.

89 Ma, Y.-D., Li, K.-H., Chen, Y.-H., et al. (2019). A sample-to-answer, portable platform for rapid detection of pathogens with a smartphone interface. *Lab on a Chip* 19 (22): 3804–3814.

90 Valizadeh, A., Sohrabi, N., and Badrzadeh, F. (2017). Electrochemical detection of HIV-1 by nanomaterials. *Artificial Cells, Nanomedicine, Biotechnology* 45 (8): 1467–1477.

91 Esseghaier, C., Ng, A., and Zourob, M. (2013). A novel and rapid assay for HIV-1 protease detection using magnetic bead mediation. *Biosensors and Bioelectronics* 41: 335–341.

92 Hassen, W.M., Chaix, C., Abdelghani, A., et al. (2008). An impedimetric DNA sensor based on functionalized magnetic nanoparticles for HIV and HBV detection. *Sensors and Actuators B: Chemical* 134 (2): 755–760.

93 Ngamdee, T., Yin, L.S., Vongpunsawad, S., et al. (2020). Target induced-DNA strand displacement reaction using gold nanoparticle labeling for hepatitis E virus detection. *Analytica Chimica Acta* 1134: 10–17.

94 Zheng, J., Hu, L., Zheng, M., et al. (2015). An electrochemical sensing strategy for the detection of the hepatitis B virus sequence with homogenous hybridization based on host–guest recognition. *RSC Advances* 5 (112): 92025–92032.

95 Aytur, T., Foley, J., Anwar, M., et al. (2006). A novel magnetic bead bioassay platform using a microchip-based sensor for infectious disease diagnosis. *Journal of Immunological Methods* 314 (1–2): 21–29.

96 Alipour, E., Shariatpanahi, S.P., Ghourchian, H., et al. (2020). Designing a magnetic inductive micro-electrode for virus monitoring: modelling and feasibility for hepatitis B virus. *Microchimica Acta* 187 (8): 1–12.

97 Li, X., Scida, K., and Crooks, R.M. (2015). Detection of hepatitis B virus DNA with a paper electrochemical sensor. *Analytical Chemistry* 87 (17): 9009–9015.

98 Razo, S.C., Panferov, V.G., Safenkova, I.V., et al. (2018). Double-enhanced lateral flow immunoassay for potato virus X based on a combination of magnetic and gold nanoparticles. *Analytica Chimica Acta* 1007: 50–60.

99 Eissa, S. and Zourob, M.J.A. (2020). A dual electrochemical/colorimetric magnetic nanoparticle/peptide-based platform for the detection of *Staphylococcus aureus*. *Analyst* 145 (13): 4606–4614.

100 Yang, Z., Sun, X.-c., Wang, T., et al. (2015). A giant magnetoimpedance-based biosensor for sensitive detection of *Escherichia coli* O157: H7. *Biomedical Microdevices* 17 (1): 5.

101 Zhang, D., Huarng, M.C., and Alocilja, E.C. (2010). A multiplex nanoparticle-based bio-barcoded DNA sensor for the simultaneous detection of multiple pathogens. *Biosensors and Bioelectronics* 26 (4): 1736–1742.

102 Shi, X., Zhang, J., and He, F. (2019). A new aptamer/polyadenylated DNA interdigitated gold electrode piezoelectric sensor for rapid detection of *Pseudomonas aeruginosa*. *Biosensors and Bioelectronics* 132: 224–229.

103 Chen, Q., Huang, F., Cai, G., et al. (2018). An optical biosensor using immunomagnetic separation, urease catalysis and pH indication for rapid and sensitive detection of Listeria monocytogenes. *Sensors and Actuators B: Chemical* 258: 447–453.

104 Watanabe, K., Kuwata, N., Sakamoto, H., et al. (2015). A smart DNA sensing system for detecting methicillin-resistant *Staphylococcus aureus* using modified nanoparticle probes. *Biosensors and Bioelectronics* 67: 419–423.

105 Lee, J., Takemura, K., Kato, C.N., et al. (2017). Binary nanoparticle graphene hybrid structure-based highly sensitive biosensing platform for norovirus-like particle detection. *ACS Applied Materials Interfaces* 9 (32): 27298–27304.

106 Wang, C. and Irudayaraj, J.J.S. (2010). Multifunctional magnetic–optical nanoparticle probes for simultaneous detection, separation, and thermal ablation of multiple pathogens. *Small* 6 (2): 283–289.

107 Park, S.-J., Taton, T.A., and Mirkin, C.A.J.S. (2002). Array-based electrical detection of DNA with nanoparticle probes. *Science* 295 (5559): 1503–1506.

108 Lee, J., Morita, M., Takemura, K., et al. (2018). A multi-functional gold/iron-oxide nanoparticle-CNT hybrid nanomaterial as virus DNA sensing platform. *Biosensors and Bioelectronics* 102: 425–431.

109 Zhang, Y., Zeng, G.-M., Tang, L., et al. (2011). An electrochemical DNA sensor based on a layers–film construction modified electrode. *Analyst* 136 (20): 4204–4210.

110 Li, C., Li, G., Liu, H., et al. (2020). A novel gold nanoparticles decorated magnetic microbead-based molecular beacon for DNA multiplexing detection by flow cytometry. *Analytica Chimica Acta* 1110: 19–25.

111 Faridi, E., Moradi, M., Ansari, N., et al. (2017). Au/NiFe/M (Au, MoS 2, graphene) trilayer magnetoplasmonics DNA-hybridized sensors with high record of sensitivity. *Journal of Biomedical Optics* 22 (12): 127001.

7

Plasmonic Sensors for Vitamin Detection

Duygu Çimen and Nilay Bereli

Department of Chemistry, Hacettepe University, Ankara, Turkey

7.1 Introduction

Vitamins are organic compounds that are not produced by body cells but essential for metabolism that occur within the body [1]. Vitamins play an important role in carbohydrate, fat, and protein metabolism, healthy development of the body, immunity against infections, digestive functions. For this reason, when vitamins taken from foods are deficient, disorders occur in metabolism. Vitamins are called water and oil-soluble vitamins according to their chemical properties and structures. Vitamins are the most essential organic compounds in terms of nutrition. Vitamins are classified in two groups as water-soluble (vitamins B and C) and fat-soluble (vitamins A, D, E, and K) (Figure 7.1). The fat-soluble vitamins are stayed for longer time in the body and stored in the body and liver. The water-soluble vitamins are unable to store in the body and are excreted in the urine [2].

7.1.1 Vitamins

Vitamins are active elements, such as enzymes and hormones, called biocatalysts that regulate reactions in the body. The bioavailability of vitamins is affected by chemical and biochemical changes that occur during the production and storage of foods. Vitamins and enzymes act as catalysts in biochemical reactions. Environmental factors such as oxygen, sunlight, interaction with metals such as copper, iron, free radicals, air pollution, low (reaction slows down) and high temperature, and temperature increase disrupt the structure of vitamins. In our country, formula, milk, and dairy products for children and babies, yogurt and yogurt-containing drinks that help lower cholesterol and digestion are offered to consumers. These foods contain ingredients that are beneficial for health as well as nutrients. These are strengthened during processing or by adding to the amount that found naturally [3–5].

In the analyzes for quality and safety in foods, it is necessary to determine whether natural components of foods, substances formed by contamination and added color, aroma,

Plasmonic Sensors and their Applications, First Edition. Edited by Adil Denizli.

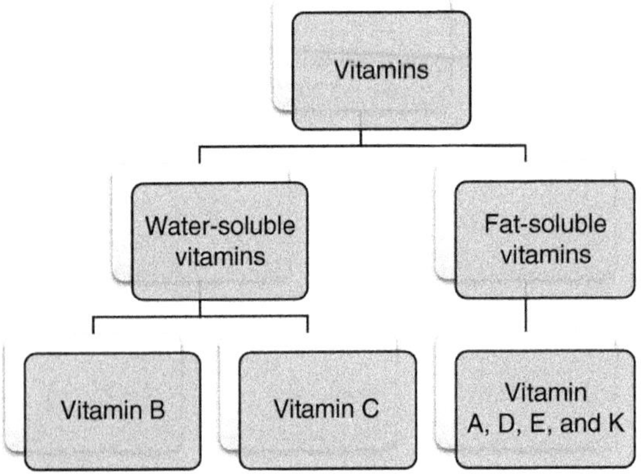

Figure 7.1 Schematic presentation of the classification of vitamins. Source: © John Wiley & Sons.

antioxidant, and antimicrobial substances are used in permitted amounts. In addition, due to the factors such as temperature, humidity, pH, light, oxygen, vitamin losses in the production and storage and low shelf life, vitamin analysis in foods should be carefully monitored. In recent years, certain standards, laws, and monitoring procedures have been established for the reliability of foods and the analysis of the components, and food analysis studies are carried out in food companies and private laboratories. In these laboratories, it is important to carry out more effective, fast, short, and reliable way for food analysis studies in order to conduct national and international proficiency tests. In food analyzes, determination of vitamin value in foods, compliance with the legislation, and label information are investigated. Certain food supplements are used to meet the nutrients required for the body, especially in patients receiving cancer and chemotherapy, and when the individual cannot get enough nutrients from the mouth in case of anorexia for various reasons. There are various vitamins necessary for growth and vitality in the products used in this form of nutrition. Many methods such as titrimetric, spectrometric, fluorometric, spectrofluorometric, chromatographic, and sensors are used for the determination of vitamins in various foods [6–8].

While the biggest advantage of classical chromatographic methods is the possibility of sensitive analysis, the biggest disadvantage is the expensive methods. More sensitive immunoassay techniques and enzymatic methods for specific analyzes require a time consuming, expensive, equipped laboratory environment, and workforce. Methods such as spectrophotometry, electrophoresis, and titration require slow extraction and pretreatment, which increase the analysis time. Sensors have advantages such as portable, fast, high sensitivity, and cheap determination, which are directly related to the measurement system and are used in food analysis [3, 9–14].

7.2 Plasmonic Sensors

The plasmonic sensors were first used in the 1950s to determine Clark's O_2 level in blood with the aid of an electrode. It was used by Clark and Lyons in the 1960s in their studies to measure the glucose level in the blood on the enzyme-electrode. The main principle in plasmonic sensors, the combination of the substance to be analyzed with the interaction of a biologically active and selective component is combined with a transmitting system and measuring these interaction products with a measurement system [15].

Sensor systems; "Biomolecule/bioreceptor" with selective recognition mechanism consists of three parts:

- Transducer and detector, which convert physicochemical signals resulting from their interaction with the substance to be determined into electrical signals.
- Bioreceptors: Biomolecules such as enzyme, tissue cultures, micro-organisms, organelles, antibodies, and nucleic acid can be used. It is also the part with biological sensitivity and selectivity in the recognition of the target molecule in the sensors.
- Converters: They are the devices that measure the physical or chemical change occurring in the environment after recognizing the target molecule of the bioreceptor and convert it into digital signals.

The plasmonic sensor techniques include mainly three types: Surface Plasmon Resonance Sensors, Localized Surface Plasmon Resonance Sensors, and Colorimetric Sensors.

7.2.1 Surface Plasmon Resonance Sensors

Surface plasmon resonance (SPR) sensors observed bright and dark bands in reflected light for the first time in 1902 when Wood reflected polarized light from a diffraction-regulating surface onto a mirror. In 1968, Otto, Kretschmann and Raether reported the formation of surface plasmones. The first use of SPR-based sensors in measuring biomolecular interactions was reported by Liedberg et al. in 1983. Surface plasmon resonance is an important method used to measure changes in refractive index close to the metal surface. In the SPR method, a thin metal film is used between two permeable and refractive index media (glass prism and solution). Metal film is usually gold or silver. Surface plasmon resonance (SPR) is an event that occurs as a result of resonance oscillation of electrons with light between two materials with positive and negative permeability. The theory of surface plasmon resonance is based on the principle of combining the energy-carrying light photon with the electrons of the metal or energy transfer (Figure 7.2) [16–19].

The angle of incidence of the light at which the junction occurs varies with each metal and the perimeter of the metal surface. Energy transfer occurs when a match or resonance is captured between light photons and electrons on the metal surface. Energy transfer can be determined by measuring the amount of light reflected from the bottom surface of the metal film. While every light is reflected at many angles, some of the light is absorbed at the resonance angle. Measurements can be made simultaneously, very quickly, while the interacting molecules can be analyzed directly by making very small chemical modifications or without the need for any chemical modifications. Also, a very low number of samples and materials are consumed. SPR sensors are used in various fields for food quality control analysis, environmental analysis, and diagnostic purposes in medicine increases [20–23]. As the validity of SPR sensor technology in food analysis increases, the number of studies on target analyte determination in this field also increases [24–27].

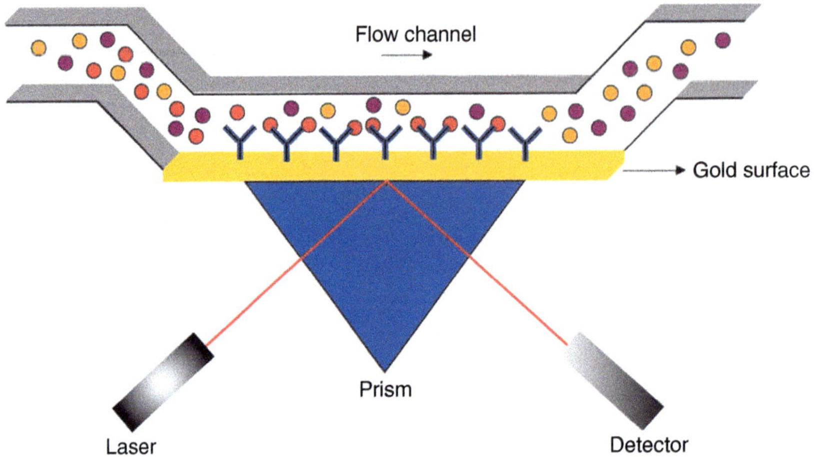

Figure 7.2 Schematic presentation of surface plasmon resonance (SPR). Source: © John Wiley & Sons.

7.2.2 Localized Surface Plasmon Resonance Sensors

The localized surface plasmon resonance (LSPR) shown by the excitation of noble metal nanoparticles with light are used in unlabeled biosensor applications and bioimaging. The fact that the molecular interactions can be measured spectrally as a result of the high sensitivity to the local refractive index, which changes as a result of the interaction of nanoparticles with biomolecules, allows numerous studies on this subject (Figure 7.3) [28–30].

LSPR, which is used as an optical based converter in biosensors, depends on the size, shape, and material properties of metal nanoparticles. Single metal nanoparticles with large surface area and strong optical properties have unique chemical and physical properties compared to bulk metals. Noble metals such as silver, copper, and gold and nanoparticles such as nano-sphere, nano-stick, nano-wire, nano-cube can be synthesized. Silver and

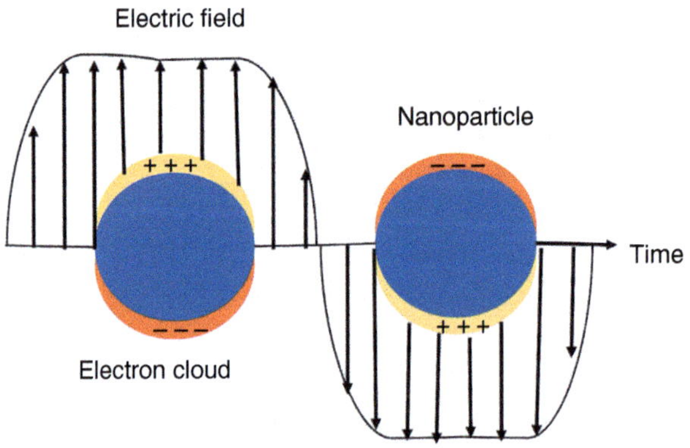

Figure 7.3 Schematic presentation of localized surface plasmon resonance (LSPR). Source: © John Wiley & Sons.

gold are less reactive and more stable metals and are also more preferred than copper. Especially golden nanoparticles (AuNP) are widely used in biotechnology with their high conductivity and chemical stability. LSPR is a unique optical property responsible for the bright colors of colloidal suspensions of nanoparticles synthesized from noble metals such as gold and silver. The preference of gold especially in plasmonic studies is due to the absorption and scattering of light at high efficiency. Strong absorption and scattering bands in the visible region (500–600 nm), which can be measured by UV-VIS spectroscopy provided by gold nanoparticle LSPRs, allow visible colorimetric perception of changes in the LSPR band of biomolecules [31, 32].

7.2.3 Colorimetric Sensors

They are identified by different names such as colorimetric sensors, thermal enzyme sensors, enzyme thermistors, or enthalpimetric enzyme sensors. Its basic principles consist in determining the substrate concentration by utilizing the enthalpy change in an enzymatic reaction. In general, the exothermic nature of enzymatic reactions is used. The linear relationship between the temperature change resulting from the enzymatic reaction and the substrate concentration is reached. Very small temperature changes such as °C are monitored with the help of thermistors or thermophiles in thermally insulated environments. For the first time, heat exchange was measured during the enzyme substrate interaction by the use of sensitive thermistors bound by enzymes by Danielson and Mosbach. Colorimetric devices operate on the basis of temperature measurement absorbed or evaluated during chemical reactions between biomolecules. The colorimetric sensors determine the temperature measurement produced in an enzymatic reaction. It was originally used for colorimetric transduction enzyme-based sensors and then applied to the cell and immunosensor. Thermometers are used to convert colorimetric sensors, heat generated (or lost) into an electrical signal. The analyte is the substrate of the enzyme. Recently, the colorimetric sensor method has been used frequently in the proteomic studies, food industry, and environmental monitoring [33–36].

7.3 Vitamin Applications of Plasmonic Sensors

Shrivastav et al. [37] developed a fiber optic sensor based on surface plasmon resonance for the detection of ascorbic acid. In this study, molecularly imprinted polyaniline film was prepared using ascorbic acid as template molecule. A shift of 22 nm in resonance wavelength has been obtained with concentration solution in the range from 10^{-8} to 10^{-4} M. The concentration of ascorbic acid is optimized to maximize the sensitivity and selectivity of designed SPR sensor. This new SPR sensor system has a characteristic such as response, high selectivity, sensitivity, and low cost. Also, the effect of pH of ascorbic acid solution was investigated in SPR sensor performance and it have been found most appropriate pH 7.0.

Prakashan et al. [38] designed SPR based vitamin A sensors using Au@Ag core–shell nanoparticles. The synthesized of Au@Ag core–shell nanoparticles were characterized with TEM analysis. The average size of the Au@Ag core–shell nanoparticles was 30 nm. The sensing selectivity and capability of Au@Ag core–shell nanoparticles were performed using adsorption spectroscopy. In this study, the linear range was between 10 and 1000 μM. For the proposed sensor, a linear response was observed in the range from 10 to 1000 μM. The designed SPR sensors are recommended as a sensitive, economical, and low-toxic biosensor for vitamin A detection (Figure 7.4).

Alsager et al. [39] were prepared a novel colorimetric aptasensors based on gold nanoparticles (AuNP). In this study, the performance of two different vitamin D3 (VTD3) and vitamin D3 (VTD2) aptamers were tested and the limit of detection was 1 μM. The most important feature of this proposed method is analyzed in nanomolar detection level in a non-functioning colorimetric sensor. It is also a highly sensitive sensor to potentially interfering molecules, including VTD2, when centrifugation and resuspension is applied. The level of vitamin 3 in human blood was determined colorimetrically after extraction with *n*-hexane. The experimental results were in agreement with those obtained by HPLC.

Zarei et al. [40] prepared a sensitive and simple procedure using silver nanoparticles (Ag-NPs) for detection of ascorbic acid. Silver nanoparticles were synthesized using a stabilizing agent polyvinylpyrrolidone (PVP). The surface plasmon absorbance of Ag-NPs was at $\lambda = 440$ nm for the quantitative spectrophotometric detection of the ascorbic acid. As a result of experiments, the detection limit was obtained as 0.08 μM in the linear range of 0.5–60 μM. The prepared method was applied for the detection of vitamin C from commercial and natural orange juice, powdered drink mixtures, multivitamin, and effervescent tablets.

Verma et al. [41] developed using surface plasmon resonance (SPR) based optical fiber biosensor by combining colloidal crystal and molecular-imprinted hydrogel for the detection of 3-pyridinecarboxamide (PA) (vitamin B3). The limit of detection was found to be 1.483 mg ml^{-1} with different solutions of vitamin B3 with concentration range 0–10 mg ml^{-1}. The new present study has very advantage such as reusable and small probe, low cost, and fast SPR sensor response (Figure 7.5).

Caelen et al. [42] reported a surface plasmon resonance with on-chip measurement for quantification of riboflavin in milk samples. SPR chip surfaces were modified with covalently immobilized riboflavin for binding the riboflavin binding protein. Riboflavin calibration solution was prepared mixing by riboflavin binding protein and riboflavin standard solution. Riboflavin in milk samples was measured by comparing the response to calibration. Consequently, SPR sensor reached the detection limit of 70 μg l^{-1} and the limit of quantification of 234 μg l^{-1} with injection volume of 160 μl.

Indyk et al. [43] produced an optical surface plasmon resonance biosensor for folate detection in milk samples. In this study, milk samples were prepared from supplemented and non-supplemented foods for folate detection. This method is a highly efficient inhibition assay method using folate-binding protein. Certified reference food material panel was used to determine the sensitivity (HorRatr: 0.3–0.8), recovery and method accuracy (72–112%) of the prepared SPR biosensor. Various food samples such as milk, cereal, flour, broccoli, eggs, fish meal, liver were analyzed with this method. Experimental results were determined to be a highly sensitive method for determining folate content in foods.

Çimen et al. [44] prepared a different vitamin imprinted surface plasmon resonance (SPR) sensor for the detection of vitamin B12, B9, and B2 from milk and infant formula samples (Figure 7.6). The characterization studies of prepared SPR sensors carried out atomic force microscopy, contact angle measurements, ellipsometry and Fourier transform

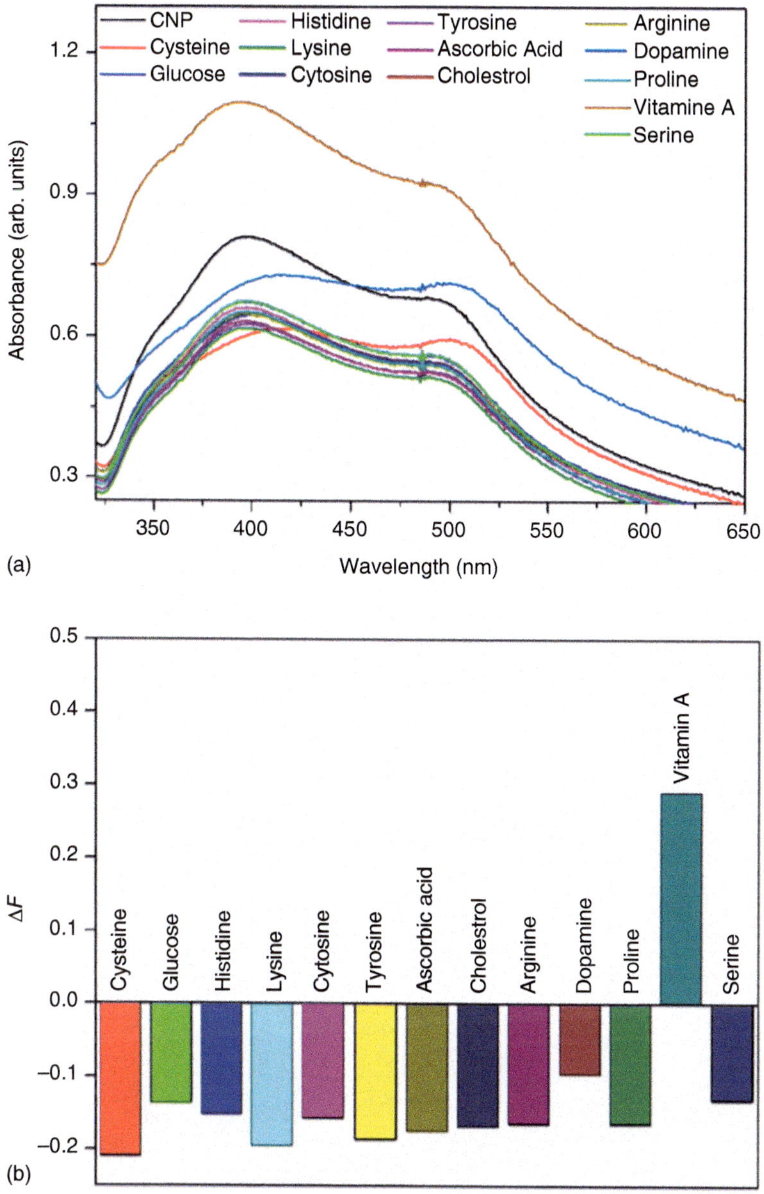

Figure 7.4 (a) Absorption spectra of Au@Ag CNPs in the presence of different biomolecules (1000 μM) and (b) selectivity Au@Ag CNPs toward vitamin A and other biomolecules. Source: Reprinted with permission from Prakashan et al. [38]. © 2019, Elsevier.

infrared-attenuated total reflectance. Vitamins B6 and B1 were used for selectivity experiments. The most important feature of SPR sensors is that they can be used repeatedly. The reproducibility of the prepared vitamins was investigated on the same day and on different days for five times. In this study, validation studies from real food samples were performed using liquid chromatography-tandem mass spectrometry (LC-MS/MS).

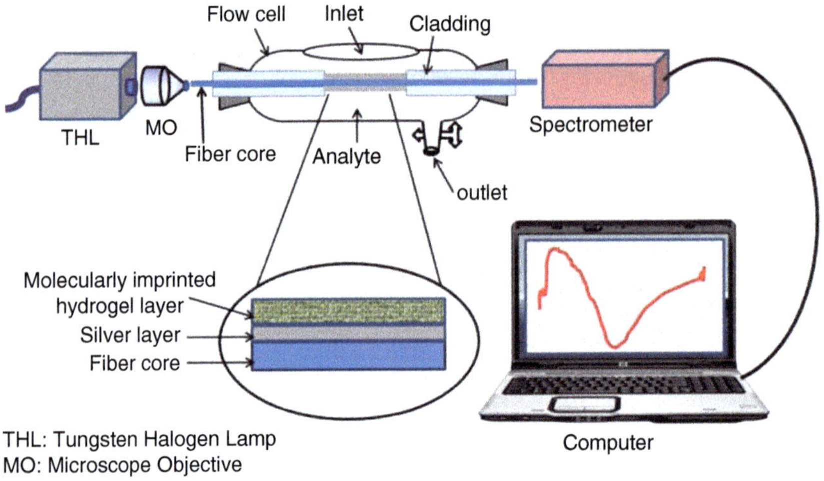

Figure 7.5 Schematic presentation of experimental setup for the characterization of the optical fiber sensor. Source: Reprinted with permission from Verma et al. [41]. © 2013, Elsevier.

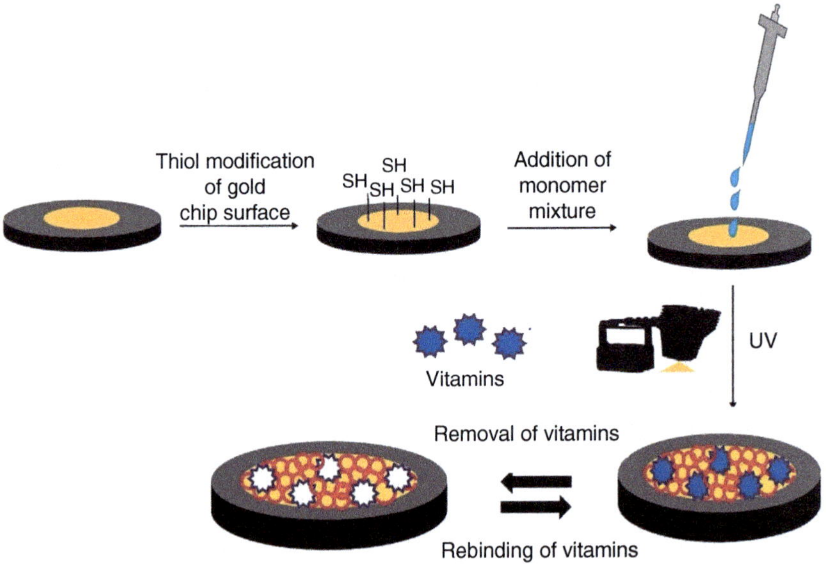

Figure 7.6 Schematic representation of the preparation of vitamin imprinted chip surface. Source: Reprinted with permission from Çimen and Denizli [44]. © 2020, Springer Nature. Licensed under CC BY 4.0.

Vitamin D plays an important role in the treatment of many diseases such as skeletal damage and pathogenesis. In this study, different approaches for the detection of vitamin D were developed using affinity-based biosensors. Carlucci et al. [45] developed a novel surface plasmon resonance and electrochemical based biosensors for the determination of 25-OH vitamin D. Firstly, an immunosensor based on SPR transduction was performed for real-time detection of vitamin D in clinical analysis. The limit of detection was obtained as

$2\,\mu g\,ml^{-1}$. The second approach was modified with vitamin D gold nanoparticles to increase the sensitivity of the sensor. The limit of detection was obtained as $1\,\mu g\,ml^{-1}$. In another alternative approach, SPR sensor prepared for detection of vitamin D with vitamin D binding protein. The limit of detection was obtained as $45\,ng\,ml^{-1}$. Finally, an electrochemical biosensor was prepared based on the reaction of vitamin D with 4-ferrocenylmethyl-1,2,4-triazolin-3,5-dione (FMTAD). As a result of experiments, the limit of detection was obtained as $10\,ng\,ml^{-1}$ in the linear range of 20–200 $ng\,ml^{-1}$. According to experimental results, the proposed sensor systems were determined to be a highly sensitive method for detection of vitamin D in clinical applications.

Wang et al. [46] designed a non-aggregation colorimetric sensor for detection of vitamin C based on surface plasmon resonance of gold nanorods (Figure 7.7). Gold nanorods (AuNRs) were characterized by transmission electron microscope (TEM). The aspect ratio of AuNRs has been reduced in length and width. The morphology of the AuNRs has been changed. In the concentration (0.11–85.0 μM) of vitamin C was shifted the longitudinal plasmon absorption wavelength (LPAW, λ_L) blue ($\Delta\lambda_L$). In the concentration (85.0–385 μM) of vitamin C was shifted the transversal plasmon absorption wavelength (TPAW, λ_T) red shifted ($\Delta\lambda_T$) gradually. The limit of detection was calculated as $3.5\times10^{-9}\,gVc\,ml^{-1}$ for λ_L and $5.1\times10^{-6}\,gVc\,ml^{-1}$ for λ_T, respectively. It was observed that the color of the solutions clearly changed.

Tabassum et al. [47] investigated a simultaneous interactions of vitamin K1 (VK1) and heparin with the step channel multianalyte sensing probe using the fiber optic surface plasmon resonance technique. The fiber, which is divided into two parts, is covered with thin silver and copper films. The multi-walled carbon nanotube nanohydride in chitosan was produced on the silver layer to detect vitamin K1. The core–shell nanostructure of polybren@ZnO was coated on the copper layer to detect heparin. Characterization of the SPR sensor was done using the spectral interrogation method. As a result of experiments, the limit of detection was obtained as $2.66\times10^{-4}\,mg\,l^{-1}$ and $2.88\times10^{-4}\,mg\,l^{-1}$ for vitamin K1 and heparin, respectively. According to experimental results, the important advantages of the present SPR sensor systems are possibility of online monitoring and low cost for detection of vitamin K1.

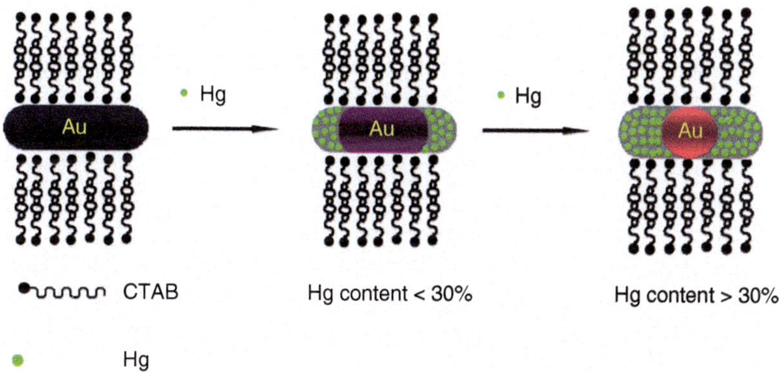

Figure 7.7 Morphology change of AuNRs during the formation of gold amalgamation. Source: Reprinted with permission from Wang et al. [46]. © 2013, Elsevier.

Khalkho et al. [48] prepared a new L-cysteine modified silver nanoparticles as a colorimetric probe for detection of vitamin B1 (thiamine). The characterization of Cys-capped AgNPs was performed using Fourier transform infrared spectroscopy (FTIR), transmission electron microscope (TEM), UV-Visible spectrophotometry, and dynamic light scattering (DLS) techniques. The analysis of L-cysteine modified silver nanoparticles (Cys-capped AgNPs) in the 200–800 nm was performed by measuring the red shift of the localized surface plasmon resonance (LSPR) band. After the interaction of Cys-capped AgNPs with vitamin B1, the color of the gold nanoparticles has become colorless from yellow. The limit of detection was obtained as 7 μg ml^{-1} in the linear range of 25–500 μg ml^{-1} (Figure 7.8).

Duenchay et al. [49] designed a colorimetric sensor for the determination of vitamin B1 in urine samples using unmodified gold nanoparticles (AuNPs). In the kinetic analysis, strong electrostatic interactions were occurred between negatively charged AuNPs and positively charged vitamin B1. In this study, a color change from red to blue was observed within 10 minutes in the interaction of different concentrations of B1 vitamins with gold nanoparticles. In the vitamin B1 linear range of 40–200 ppb, the limit of detection and limit of quantitative were found to be 3.002 and 10.006 ppb, respectively. The accuracy and sensitivity properties of the developed sensors were investigated and validation studies were carried out with classical methods. As experimental results, it has been proven that the developed colorimetric sensors can be used effectively for the analysis of vitamin B1 in urine samples.

Yao et al. [50] prepared sensitive and rapid colorimetric sensors for the detection of ascorbic acid. In this study, copper metal–organic framework nanoparticles functionalized with amino acid with peroxidase-like properties were used. The prepared nanoparticles were catalyzed by H_2O_2 oxidation of 3,3,5,5-tetramethylbenzidine (TMB) producing (Figure 7.9). The change of blue color at 655 nm in UV-VIS was indicated a decrease in absorbance. The limit of detection was obtained as 0.15 mM in the linear range of 0.5–60 mM. It has been proven that this designed colorimetric sensor can be successfully applied for colorimetric detection of ascorbic acid in food, pharmaceutical, and human serum samples.

Yang et al. [51] proposed a simple colorimetric and spectrophotometric method for detection of ascorbic acid. In this study, silver nanoparticles were prepared from silver (Ag) and bovine serum albumin protected-silver nanosets (BSA-AgNCs) mixture. After the catalysis of BSA-AgNCs, the color of the mixture changed from colorless to yellow at room temperature. A band at 420 nm for the produced silver nanoparticles were observed by localized surface plasmon resonance (LSPR). BSA protected silver nanoscopes were reduced to Ag^0 of Ag^+ with ascorbic acid at room temperature. As a result of experiments, the detection limit was obtained as 0.16 μM in the linear range of 2.0–50 μM (Figure 7.10).

Wang et al. [52] prepared a seed-mediated growth bimetallic nanoparticles for a simple and sensitive detection method of vitamin C. The analysis of Au nanocages was performed by localized surface plasmon resonance (LSPR). Thanks to the morphology and transformations of the au nanocages, a shorter wavelength passed through the infrared region. As a result of a linear relationship between wavelength changes of the LSPR peak for detection of vitamin C, the limit of detection was obtained as 24 nM in the linear range of 0.05–7.5 μM. The results of this designed LSPR method showed that the determination of

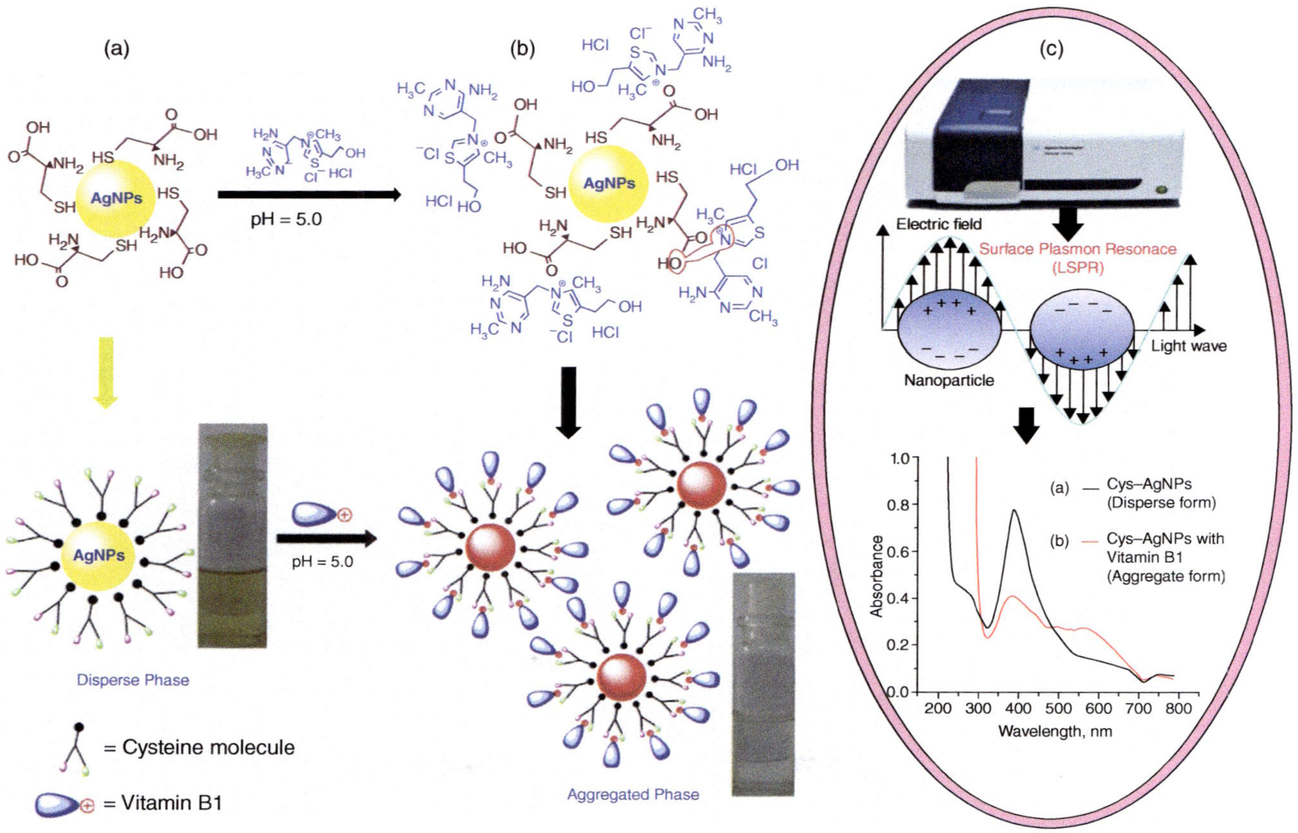

Figure 7.8 The schematic illustration of Cys-capped AgNPs based sensing strategy for determination of vitamin B1 using colorimetric probe. Source: Reprinted with permission from Khalkho et al. [48]. © 2020, Elsevier. Licensed under CC BY 4.0.

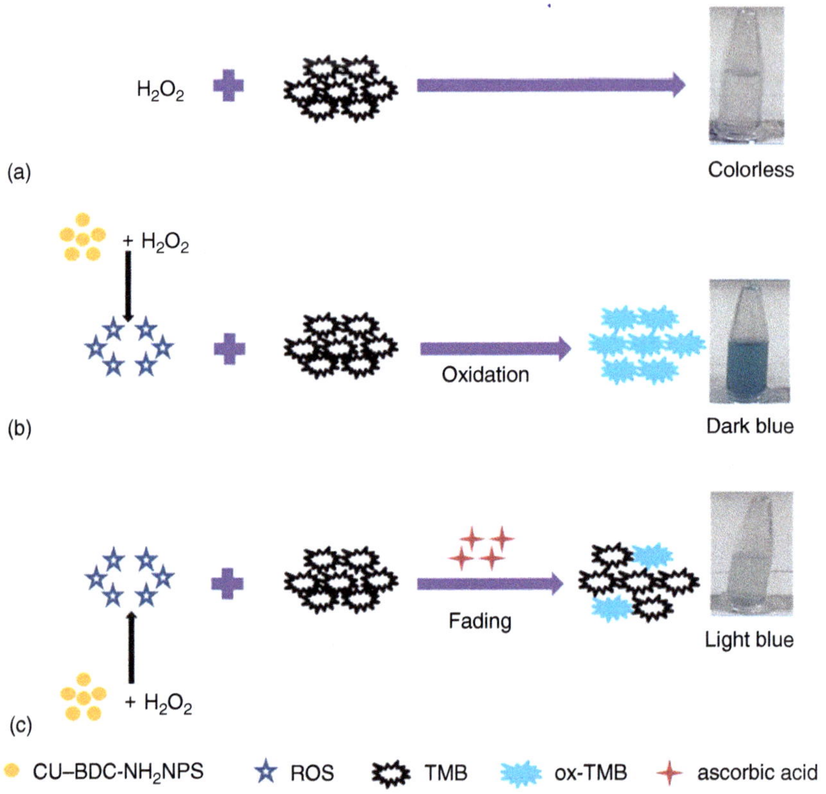

Figure 7.9 Schematic illustration of the proposed colorimetric assay. Source: Reprinted with permission from Yao et al. [50]. © 2019, Royal Society of Chemistry.

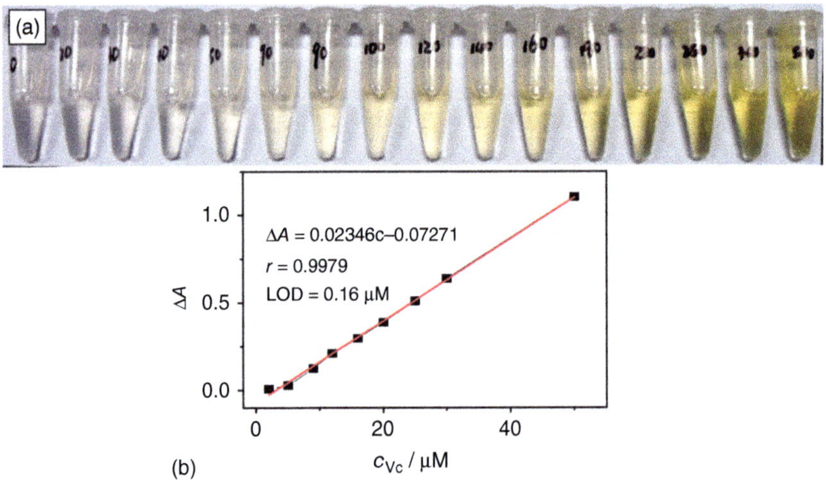

Figure 7.10 (a) Colorimetric detection of ascorbic acid with the common photograph of silver NPs solution and (b) linear relationship between absorption changes of silver NPs suspension at 416 nm and the concentration of ascorbic acid. Source: Reprinted with permission from Ling et al. [51]. © 2013, Elsevier.

vitamin C in pharmaceutical products would be successful in high selectivity for the determination of vitamin C.

Shrivas et al. [53] investigated a plasmonic chemical sensor for selective detection of vitamins B1 and B6 in white and brown rice food samples. Sucrose-capped gold nanoparticles, a simple and selective method, localized surface plasmon resonance (LSPR) was used for the detection of vitamin B1 and B6. It was observed that LSPR absorption band changed color from pink to blue in UV-VIS region by adding vitamin B1 and B6 to gold nanoparticle solutions. The limit of detection was obtained as $8\,ng\,ml^{-1}$ in the linear range of $25–1000\,ng\,ml^{-1}$ for vitamin B1. The limit of detection was obtained as $15\,ng\,ml^{-1}$ in the linear range of $50–1000\,ng\,ml^{-1}$ for vitamin B6. They showed that they prepared sensors with high sensitivity, simple, low cost, and repeatability properties using gold nanoparticles for the detection of vitamin B1 and B6 in different rice food samples.

7.4 Conclusions and Prospects

Over the past several decades, significant effort has been invested with the aim of developing plasmonic sensors technologies that will impact the practice of vitamin detection. The use of plasmonic sensors in vitamin detection is important especially by chemical and biochemical changes that occur during the production and storage of foods. Continued development is warranted because the plasmonic sensors have advantages over more complex analytical techniques such as mass spectrometry and optical techniques. The integration of all analytical process, from sample collection to analyte detection, on the same device, lab-on-a-chip system, can be sought. Plasmonic sensors platform able to perform sample collection and analysis on the same device could be designed. These techniques include time-consuming procedures, expensive instruments, and hazardous labels. The development of rapid, easy to use, and cost-effective vitamin detection based on different plasmonic sensor systems has been reported in the literature. Plasmonic sensors are a highly sensitive technique and therefore very convenient for use in vitamins detection.

References

1 Ball, G.M.F. (2006). *International Journal of Food Properties* 9: 927–928.

2 Santos, J., Mendiola, J.A., Oliveira, M.B. et al. (2012). *Journal of Chromatography A* 1261: 179.

3 Babakhanian, A., Kaki, S., Ahmadi, M. et al. (2014). *Biosensors and Bioelectronics* 60: 185–190.

4 Eggersdorfer, M., Laudert, D., Létinois, U. et al. (2012). *Angewandte International Edition Chemie* 51: 12960–12990.

5 Phillips, M.M. (2015). *Analytical and Bioanalytical Chemistry* 407: 2965–2974.

6 A.C. Ortega, D.C. Silva, J.V. Visentainer, N.E. Souza, V.C. Almeida, C.C. Oliveira, *Analytical Letters,* 2011, 44, 778–786.

7 Mirmoghtadaie, L., Ensafi, A.A., Kadivar, M. et al. (2013). *International Journal of Electrochemical Science* 8: 3755–3767.

8 Petteys, B.J. and Frank, E.L. (2011). *Clinica Chimica Acta* 412: 38–43.

9 Schmidt, A., Schreiner, M.G., and Mayer, H.K. (2017). *Journal of Chromatography A* 1500: 89–95.
10 Campos-Giménez, E., Bénet, S., Oguey, Y. et al. (2018). *Food Chemistry* 249: 91–97.
11 Nakos, M., Pepelanova, I., Beutel, S. et al. (2017). *Food Chemistry* 216: 301–308.
12 He, X., Meng, M., and Xi, R. (2013). *Journal of Agricultural and Food Chemistry* 61: 7048–7054.
13 Hampel, D., Shahab-Ferdows, S., Domek, J.M. et al. (2014). *Food Chemistry* 153: 60–65.
14 Luo, X., Chen, B., Ding, L. et al. (2006). *Analytica Chimica Acta* 562: 185–189.
15 Clark, L.C. and Lyons, C. (1962). *Annals of the New York Academy of Sciences* 102: 29–45.
16 Wood, R., XlII (1902). *The London, Edinburgh, and Dublin Philosophical Magazine and Journal of Science* 4: 396–402.
17 Wood, R., XXVII (1912). *The London, Edinburgh, and Dublin Philosophical Magazine and Journal of Science* 23: 310–317.
18 Otto, A. (1968). *Zeitschrift für Physik* 216: 398–410.
19 Liedberg, B., Nylander, C., and Lunström, I. (1983). *Sensors and Actuators* 4: 299–304.
20 Faalnouri, S., Çimen, D., Bereli, N., and Denizli, A. (2020). *ChemistrySelect* 5: 4761–4769.
21 Saylan, Y., Akgönüllü, S., Çimen, D. et al. (2017). *Sensors and Actuators B* 241: 446–454.
22 Çimen, D., Bereli, N., Günaydın, S., and Denizli, A. (2020). *Talanta* 219: 121259.
23 Jalilzadeh, M., Çimen, D., Özgür, E. et al. (2019). *Journal of Macromolecular Science, Part A* 9: 877–886.
24 Chaubey, A. and Malhotra, B. (2002). *Biosensors and Bioelectronics* 17: 441–456.
25 Nguyen, H.H., Park, J., Kang, S., and Kim, M. (2015). *Sensors* 15: 10481–10510.
26 Karlsson, R. (2004). *Journal of Molecular Recognition* 17: 151–161.
27 Englebienne, P., Hoonacker, A.V., and Verhas, M. (2003). *Spectroscopy* 17: 255–273.
28 Fong, K.E. and Yung, L.Y.L. (2013). *Nanoscale* 5: 12043–12071.
29 Mayer, K.M. and Hafner, J.H. (2011). *Chemical Reviews* 111: 3828–3857.
30 Hutter, E. and Fendler, J.H. (2004). *Advances Materials* 16 (19): 1685–1706.
31 Unser, S., Bruzas, I., He, J., and Sagle, L. (2015). *Sensors* 15: 15684–15716.
32 Ahn, H., Song, H., Choi, J., and Kim, K. (2018). *Sensors* 18: 98.
33 Mosbach, K. (1991). *Biosensors and Bioelectronics* 6: 179–182.
34 Kumar, V.V. and Anthony, S.P. (2014). *Sensors and Actuators* 191: 31–36.
35 Xie, B., Ramanathan, K., and Danielsson, B. (2000). *TrAC, Trends in Analytical Chemistry* 19: 340–349.
36 Ramanathan, K. and Danielsson, B. (2001). *Biosensors and Bioelectronics* 16: 417–423.
37 Shrivastav, A.M., Mishra, S.K., and Gupta, B.D. (2015). *Plasmonics* 10: 1853–1861.
38 Prakashan, V.P., Gejo, G., Sanu, M.S. et al. (2019). *Applied Surface Science* 484: 219–227.
39 Alsager, O.A., Alotaibi, K.M., Alswieleh, A.M., and Alyamani, B.J. (2018). *Scientific Reports* 8: 12947.
40 Zarei, K. and Moghaddary, S. (2015). *Cogent Chemistry* 1: 1109172.
41 Verma, R., Banshi, D., and Gupta, B.D. (2013). *Sensors and Actuators B* 177: 279–285.
42 Caelen, I., Kalman, A., and Wahlstro, L. (2004). *Analytical Chemistry* 76: 137–143.
43 Indyk, H.E. and Woollard, D.C. (2013). *Journal of Food Composition and Analysis* 29: 87–93.
44 Çimen, D. and Denizli, A. (2020). *Photonic Sensors*: 1–17.
45 Carlucci, L., Favero, G., Tortolini, C. et al. (2013). *Biosensors and Bioelectronics* 40: 350–355.

46 Wang, X.X., Liu, J.M., Jiang, S.L. et al. (2013). *Sensors and Actuators B* 182: 205–210.
47 Tabassum, R., Banshi, D., and Gupta, B.D. (2016). *Biosensors and Bioelectronics* 86: 48–55.
48 Khalkho, B.R., Kurrey, R., Deb, M.K. et al. (2020). *Heliyon* 6: e03423.
49 Duenchay, P., Kaewjua, K., Chailapakul, O., and Siangproh, W. (2020). *New Journal of Chemistry* 44: 9223–9229.
50 Yao, Z., Li, Z., Liu, H. et al. (2019). *Analytical Methods* 11: 1697–1706.
51 Ling, X.H., Ling, J., Peng, J. et al. (2013). *Spectrochimica Acta Part A: Molecular and Biomolecular Spectroscopy* 106: 224–230.
52 Wang, Y., Zhang, P., Mao, X. et al. (2016). *Sensors and Actuators B* 231: 95–101.
53 Shrivas, K., Nirmalkar, N., Thakur, S.S. et al. (2018). *Food Chemistry* 250: 14–21.

8

Proteomic Applications of Plasmonic Sensors

Duygu Çimen, Merve Asena Özbek, Nilay Bereli, and Adil Denizli

Department of Chemistry, Hacettepe University, Ankara, Turkey

8.1 Introduction

The definition of the proteome is the set of proteins produced by a genome. The proteomics study covers a wide area and includes simultaneous analysis of all the proteins that in a cell, tissue or fluid. Protein characterization such as identification, post-translational modifications and protein–protein interactions are part of the protein level proteomic methods in terms of disease mechanism and cellular processes [1–3].

In the recent years, DNA sequence analysis, mutation, polymorphism determination and expression imaging using genomic techniques in the pharmaceutical industry has increased the number of studies for the diagnosis of diseases [4, 5]. The main objectives used in the vast majority of these methods are to define the functions and structures of proteins. The developed methods to define the functions of proteins by genomic technology is increasing day by day.Proteomic approaches uses mass spectrometry (MS) and sensor systems for protein identification [6, 7] (Figure 8.1).

Proteome provides access to instant information of the protein composition at a given time in a particular cell or tissue. Thanks to proteomics, known as large-scale analysis of proteins, new protein biomarkers can be identified for disease diagnosis, as well as drug analysis by means of quantitative analysis at the level of protein–protein interactions and cellular protein expression within the communication network of cells [8]. Proteomics on the whole can be divided into three kinds as described below (Figure 8.2).

The extraction of protein from a cell is the first step of the proteomics. At the second step, separation of the extracted proteins via two-dimensional electrophoresis takes place. Detection of peptides are relatively easy, therefore in the third step proteins are cut into peptides. The next step is the detection of those peptides and peptide fragments usually by mass spectrometry. The sequence of the protein is obtained at the final step with all data from interpreting (Figure 8.3).

Studies on the human proteome report the human plasma proteome, human liver, human lung fibroblasts, tumor cells, and different diseases [9, 10]. The human plasma

Plasmonic Sensors and their Applications, First Edition. Edited by Adil Denizli.

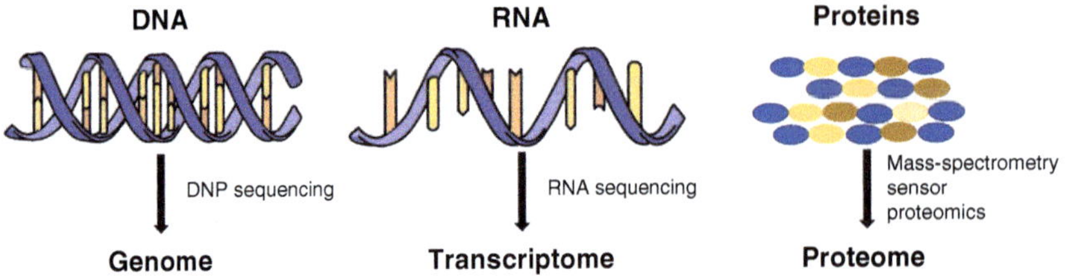

Figure 8.1 Integrating the DNA and RNA sequencing and proteins for proteomics. Source: Reprinted with permission from Westbrook et al. [6] © 2001; Roh et al. [7]. © 2010, John Wiley & Sons.

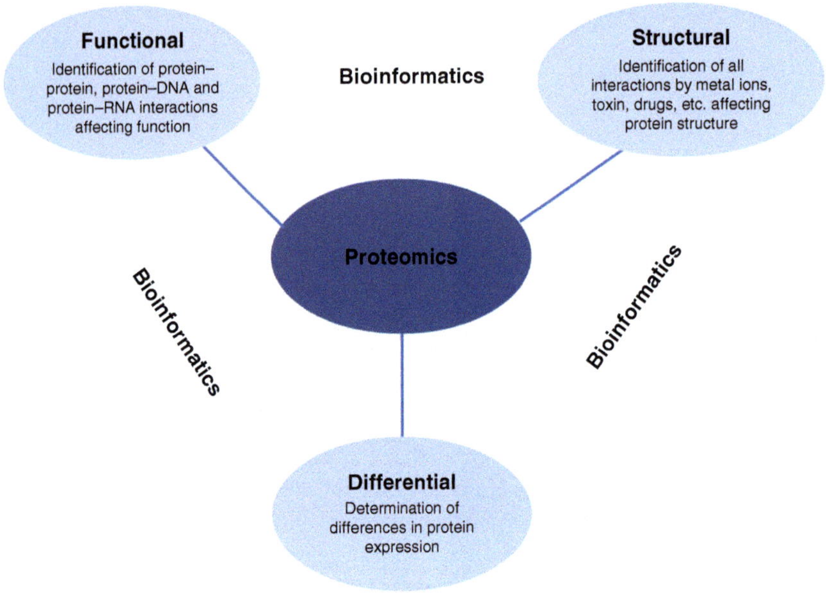

Figure 8.2 Types of proteomics (functional, structural and differential proteomics). Source: © John Wiley & Sons.

proteome is a revolutionary resource for disease diagnosis and monitoring of treatment. Because plasma is the primary clinical source and forms the largest part of the human proteome. Plasma contains many proteins from tissue and these proteins are released from tissues to plasma. They are also used as protein markers in a very large pool of immunoglobulins.

Protein characterization for integrated analysis of cellular processes and disease mechanisms at the protein level includes proteomic methods such as protein identification, post-translational modifications, protein–protein interactions, and elucidation of metabolic pathways. Protein expression levels in a wide variety of organisms are determined by mass spectrometry studies combined with different separation methods. Recently, with the realization of the potential effects of proteomic methods on biotechnology and medicine, clinicians have begun to take advantage of proteomic strategies to clarify the molecular mechanisms of diseases [11–13].

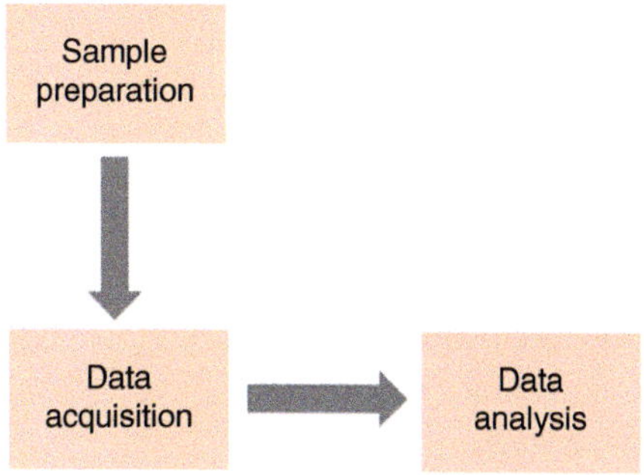

Figure 8.3 Proteomic workflow Source: © John Wiley & Sons.

Proteomic studies started with the application of two-dimensional gel electrophoresis, chromatographic methods and sensor , and the difficulties in protein identification restricted the wide range of applications in this area [1, 14–16]. Much of these technological obstacles have been overcome by the use of developing genomic databases and mass spectrometers. There are many features that can be used to identify a protein. A few of them are expressions, placement, interactions, domain structure, modifications, and activity. If the proteome of an organism is considered as all expressed protein forms, including spliced forms and post-translational modifications, the size and complexity of the amount of information to be considered can be understood. Proteomics is widely used in oncology, bio-medicine, agriculture, and food microbiology with the plasmonic sensor technique [16, 17].

In this review, we have discussed several examples of recent advancements in the application of proteomics to a number of oncology, biomedicine, agriculture, and food microbiology.

8.2 Plasmonic Sensors

All living systems in nature perceive changes in their environment and try to adapt to changing environmental conditions in order to sustain their vitality by reacting quickly to the conditions caused by these environmental changes. Plasmonic sensor systems are designed exactly with this biosensing mechanism. Plasmonic sensors are an analytical device consisting of a biologically active recognition part that specifically interacts with the target analyte and a transducer that can convert this interaction result into a measurable signal. Biological elements (bioreceptors) used in sensors can be enzymes, antibodies, antigens, microorganisms, nucleic acids, aptamers, cells, bacteria, or organelles. The only condition for the use of any biomaterial for bioreceptor purposes is that the material has the ability to uniquely identify the desired analyte. Plasmonic sensors play an important role in many fields such as medicine, chemistry, biology, physics, pharmacy, agriculture, food, environment, defense industry, and space [18–22]. Today, there are many commercialized plasmonic sensor applications.

8.2.1 Surface Plasmon Resonance Sensors

Surface plasmon resonance (SPR) sensors were first observed by Wood in 1902. When Wood reflected polarized light onto a mirror from a diffraction-regulating surface, he observed bright and dark bands in the reflected light [22, 23]. In 1968, Otto, Kretschmann and Raether reported the formation of surface plasmon [24]. The first use of SPR-based sensors in measuring biomolecular interactions was reported by Liedberg et al. in 1983 [25]. The SPR technique is considered as an important analytical tool thanks to its remarkable features such as the fact that it does not require labeling when examining the interactions of biomolecules, and it is real-time and highly sensitive.

When a photon of incoming light hits a metal surface (usually a gold surface), the surface plasmon resonance occurs. Some of the light energy coming at a certain angle is covered with electrons of the metal layer moving due to the stimulation on the surface. This electron movement that radiates parallel to the metal surface is called plasmon. In the presence of a fixed wavelength (from the light source) and the thin metal layer on the surface, the SPR angle at which the resonance takes place depends on the refractive index of the material near the metal surface. As a result, a small change in the analysis environment will change the reflection index, creating a change in plasmon (Figure 8.4) [26–29].

SPR-based sensors can be used to measure interactions between biomolecules without the need for any labeling. Measuring interactions simultaneously and directly enables the determination of kinetic or thermodynamic parameters, concentration, or interactions between ligand–analyte. SPR-based sensors can be used to understand a wide variety of biomolecular mechanisms, from protein–protein, antibody–antigen, receptor–ligand interactions to the identification of low molecular weight compounds, compared to other techniques such as enzyme or radiolabeling methods, as it has fast response time and high selectivity [30, 31].

8.2.2 Localized Surface Plasmon Resonance

Collective oscillations caused by the excitation of free electrons on metallic surfaces by electromagnetic waves are called localized surface plasmon resonance (LSPR). Electrons on the surface of metal nanoparticles oscillate when they are stimulated with a larger

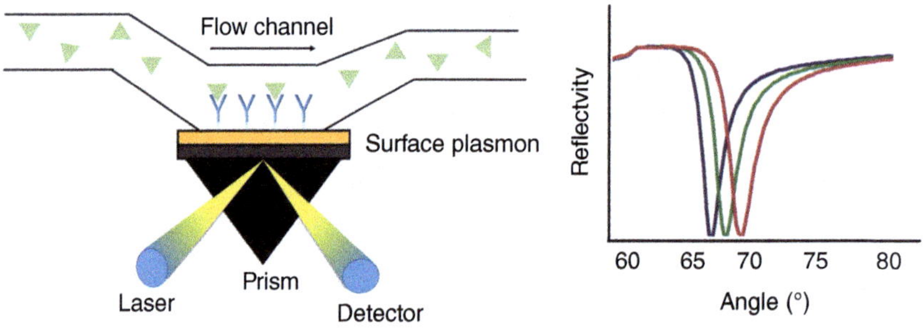

Figure 8.4 Schematic of surface plasmon resonance (SPR). Source: © John Wiley & Sons.

wavelength than their size, and a localized electromagnetic field is formed on the surface of the nanoparticles. When the oscillations of the incoming electromagnetic wave resonate with the electromagnetic oscillations of the nanoparticles, localized surface plasmon resonances are characterized by the resonance oscillation frequency. Thanks to LSPR, the local electric field on the surface of the particle increases considerably compared to the stimulated electric field. This occurs as a strong optical quenching and scattering signal in metal nanoparticles. The plasmon signal is maximum at the plasmon resonance frequency and can be measured by separating from the background signal with high contrast at certain wavelengths. Resonance occurs in visible metal wavelengths in noble metal nanoparticles such as gold and silver, while maximum absorption occurs in near-infrared and mid-infrared in semiconductor nanoparticles.

This feature that is frequently used in biosensor applications in biology, medicine, and biochemistry fields, can be optimized by controlling the shape, size, homogeneity, and surface coatings of metallic nanoparticles (Figure 8.5) [32, 33].

LSPR is a unique optical property responsible for the bright colors of colloidal suspensions of nanoparticles synthesized from noble metals such as gold and silver. The preference of gold especially in plasmonic studies is due to the absorption and scattering of light at high efficiency. Absorption and scattering bands in the visible region (500–600 nm) can be measured by UV–Vis spectroscopy provided by gold nanoparticle LSPRs. The location of LSPR bands depends on the dimensions, homogeneities, shapes, size distributions of AuNPs, as well as the dielectric constant of the medium. That is, modifying any of these parameters will cause shifts in absorption wavelength. For example, the growth of AuNP size or clustering of AuNP shifts the LSPR band to high wavelengths. Looking at the plasmon spectrum of spherical gold nanoparticles, whose synthesis and analytical solution is easier than others. It is observed that they have a single peak in the visible region. The optical properties of these particles largely depend on the diameter of the nanoparticle. As the diameter grows, these spheres emit more light. Therefore, gold nanospheres with a diameter greater than 40–50 nm are often used as bioimaging tags in the dark field microscopy technique. Compared to other gold nanoparticles, gold nano-sticks attract special attention by researchers because of their large surface to interact with light according to their volume [34, 35].

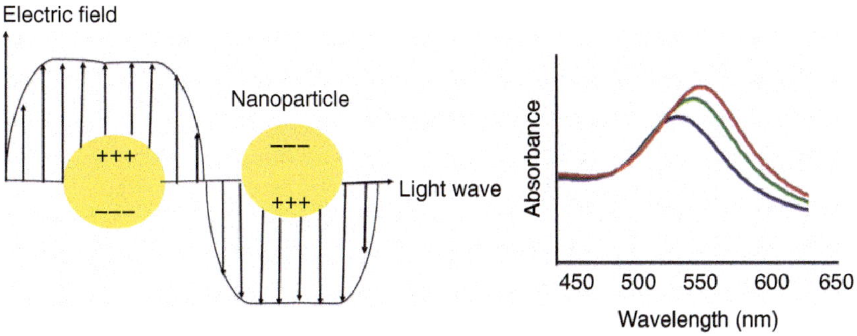

Figure 8.5 Schematic of localized surface plasmon resonance (LSPR). Source: © John Wiley & Sons.

8.2.3 Colorimetric Sensors

The colorimetric sensors are based on the measurement of the temperature exchange released as a result of the enzymatic reaction between the analyte and the biorecognitory molecule. Analyte concentration is measured depending on the enthalpy change. Colorimetric sensors determine the temperature measurement produced in an enzymatic reaction. It was originally used for colorimetric transduction enzyme-based sensors. It was then applied to cells and immunosensors. Colorimetric sensors use thermistors that convert the heat generated during the reaction into an electrical signal. The analyte is the substrate of the enzyme. Recently, the colorimetric sensor method has been used frequently in the food industry, and environmental monitoring [36–40].

8.3 Proteome Applications with Plasmonic Sensors

8.3.1 Food Applications

Proteomics has an important place among the technological developments in the food industry. This innovative area makes it possible to analyze largely processed food matrices and to identify low quantities of analytes with high selectivity. Thus, proteomics provides important contributions in terms of food quality control.

Zhu et al. [41] fabricated a SPR sensor for determination of Ochratoxin A (OTA) which has negative effects on human health using an anti-OTA aptamer immobilized sensor chip. In the study performed with OTA, which is a kind of chlorophenolic mycotoxin, streptavidin protein was used as a cross-linker and optimum conditions were defined. With the biosensor prepared in the following process, OTA was determined with a detection limit of 0.005 $ng\,ml^{-1}$ in the linear range of 0.094–10 $ng\,ml^{-1}$. In real sample experiments with red wine and peanut oil samples, the recovery of Ocratoxin A values between 86.9 and 116.5% were obtained.

Homola et al. [42] designed a SPR sensor based on wavelength modulation to detect of *staphylococcal enterotoxin B* (SEB) in milk sample. Easy and rapid analysis of SEs that cause gastroenteritis is important for early intervention in food poisoning. For this purpose, they conducted their studies with two different survey methodologies: direct detection of SEB and sandwich analysis. In line with the experiments, the SPR biosensor was determined as 5 $ng\,ml^{-1}$ in the buffer for direct detection, and in the sandwich analysis, the detection limit in both the buffer and milk sample was 0.5 $ng\,ml^{-1}$.

Joshi et al. [43] developed a portable nanostructured imaging surface plasmon resonance (iSPR) biosensor to investigate the presence of deoxynivalenol (DON) and OTA in beer samples. DON and OTA, known as mycotoxins, are mutagenic and carcinogenic secondary metabolites of fungi that are common in various beverages and foods. In the study based on competitive inhibition immunoassay, detection limits (LODs) in the beer samples of the obtained portable biosensor were observed as 17 $ng\,ml^{-1}$ for DON and 7 $ng\,ml^{-1}$ for OTA.

By utilizing molecular imprinting technology, Faalnouri et al. [44] reported SPR nanosensors to identify amoxicillin (AMOX) in milk samples. AMOX, an antibiotic used in the treatment of various bacterial infections, causes residues in many foods consumed daily in improper use. In accordance with this purpose, amoxicillin imprinted polymeric film and nanoparticles containing poly (hydroxyethyl methacrylate-*N*-methacryloyl-(L)-glutamic acid) poly (HEMAGA) were produced thanks to the ultraviolet polymerization technique. After various characterization processes, the detection limit for nanoparticles and polymeric film SPR nanosensors was calculated as 0.0009 and 0.0012 $ng\,ml^{-1}$ in the linear range of 0.1–200 $ng\,ml^{-1}$, respectively.

Based on the advantage of the molecular imprinting method, Çimen et al. [20] designed a SPR sensor to identify vitamin B2, vitamin B9 and vitamin B12 in milk samples and infant formula. The need to take vitamins from diet and supplements, as they cannot be synthesized in the human body, has made it imperative to monitor food quality and content of the food. After the characterization studies was successfully performed in detail, the detection limit values were reached as $1.6 \times 10^{-4}\,ng\,ml^{-1}$ for vitamin B2, $13.5 \times 10^{-4}\,ng\,ml^{-1}$ for vitamin B9 and $2.5 \times 10^{-4}\,ng\,ml^{-1}$ for vitamin B12. In addition to selectivity studies, validation studies of the developed SPR sensors were carried out via liquid chromatography–tandem mass spectrometry (LC-MS/MS). In reproducibility experiments, a negligible decrease was observed in the performance of the SPR sensors obtained (Figure 8.6).

Alzahrani et al.[45] aimed a colorimetric detection study based on LSPR, which is sensitive to Hg(II) ions. Mercury ions, which are highly harmful in terms of environment and human health, can leave permanent effects even at low concentrations. An environmentally friendly method was preferred in the production of green-fabricated silver nanoparticles (AgNPs) used in the study, thus avoiding the use of harmful chemicals. The characterization of the nanoparticles produced was successfully accomplished with various steps. In the experimental study, detection of Hg(II) ions was achieved by UV–Vis spectrophotometer, and it was also observed that the color of the solution changed from yellow to colorless. In the real sample study using tap and groundwater, recovery rates were obtained between 92 and 97%.

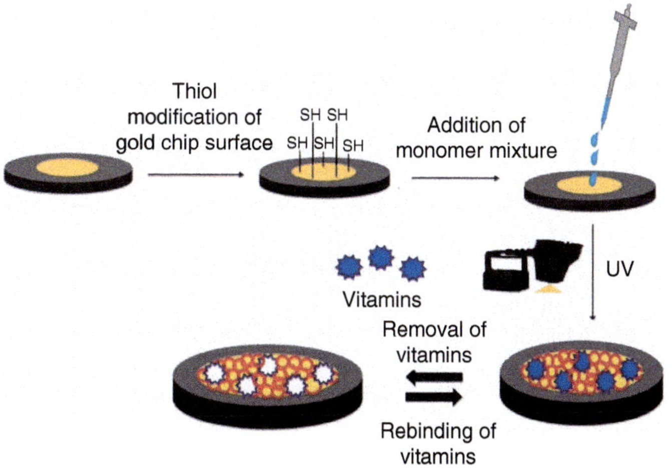

Figure 8.6 Schematic representation of the preparation of vitamin imprinted chip surface. Source: Reprinted with permission from Çimen and Denizli [20]. © 2020, Springer Nature.

Şener et al. [46] prepared a simple, rapid, sensitive, and selective colorimetric method for the detection of toxic mercury (Hg^{2+}) ions in drinking water. The proposed method of this sensitive colorimetric assay is the spontaneous reduction of Hg^{2+} ions onto the gold nanoparticles surface, which causes the gold nanoparticle aggregation in the presence of lysine. Using this proposed colorimetric assay, Hg^{2+} ions can be detected with naked eye within a few minutes. The detection limit of this colorimetric assay is 2.9 nM. In addition, selectivity of the colorimetric assay was tested by using several competing metal ions or a mixture of competing metal ions.

Detection of rare uranyl ions is important in terms of both for health and industrial reasons. Lee et al. [47] reported a colorimetric assay for the detection of uranyl (UO_2^{2+}) ions with a detection limit of 50 nM, which is below the maximum contamination limit have defined by the U.S. Environmental Protection Agency (EPA). Firstly, they used DNAyzmes to aggregate gold nanoparticles, which can specifically bind to UO_2^{2+}. Then, the bond between gold nanoparticle aggregates was cleaved by uranyl ions and the color of the solution changed from purple to red.

Liu et al. [48] demonstrated how they prepared a SPR immunosensor system, which combines the antibody-functionalized Fe_3O_4 magnetic nanoparticles (immunoMNPs) separation used in the determination of *Salmonella enteritidis(S. enteritidis).Salmonella,* which is a gram-negative type of bacteria generally transmitted from meat and eggs, can result in food poisoning cases that show symptoms such as fever and vomiting. In the study utilizing the sandwich immunoassay technique, MNPs also served as amplification reagent to increase the SPR signal in addition to helping specific recognition of *S. enteritidis*. With the developed system, the detection limit for *S. enteritidis* in the linear range of 1.4×10^1–1.4×10^9 CFU ml^{-1} was achieved as 14 CFU ml^{-1}. After the selectivity study was successfully performed with *Escherichia coli K12 ER2738 and Lactobacillus LJ-3*, recovery rates in the range of 92.76–113.25% were observed in real sample experiments with egg shell (Figure 8.7).

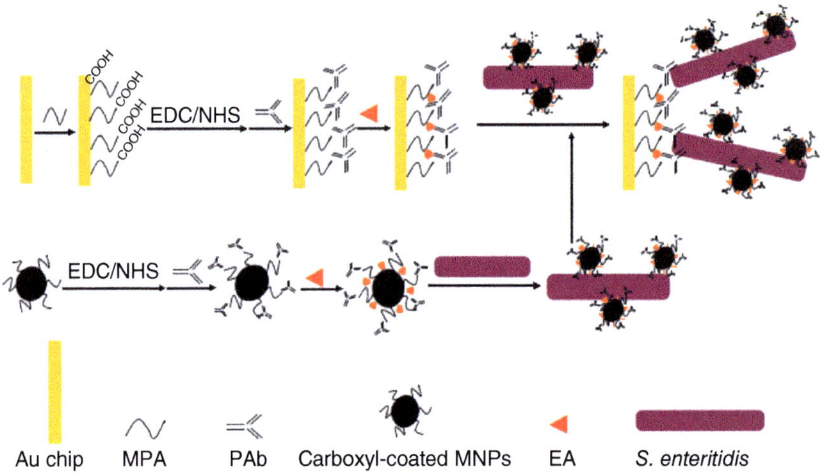

Figure 8.7 Schematic representation for the detection of *S. enteritidis* by MNPs-enhanced SPR sandwich assay. Source: Reprinted with permission from Liu et al. [48]. © 2016, Elsevier.

8.3.2 Biomedical Applications

Proteins are known as a significant step in the interaction between biomaterials and cells or tissues. Thanks to proteomics, conclusions can be made about protein adsorption with biomaterials, protein production by cells cultured on various materials, the applicability of new materials as a result of nanoparticle synthesis, and specific biomarkers for the progression of a disease.

Xu et al. [49] presented a new technique by combining the protein microarray and SPR sensor system for use in hepatitis B determination. Hepatitis B virus (HBV) is a dangerous virus type that carries great risk for human health, resulting in death in many people every year. In optimization studies, parameters such as immobilization buffer, reaction time, and concentration were discussed for five different protein probes. The results showed that other methods such as ELISA and PCR could detect only one marker within four hours, while five hepatitis B markers could be detected simultaneously within 30 minutes with the developed technology. In addition, the results obtained were more than 85% compatible with the ELISA test and showed a sensitivity of 0.1 $ng\,ml^{-1}$ for HBsAg (the hepatitis B surface antigen).

Osman et al. [50] produced a SPR sensor for sensing myoglobin in human serum by utilizing the molecular imprinting approach. Myoglobin shows a marked increase in the diagnosis of acute myocardial infarction (MI) as early as one hour; therefore, it plays a role in the diagnosis of this pathology. They first prepared two different types of nanofilm, namely myoglobin-imprinted poly (hydroxyethylmethacrylate-*N*-methacryloyl-L-tryptophan methyl ester) [poly (HEMA-MATrp)] nanofilm and non-imprinted poly (HEMA-MATrp) nanofilm without myoglobin to compare the results. After a wide range of characterization studies, the limit of detection (LOD) and limit of quantitation (LOQ) values of the developed sensor were obtained as 26.3 and 87.6 $ng\,ml^{-1}$, respectively. In studies conducted to evaluate the specificity of the prepared sensor, competitive agents such as lysozyme, cytochrome c, and bovine serum albumin were used, and it was observed that the designed sensor was much more sensitive to myoglobin.

Saylan et al. [51] suggested a nanofilm molecularly imprinted on the chip, which can sense hemoglobin utilizing a SPR sensor system. Hemoglobin, which takes an active role in the transportation of oxygen and carbon dioxide, is an iron-containing protein in its structure. The change in the amount and structural content of this protein can cause negative health problems such as anemia and thalassemia. In the study carried out via photopolymerization method, methylenebisacrylamide was used as a cross-linker and ammonium persulfate–tetramethylethylenediamine were preferred as a initiator–activator pair. In the experimental studies conducted with the prepared nanofilm, the limit of detection for hemoglobin was calculated as 0.000 35 $mg\,ml^{-1}$, and the selectivity of the sensor was evaluated with competitive agents such as transferrin (Trf), myoglobin (Myb) bovine serum albumin (BSA) and lysozyme (Lyz).

Özgür et al. [52] took advantage of the molecular imprinting strategy to identify *Escherichia coli(E. coli)* and presented an approach about a SPR-based biomimetic sensor. Urinary tract infection (UTI) caused by *E. coli* is a common disease and causes much more serious health problems when proper treatment is not applied. In the study, *N*-methacryloyl (L)-histidine methyl ester (MAH) was preferred as a functional monomer and Ag nanoparticles (AgNPs) that contribute to the sensitivity of the prepared sensor were used. With the developed system, the detection limit for *E. coli* was reached as 0.57 CFU ml^{-1}. Compared with the studies in the literature, it was observed that the sensor system created had shorter analysis time and when the reusability parameter was examined, there was no significant decrease in adsorption capacity (Figure 8.8).

Sarı et al. [53] synthesized molecularly imprinted nanoparticles (MIP/NPs) and fabricated a SPR nanosensor for use in the determination of ciprofloxacin. Ciprofloxacin (CPX) residues, a broad spectrum antibiotic belonging to the fluoroquinolone class, have been found to threaten both the environment and human health. With the system, in which nanoparticles were obtained by the method of miniemulsion polymerization, the limits of detection were found in pure water and synthetic wastewater (SWW) as 3.21 and 7.1 ppb, respectively. Additionally, in the study, the applicability of the prepared nanosensor was evaluated by collecting all relevant data including selectivity, reusability and kinetic analysis (Figure 8.9).

Koyun et al. [54] provided a SPR aptasensor that can detect human activated protein C (APC) sensitively. Protein C plays a significant role in the coagulation system as an antithrombotic and anticoagulant agent. In the study, SPR aptasensors with random DNA and HEMA-MAC polymeric films were formed and the *N*-methacryloyl-L-cysteine (MAC) monomer, an amino acid based monomer, was used. After the characterization processes of SPR aptasensors were evaluated, LOD and LOQ of the DNA-Apt SPR aptasensor were found to be 1.5 and 5.2 ng ml^{-1}, respectively. Other prominent parameters examined in the study were kinetic analyses, selectivity, and reusability studies (Figure 8.10).

Liu et al. [55] demonstrated an integrated sensor system for the quantitative determination and qualitative recognition of proteins by taking advantage of complementary metal-oxide-semiconductor (CMOS) image sensors and capillary LSPR sensors. They conducted the experimental study with transferrin, which carries iron to tissues in the body, and IgG, which acts as the most common type of antibody in the bloodstream. The ability to functionalize the surface of the gold nanoparticles used in the study by various proteins provided an advantage in detecting a wide variety of biochemical analytes. They also

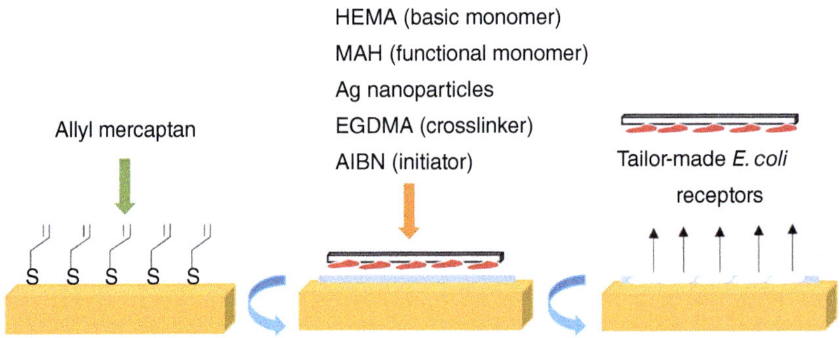

Figure 8.8 Schematic representation of *E. coli* imprinted polymeric film synthesis via microcontact imprinting technique. Source: Reprinted with permission from Özgür et al. [52]. © 2020, Elsevier.

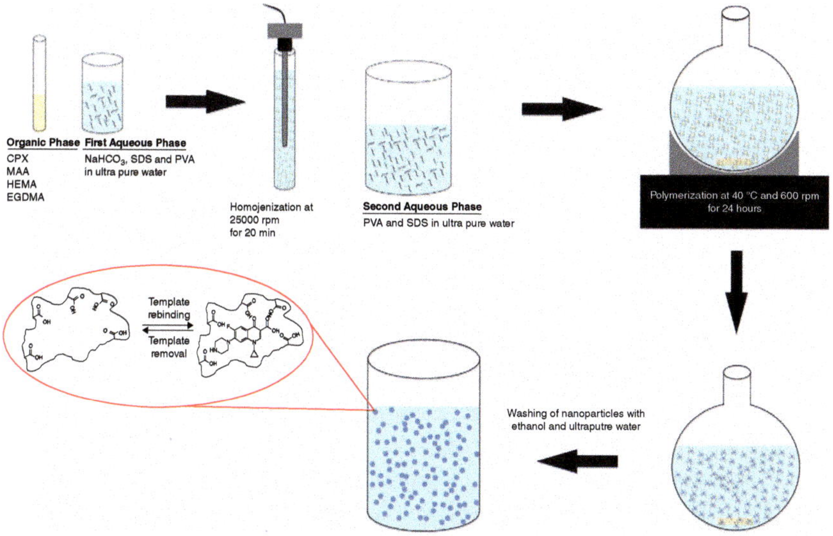

Figure 8.9 Synthesis process of CPX imprinted polymer. Source: Reprinted with permission from Sari et al. [53]. © 2018, Taylor & Francis.

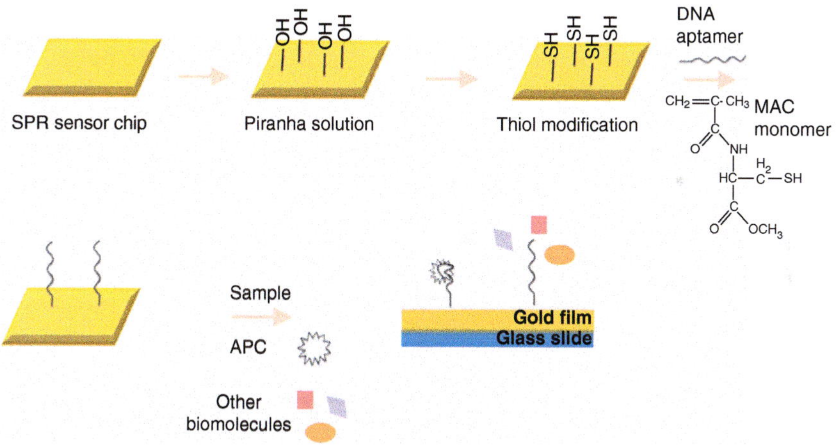

Figure 8.10 Schematic diagram of SPR aptasensor for APC detection. Source: Reprinted with permission from Koyun et al. [54]. © 2019, Elsevier.

emphasized that the technology they developed was accessible, inexpensive, and easy, stating that many methods of analysis used in the identification of proteins in the human body are very costly, and these applications are limited only to institutional locations.

Kim et al. [56] described a LSPR biosensor with a new protocol for immobilization of nanoparticles on an optical fiber, thanks to an Au capping procedure based on seed-mediated growth. In the study, citrate reduction technique was preferred and AuNPs were better adhered to the optical fiber surface without losing its precious properties. The obtained LSPR sensor was applied in the determination of thyroglobulin (Tg), and the detection limit for Tg was calculated as 0.19 pg ml^{-1}. Additionally, the reliability of the prepared sensor was tested by identifying Tg levels in serum samples taken from patients (Figures 8.11 and 8.12).

Çimen et al. [57] fabricated an anti-cardiac troponin I monoclonal antibody immobilized SPR sensor system, which is quite sensitive and specific to cardiac troponin I. Nowadays, deaths from ischemic heart diseases are frequently encountered. Rapid and early diagnosis of this pathology is very important in terms of effective treatment, and one of the most important biomarkers used in this regard is cardiac troponin I (cTnI). Kinetic analysis was carried out after the immobilization procedure was successfully verified by various characterization processes. In addition, selectivity studies with immunoglobulin G (IgG), prostate-specific antigen (PSA) and myoglobin (Mb), and reproducibility of the biosensor were examined. The detection limit and quantification limit of the developed biosensor were obtained as 0.000 12 and 0.00041 ng ml^{-1}, respectively. Validation studies were carried out via enzyme-linked immunosorbent analysis (ELISA) technique in order to demonstrate the applicability and superiority of the prepared biosensor (Figure 8.13).

Jalilzadeh et al. [18] presented a SPR nanosensor capable of detecting Zn(II) ions in an artificial plasma and aqueous solution medium. Zinc plays an important role in the immune system and nervous system and the lack or excessive amount of zinc ions in the human body causes negative health effects. First, both Zn (II) ions imprinted poly (2-hydroxyethyl methacrylate-*N*-methacryloyl-(L)-histidine methyl ester) [poly (HEMAH)] nanofilm and non-imprinted nanofilm were synthesized to demonstrate the advantage of the molecular imprinting technique. They also conducted kinetic analysis and selectivity studies. In the

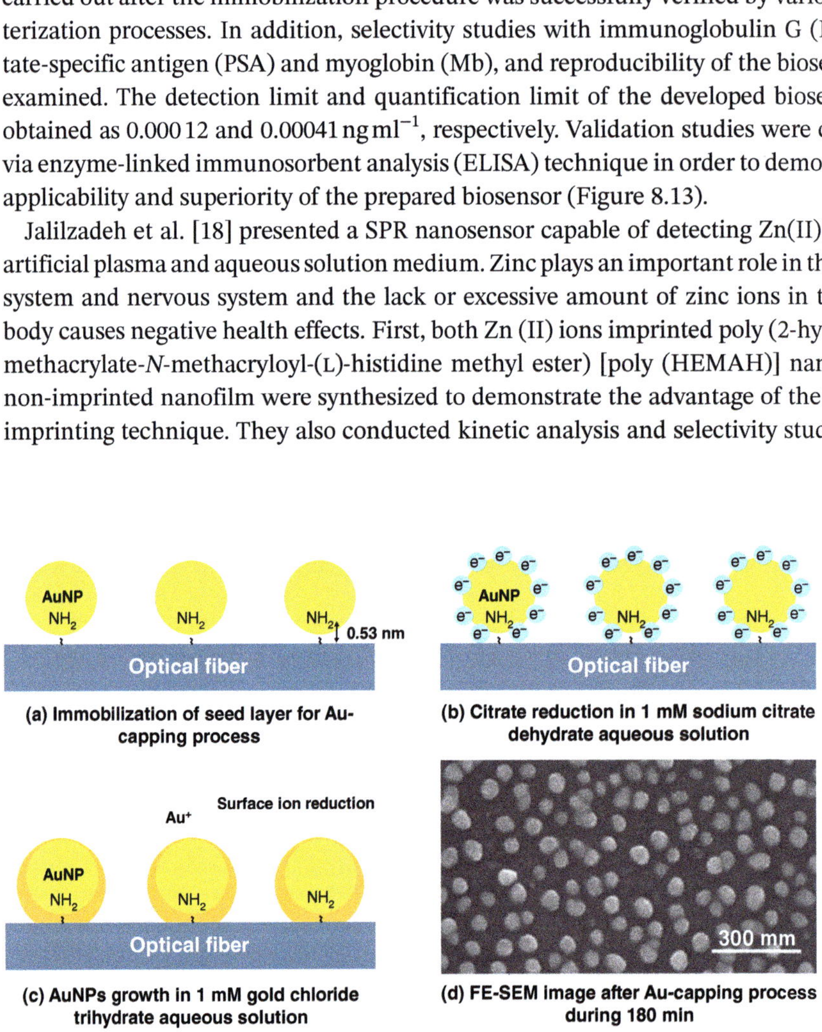

Figure 8.11 Schematic diagram of the Au-capping process on a surface of an optical fiber. (a) Firstly, AuNPs are immobilized on the optical fiber. (b) Secondly, citrate ions are attached on AuNPs to reduce Au$^+$ ions. (c) Finally, AuNPs are physically absorbed to optical fiber surface owing to the growth of AuNPs by citrate reduction. (d) The FE-SEM image shows that there is no aggregation of the nanoparticles following the Au-capping process of 180 min. Source: Reprinted with permission from Kim et al. [56]. © 2019, Elsevier.

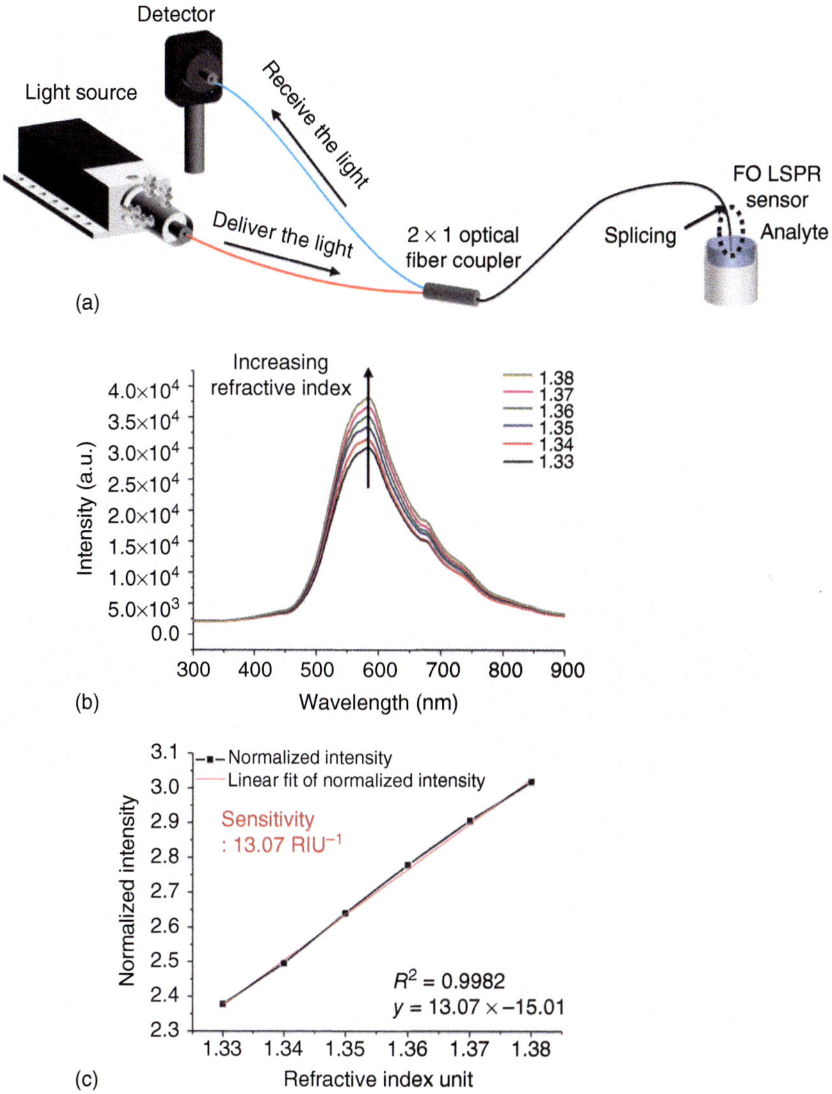

Figure 8.12 Optical measurement setup and changes in LSPR intensities from the FO LSPR sensor by varying refractive indices. (a) Schematic diagram of the optical measurement system with the FO LSPR sensor is shown. 2×1 coupler is used, which has a Y connector on one side. One end is connected to a light source, while the other end is connected to a detector. One end of the opposite side is connected to the FO LSPR sensor. (b) The spectra are measured with various refractive indices. (c) The changes in refractive index solution intensities are obtained and plotted at peak resonance intensities. The LSPR intensities show a linear response to increases of refractive indices. Source: Reprinted with permission from Kim et al. [56]. © 2019, Elsevier.

study, the Zn (II) ions detection limit for artificial plasma was reported as 0.31 ng ml^{-1} and for aqueous solution as 0.19 ng ml^{-1}.

Shahbazi et al. [58] suggested a novel spectrophotometric and colorimetric sensors for the detection of metformin in human serum. Firstly, they were prepared citrate-capped

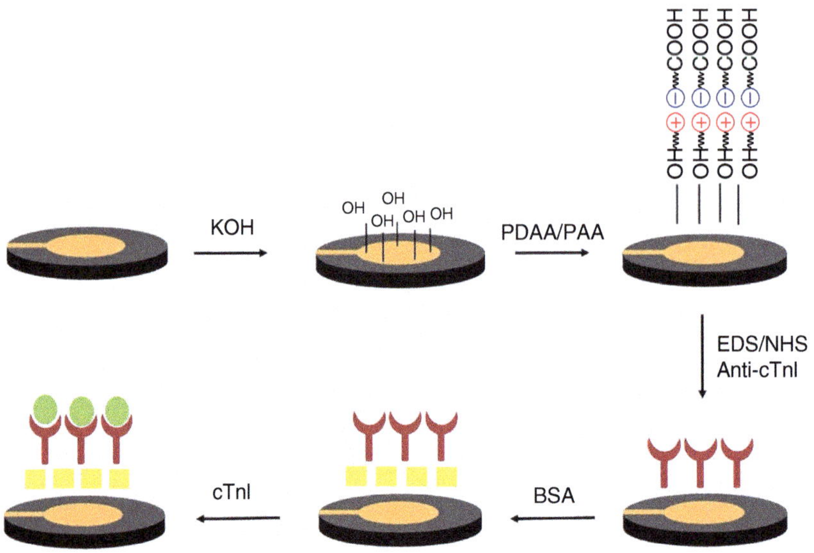

Figure 8.13 Schematic illustration of SPR biosensor surface modification for detection of cardiac troponin I (cTnI). Source: Reprinted with permission from Çimen et al. [57]. © 2020, Elsevier.

gold nanoparticles (citrate-GNPs). In this probe, a color change was observed from the accumulation of gold nanoparticles by adding metformin to the gold nanoparticle solution. These changes of color and analyte quantity analysis were performed spectrophotometrically. In addition, parameters such as pH, time, and GNP ratio have been optimized. The limit of detection was 1.79 ppm for metformin in a linear range of 6.25–133.3 ppm.

Antibiotics are frequently used in pharmaceutical, agriculture, livestock production, and aquaculture industry. Pourreza et al. [59] used the colorimetric detection method of ceftriaxone from serum and urine samples. They synthesized a hydrogel network made from poly (vinyl alcohol) and borax and then was modified with gold nanoparticles. The characterization of the modified hydrogel network used transmission electron microscopy, energy dispersive X-ray analysis, zeta-sizing, and viscosimetry. Maximum absorption in the UV–vis spectrum was observed at 517 nm to confirm the nanosynthesis of AuNPs. The detection limit of this colorimetric method is $0.33\,\mu g\,ml^{-1}$ in linear range of $1–90\,\mu g\,ml^{-1}$.

Wang et al. [60] proposed a rapid, easy-to-use, cheap, and hand-held colorimetric sensor array for multiple and discrimination of protein detection. In this study, they proposed a sensitive, simple, visible, and smartphone readable method for the determination of multiple proteins. Different aggregation behaviors were observed for different proteins with the effect of differential ion strength, which led to different color changes by using NaCl salt in five different concentrations of gold nanoparticles. This colorimetric sensor assay system did not distinguish 12 proteins in aqueous solution at a concentration of only 50 nM. It also showed that these proteins can be distinguished at 100 nM with 100% accuracy in the presence of human urine.

Although proteins play an important role in the diagnosis of the disease, protein separation is quite difficult. Jia et al. [61] prepared a colorimetric sensor for the determination of proteins based on the aggregation of DNA-functionalized gold nanoparticles with exonuclease I. The gold nanoparticles were formed a different DNA–protein binding without aggregation of proteins in high concentration NaCl solution. Firstly, it is obtained on a array of sensors called Fingerprints, and then determined by linear discriminant analysis (LDA). With this sensor array, 15 proteins at 10 nM concentration were separated correctly in the buffer solution and real serum samples. In addition, these colorimetric sensors have the ability to distinguish both individual proteins and their mixtures.

8.3.3 Agricultural Applications

Proteomic, a developing branch of science in terms of plant science, enables identification of proteins in physiological conditions and interactions. However, plant proteomic research has not become a focus as much as humans and cancer studies.

Song et al. [62] published an experimental research on developing a SPR-based immunosensor so that phosmet in fruits and vegetables can be selectively detected. Phosmet, which is involved in applications for pest control in agriculture, is known as an insecticide and its residues have been identified in vegetables and fruits. With the biosensor prepared in the studies conducted in this direction, the detection limit was calculated as $1.6\,\mathrm{ng\,l^{-1}}$ ($S/N = 3$) in the linear range of $8.0–60.0\,\mathrm{ng\,l^{-1}}$. In the recovery studies carried out with various fruit and vegetable types such as peaches and cabbage, the rates were in the range of 86.4–102.8%.

By combining molecular imprinting technique with SPR sensor technology, Saylan et al. [19] prepared nanofilms used to detect pesticides such as simazine (SMZ), atrazine (ATZ), and cyanazine (SNZ) in aqueous solution. Pesticides which are used to prevent and control harmful organisms cause nonspecific toxicological effects. With the pesticide imprinted nanofilms on the SPR surface obtained by UV polymerization reactions, LOD values of 0.031, 0.091, and 0.095 nM for SMZ, ATZ, and SNZ were reached in the range of 0.10–6.64 nM, respectively. In addition, it was concluded that Langmuir adsorption isotherm model was the most suitable isotherm model for all pesticide imprinted sensor systems and as a result of experiments, the prepared sensor was highly selective and reusable.

Masdor et al. [63] aimed to create a SPR sensor system using three immunoassay methods, namely direct, sandwich and nanomaterial-modified sandwich assays, to quickly determine *Campylobacter jejuni*. *Campylobacter* spp. causes gastroenteritis in humans, and *Campylobacter jejuni*, a member of this family, is frequently encountered in poultry-based foods such as chicken. As a result of the experiments carried out under optimum conditions, LOD values were obtained as $8\times10^{6}\,\mathrm{CFU\,ml^{-1}}$ for the direct assay method and $4\times10^{4}\,\mathrm{CFU\,ml^{-1}}$ for the sandwich assay method. Compared to enzyme-linked immunosorbent assay (ELISA) (LOD value was $10^{6}–10^{7}\,\mathrm{CFU\,ml^{-1}}$), it was observed that much more promising results were achieved by using the highly selective sandwich assay method.

Chen et al. [64] investigated a label-free colorimetric method for the detection of Pb^{2+} ions in aqueous solution based on gold nanoparticles. Thiosulfate was used to form a complex with gold nanoparticles. In the presence of Pb^{2+} ions and 2-mercaptoethanol, Pb–Au alloy was formed on the surface of gold nanoparticles, which results the aggregation of gold nanoparticles. The limit of detection for this selective colorimetric method was 0.5 nM. In addition, they were analyzed using colorimetric method for the detection Pb^{2+} ions in soil and river water samples.

8.3.4 Oncology Applications

Proteins are responsible for many cellular functions; therefore, they play a decisive role in the pathogenesis of diseases. Proteomics examines the functions, structures, and interactions of proteins in complex biological systems and provides guidance in the diagnosis and treatment of the related disease.

Sankiewicz et al. [65] proposed a biosensor for serum collagen type IV (COLIV) determination thanks to surface plasmon resonance imaging (SPRI) technology based on the interaction of collagen IV with a specific antibody. COLIV is considered to be a tumor biomarker due to the detection of high concentrations of COLIV in the serum of patients in various cancer conditions such as metastatic breast cancers, lung, liver, and stomach. With the developed biosensor in a response range of 10–300 $ng\,ml^{-1}$, LOD was calculated as 2.4 $ng\,ml^{-1}$ and LOQ was 8 $ng\,ml^{-1}$. In the selectivity study conducted with albumin, laminin, fibronectin, and collagen I, the prepared biosensor was observed to be highly selective against COLIV (Figure 8.14).

Cennamo et al. [66] prepared a SPR sensor based on a plastic optical fiber (POF) for use in the detection of vascular endothelial growth factor (VEGF). It is suggested that there is a clinical correlation in the relationship between cancer tissue and VEGF protein, which is

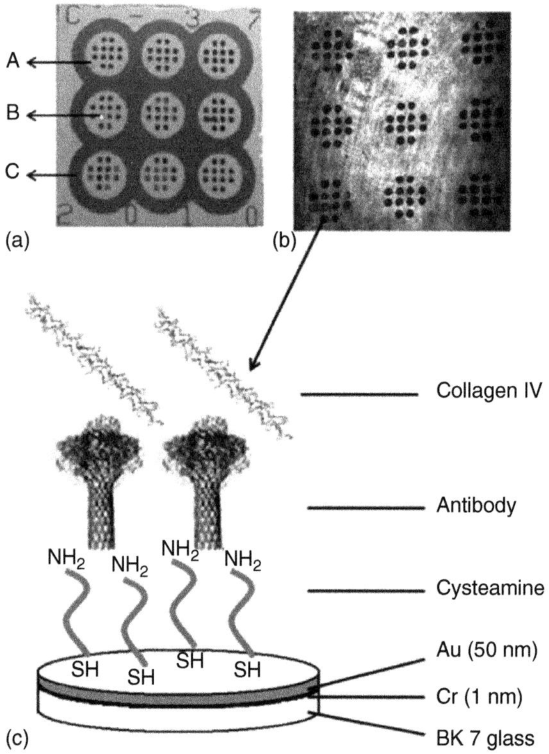

Figure 8.14 (a) Picture of chip (A-photopolymer, B-free gold surface, C-hydrophobic paint), (b) the SPR image of the chip obtained using a CCD camera, (c) the schematic illustration of the sensor active part. Source: Reprinted with permission from Sankiewicz et al. [65]. © 2016, Elsevier.

known to play a role in angiogenesis. In experiments carried out with the aptasensor system (Apt) developed by using aptamers as biological recognition agents, the detection limit of Apt, which is a sensor that has not been deactivated with mercaptoethanol, was found as approximately 3 nM, and the detection limit of the passivated sensor (Apt-MPET) was 0.8 nM.

Chen et al. [67] manufactured a SPR cytosensor for use in providing breast cancer cell detection, involving a two-marker recognition system and magnetic nanoparticle (MNP). It is very important to diagnose early in breast cancer, which is a common type of cancer in women and has high mortality mainly in developing countries. In this study, target breast cancer cells (MCF-7) were selectively captured by human mucin-1 (MUC1) aptamer, one of the most frequently used markers in monitoring metastatic breast cancer cells. In addition, folic acid conjugated monodispers MNP was used as the second selective binding reagent to create a sandwich SPR assay. As a result, the detection limit was found as low as 500 cells ml^{-1} with the sensor system created (Figure 8.15).

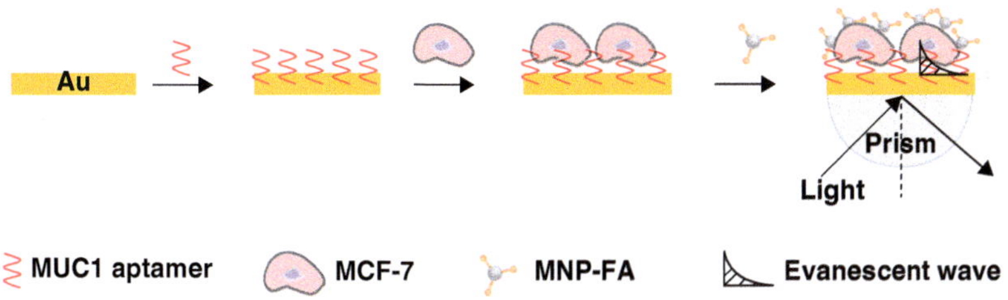

Figure 8.15 Schematic illustration of sensing strategy for SPR cytosensor. Source: Reprinted with permission from Chen et al. [67]. © 2014, Elsevier.

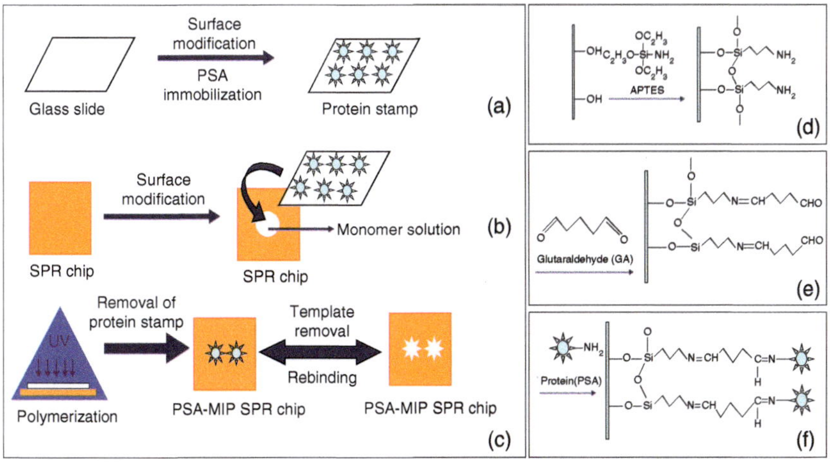

Figure 8.16 Schematic representation of microcontact imprinting of PSA onto the SPR biosensor and surface modification of glass cover slips; (a) preparation of glass cover slips (protein stamps), (b) preparation of SPR chips, (c) microcontact imprinting of PSA via UV polymerization, (d) surface modification of glass cover slips with APTES, (e) activation of amino groups on glass cover slips with glutaraldehyde, and (f) PSA immobilization onto the glass cover slips. Source: Reprinted with permission from Ertürk et al. [68]. © 2016, Elsevier.

Ertürk et al. [68] published an experimental research on the design of microcontact PSA imprinted SPR sensor chip for the determination of prostate-specific antigen (PSA). PSA, which plays a critical role in the diagnosis and treatment of prostate cancer, is a crucial biomarker. The characterization of the chip, which was imprinted using the UV polymerization technique, was performed with various instruments. As a result of experimental studies, the limit of detection for PSA was about 91 $pg\,ml^{-1}$ (18×10^{-14} M) in the concentration range of 0.1–50 $ng\,ml^{-1}$. The related isotherm models and the specificity of the SPR sensor were investigated with competitive agents such as lysozyme (Lyz) and human serum albumin (HSA) (Figure 8.16).

8.4 Conclusions and Prospects

In recent years, interest in proteomics studies has been increasing day by day. Proteomic plays an important role in disease diagnosis and drug analysis. With the realization of the potential effects of proteomic methods on biotechnology and medicine, clinicians have begun to take advantage of proteomic strategies to clarify the molecular mechanisms of diseases. The methods of proteomic analysis include inductively two-dimensional gel electrophoresis, chromatographic methods, and mass spectrometers. These methods are costly methods require intense technical training because of their time consuming and complicated procedures. The development of plasmonic sensors for monitoring of proteomic studies is very important. Plasmonic sensors have advantages such as a portability, high sensitive (low LOD), speed (fast detection) and easy-to-use (portable devices), on-site sensing capability, and improved performance of the devices. In this chapter, we discussed some examples of recent developments in proteomics applications in different fields such as agriculture, medical, diagnosis, and food.

References

1 Williams, C. and Addona, T.A. (2000). *Trends in Biotechnology* 18: 45–48.
2 Scobioala, S., Klocke, R., Michel, G. et al. (2004). *Chem* 11 (24): 3203–3218.
3 Pandey, M.M. (2000). *ProNature* 6788: 837–846.
4 Yuk, J.S. and Ha, K.S. (2005). *Experimental & Molecular Medicine* 37: 1–10.
5 Zhu, H. and Snyder, M. (2003). *Current Opinion in Chemical Biology* 7: 55–63.
6 Westbrook, J.A., Jun, X.Y., Wait, R., and Dunn, M.J. (2001). *Proteomics* 1: 370–376.
7 Roh, S., Chung, T., and Lee, B. (2010). *Advanced Sensor Systems and Applications IV* 7853: 785303.
8 Hanash, S. (2003). *Nature* 422: 226–232.
9 Zhang, C.C. and Kast, J. (2010). *Computer-Aided Drug Design* 6: 147–164.
10 Aebersold, R. and Mann, M. (2003). *Nature* 422: 198–207.
11 Petricoin, E.F. and Liotta, L.A. (2003). *Journal of Nutrition* 133: 2476S–2484S.
12 Ray, S., Reddy, P.J., Jain, R. et al. (2011). *Proteomics* 11: 2139–2161.
13 Yuk, J.S., Jung, J.M., Jung, S.H. et al. (2004a). *Proteomics* 4: 3468–3476.
14 Yan, W., Lee, H., Deutsch, E.W. et al. (2004). *Molecular & Cellular Proteomics* 3: 1039–1041.

15 Gou, M.J., Nys, G., Cobraiville, G. et al. (2020). *Journal of Chromatography A* 1618: 460873.
16 Cristea, I.M., Gaskell, S.J., and Whetton, A.D. (2004). *Blood* 103: 3624–3634.
17 Aslam, B., Basit, M., Nisar, M.A. et al. (2017). *Journal of Chromatographic Science* 2: 182–196.
18 Jalilzadeh, M., Çimen, D., Özgür, E. et al. (2019). *Journal of Macromolecular Science, Part A* 56: 877–886.
19 Saylan, Y., Akgönüllü, S., Çimen, D. et al. (2017). *Sensors and Actuators B: Chemical* 241: 446–454.
20 Çimen, D. and Denizli, A. (2020). *Photonic Sensors*: 1–17.
21 Faalnouri, S., Çimen, D., Bereli, N., and Denizli, A. (2020). *ChemistrySelect* 15: 4761–4769.
22 Wood, R. (1902). *The London, Edinburgh, and Dublin Philosophical Magazine and Journal of Science* 21: 396–402.
23 Wood, R. (1912). *The London, Edinburgh, and Dublin Philosophical Magazine and Journal of Science* 23: 310–317.
24 Otto, A. (1968). *Zeitschrift für Physik* 216: 398–410.
25 Liedberg, A., Nylander, C., and Lunström, I. (1983). *Sensors and Actuators* 4: 299–304.
26 Ladd, J., Taylor, A.D., Piliarik, M., and Homola, J. (2009). *Analytical and Bioanalytical Chemistry* 393: 1157–1163.
27 Mohanty, S.P. and Kougianos, E. (2006). *IEEE Potentials* 25: 35–40.
28 R. Birla (ed.), Springer, 2016, Switzerland, vol. 1 31–57.
29 Chaubey, A. and Malhotra, B. (2002). *Biosensors and Bioelectronics* 17: 441–456.
30 Nguyen, H.H., Park, J., Kang, S., and Kim, M. (2015). *Sensors* 15: 10481–10510.
31 Karlsson, R. (2004). *Journal of Molecular Recognition* 17: 151–161.
32 Jatschka, J., Dathe, A., Csáki, A. et al. (2016). *Sensing and Bio-Sensing Research* 7: 62–70.
33 Teranishi, T., Eguchi, M., Kanehara, M., and Gwo, S. (2011). *Journal of Materials Chemistry A* 21: 10238–10242.
34 Chang, K., Wang, S., Zhang, H. et al. (2017). *PLoS One* 12: e0177131.
35 Liu, B., Huang, R., Yu, Y. et al. (2018). *Frontiers in Chemistry* 6: 94.
36 Mosbach, K. (1991). *Biosensors and Bioelectronics* 6: 179–182.
37 Jabariyan, S. and Zanjanchi, M.A. (2019). *Applied Physics A* 125: 872.
38 Kumar, V.V. and Anthony, S.P. (2014). *Sensors and Actuators* 191: 31–36.
39 Xie, B., Ramanathan, K., and Danielsson, B. (2000). *TrAC, Trends in Analytical Chemistry* 19: 340–349.
40 Ramanathan, K. and Danielsson, B. (2001). *Biosensors and Bioelectronics* 16: 417–423.
41 Zhu, Z., Feng, M., Zuo, L. et al. (2015). *Biosensors and Bioelectronics* 65: 320–326.
42 Homola, J., Dostalek, J., Chen, S. et al. (2002). *International Journal of Food Microbiology* 75: 61–69.
43 Joshi, S., Annida, R.M., Zuilhof, H. et al. (2016). *Journal of Agricultural and Food Chemistry* 64: 8263–8271.
44 Faalnouri, S., Çimen, D., Bereli, N., and Denizli, A. (2020). *Chemistry Select* 5: 4761–4769.
45 Alzahrani, E. (2020). *Journal of Analytical Methods in Chemistry* 2020: 6026312, 1–14.
46 Şener, G., Uzun, L., and Denizli, A. (2014). *Analytical Chemistry* 86: 514–520.
47 Lee, J.H., Wang, Z.D., Liu, J.W., and Lu, Y. (2008). *Journal of the American Chemical Society* 130: 14217–14226.
48 Liu, X., Hu, Y., Zheng, S. et al. (2016). *Sensors and Actuators B: Chemical* 230: 191–198.

49 Xu, H., Gu, D., He, J. et al. (2015). *Medical Materials and Engineering* 26: S2091–S2100.
50 Osman, B., Uzun, L., Beşirli, N., and Denizli, A. (2013). *Materials Science and Engineering C* 33: 3609–3614.
51 Saylan, Y. and Denizli, A. (2018). *Sensors* 18: 3016.
52 Özgür, E., Topçu, A.A., Yılmaz, E., and Denizli, A. (2020). *Talanta* 212: 120778.
53 Sari, E., Üzek, R., Duman, M., and Denizli, A. (2018). *Journal of Biomaterials Science, Polymer Edition* 29: 1302–1318.
54 Koyun, S., Akgönüllü, S., Yavuz, H. et al. (2019). *Talanta* 194: 528–533.
55 Liu, Y., Zhang, N., Li, P. et al. (2019). *Nanomaterials* 9: 1019.
56 Kim, H.M., Jeong, D.H., Lee, H.Y. et al. (2019). *Optics and Laser Technology* 114: 171–178.
57 Çimen, D., Bereli, N., Günaydın, S., and Denizli, A. (2020). *Talanta* 219: 121259.
58 Shahbazi, N. and Zare-Dorabei, R. (2019). *ACS Omega* 4: 17519–17526.
59 Pourreza, N. and Ghomi, M. (2020). *Microchimica Acta* 187: 133.
60 Wang, F., Lu, Y., Yang, J. et al. (2017). *Analyst* 142: 3177–3182.
61 Jia, F., Liu, O., Wei, W., and Chen, Z. (2019). *Analyst* 144: 4865–4870.
62 Song, Y., Liu, M., and Wang, S. (2013). *Journal of Agricultural and Food Chemistry* 61: 2625–2630.
63 Masdor, N.A., Altintas, Z., and Tothill, I.E. (2017). *Chem* 5: 16.
64 Chen, Y.Y., Chang, H.T., Shiang, Y.C. et al. (2019). *Analytical Chemistry* 81: 9433–9439.
65 Sankiewicz, A., Lukaszewski, Z., Trojanowska, K., and Gorodkiewicz, E. (2016). *Analytical Biochemistry* 515: 40–46.
66 Cennamo, N., Pesavento, M., Lunelli, L. et al. (2015). *Talanta* 140: 88–95.
67 Chen, H., Hou, Y., Ye, Z. et al. (2014). *Sensors and Actuators B: Chemical* 201: 433–438.
68 Ertürk, G., Özen, H., Tümer, M.A. et al. (2016). *Sensors and Actuators B: Chemical* 224: 823–832.

9

Cancer Cell Recognition via Sensors System

Monireh Bakhshpour[1], Melek Özsevgiç[1], Ayşe Kevser Pişkin[2], and Adil Denizli[1]

[1] *Department of Chemistry, Hacettepe University, Ankara, Turkey*
[2] *Department of Medical Biochemistry, Hacettepe University, Ankara, Turkey*

9.1 Introduction

Cancer is characterized by uncontrolled and unregulated cell growth due to epigenetic defects and accumulation of specific genetic aberrations [1]. Abnormal cell growth causes a tumor mass that eventually becomes unattached of normal homeostatic balances and checks. The tumor cells become durable to anti-growth defenses and apoptosis in the body. The tumor rapidly metastasizes to other body systems and organs with the progressing symptoms of cancer. Unfortunately, in most cases, cancer is not curable. Therefore, the development of selective sensors for the early detection of cancer cells is of particular interest with the widespread use of developed systems [2].

Sensors are very quick systems, as such, they have much greater potential to pave the way for rapid diagnosis [3–5]. Clinical measurements of analytes will be no longer conducted exclusively within the laboratory of clinical chemistry. Sensors are instruments which can be designed as unique devices for detection of the target, using bio-recognition elements immobilized in/on a suitable matrix. Sensors are composed of bio-receptor, transducer, and a read-out part. This bio-receptor plays a major role in recognizing the analyte. An effective bio-receptor must be very specific and able to bond only to the molecule of interest. The transducer converts the process of bio-recognition into an accessible output. The read-out part is an electrical device which would identify the transducer input signals. The readable output obtains via transforming eventually the electrical signal into data [6].

In the last decades, many kinds of sensors system such as piezoelectric, optical, and electrochemical, have been prepared and reported for detecting cancer cells. Optical sensors have been reported for the detection of target molecules by the measurement of a change in the optical features of the transducer surface. Optical sensors are mostly based on surface plasmon resonance (SPR) and fluorescence that are generally used for the detection of cells. Optical measurements can be achieved by monitoring the change of the optical signal that takes place between a modified nanomaterial and a target molecule [7, 8].

Plasmonic Sensors and their Applications, First Edition. Edited by Adil Denizli.

SPR and surface-enhanced Raman spectroscopy (SERS) are the two main plasmonic sensor systems. SPR is a device that is widely used in bio-detection techniques and provides direct detection. Biochemical interactions have advantages such as real-time and high-precision measurement. SPR sensor registers the bio-recognition of the target molecule by the changes in refractive index that appear upon binding of the target molecule on the metal surface of the conductor part. SERS sensor, one of these technologies, is used for sensitive detection of pathogenic micro-organisms and cells. This powerful vibrational spectroscopy system is able to detect a low concentration of target owing to the amplification of electromagnetic fields composed by the excitation of localized surface plasmons. SERS has advantages such as single molecular specificity and molecule-level sensitivity. Otherwise, electrochemical sensors have the ability to measure the change in conductance, electrical current, impedance, or potential occurring between the target sample and the interface of the electrode [9–14].

9.2 Sensors Systems in Cancer Cell Detection

Cancer is the most important of health problems. Changing dietary habits and the impact of global climate changes on geographies are recognized as some factors that trigger cancer. Although the cause of cancer is not known for certain, cancer cells are known to reproduce very rapidly. At this point, identifying these cancer cells is of great importance for early diagnosis. Many different methods for example, magnetic resonance imaging, positron emission tomography, computerized tomography have been developed for diagnostic purposes for various types of cancer. But these devices are usually expensive and have certain limitations [15]. Sometimes, these devices may induce adverse effects on patients [16, 17]. In this context, biosensors are the solution for such problems and are used in the diagnosis of many cancers such as breast, liver, lung, prostate, ovarian, and other cancers. As they are cost-effective and provide rapid processing, this makes them rather advantageous. Cytosensors have attracted great attention recently in cancer cells detection area. Cytosensors, which directly target the cancer cells, can easily and effectively determine the type and reproductive status of tumor cells. Early diagnosis is the golden standard in cancer treatment. In this context, cytosensors help detect the circulating tumor cells (CTCs) during their biological processes [18–21].

9.3 Cancer Cells

A great number of cancer cell types are detected by plasmonic sensors including human cancers, namely, breast, prostate, hepatocellular, lung, ovarian cancers, and leukemia. The target element of these systems is mostly receptors and cell membrane proteins that are aberrantly expressed in the cell membrane; some of these are conventional tumor markers and some are potential tumor markers such as folic acid and mannose receptors or cell adhesion molecules like CD166. Selectivity is determined using normal cells or different cells like fibroblasts. Actually, to discriminate a cancer cell from its normal counterpart with high selectivity, the selection of the surface marker is of prime importance. As an example, human breast cancer cells have different surface properties. MDA-MB 231 are highly metastatic triple-negative cells without estrogen, progesterone, and Her2 receptor [22]. These poorly differentiated cells express a significant number of transferrin receptors due to their high demand for iron. MCF 7 cells, in turn, have less metastatic potential with estrogen and progesterone receptors but without Her2 receptor as such more differentiated character, they also have very low or no transferrin receptor expression. For sensitive detection of cells, antibodies with high affinity or highly selective aptamers are used. These tumor-associated membrane proteins that are secreted into the blood circulation are also targeted in some sensor studies reported here, again with the help of conventional ones like carcinoembryonic antigen (CEA) [23] as well as potential tumor markers like transgellin-2 (TAGLN2). Table 9.1 summarizes the common biomarkers used for cancer cell detection [24].

9.3.1 Prostate Cancer

As the first biomarker, prostate-specific antigen (PSA) is identified and routinely used in clinical diagnosis and screening of prostate cancer [24]. The levels of PSA biomarkers directly correlate with prostate cancer. 4.0 ng ml^{-1} levels of PSA is a normal level in a healthy person. Smith reported a study that approximately 30% of men with 4.1–9.9 ng ml^{-1} levels of PSA biomarker had prostate cancer. A study conducted by Erturk reported that a real-time and sensitive PSA detection from the clinical sample by using SPR sensor system. They used micro-contact imprinted technology for preparation sensor chip. They obtained SPR chip in the presence of ethylene glycol dimethacrylate and methacrylic acid by UV polymerization. They reported 0.1–50 ng ml^{-1} of PSA concentration with 91 pg ml^{-1} the limit of detection value [25].

In another study, they developed an ultrasensitive PSA micro-contact imprinted capacitive sensor for selective and sensitive detection of PSA. The electrodes were prepared with the same monomers via UV polymerization. They compared micro-contact technology with immobilizing anti-PSA antibodies on the electrodes of the capacitive sensor and compared the detection values of antibody immobilization and micro-contact methods. The

Table 9.1 The common cancer biomarkers.

Cancer type	Cancer biomarker
Esophageal	SCC
Lung	CA 19-9, NY-ESO-1, CEA, NSE, SCC
Ovarian	HCG, CA 549, CA 125, p53, MOV-1, CEA, CA 19-9, CASA, TAG72, CA 15-3, MCA
Prostate	PSA
Breast	ER/PR, BRCA1, CA 15-3, CA 27.29, CA 125, NY-BR-1, CEA, HER2-NEU, BRCA2, ING-1
Liver	CEA, AFP
Melanoma	NY-ESO-1, Tyrosinase
Colon	CEA, p53, EGF

detection limits were 6.0×10^{-4} ng ml^{-1} for antibody immobilization method and 8.0×10^{-5} ng m^{-1} for micro-contact imprinting method [26].

It is well known that early diagnosis is vital in cancer and its derivative diseases. It greatly increases the chances of survival. Based on this principle, Moscovici et al. developed an electrochemical sensor that enables the detection of prostate cancer cells quickly and with high selectivity. They developed a micro-fabricated glass chip with light gold holes in this sensor. Cells were counted using differential pulse voltammetry using the epithelial cell adhesion molecule (EpCAM) antibody in the chip. This separate sensor features a micro-fabricated glass chip with clearly golden holes. With this sensor, they managed to count 125 prostate cancer cells in 15 minutes in a complex medium containing serum and mixed media with many different cells. They stated that the cells were alive during the counting process and that they applied this process in a sensitive and selective manner [27].

9.3.2 Liver Cancer

Sun et al. produced aptamer-based electrochemical cytosensor for detection of human liver hepatocellular carcinoma cells (HepG2). The cytosensor was based on a dual recognition and signal amplification strategy and also a repeatable assembling and dissembling. A high-affinity thiolated TLS11, an aptamer, was covalently attached to a gold electrode through Au-thiol interactions for detection of the HepG2 cells. They designed the G-quadruplex/hemin/aptamer and horseradish peroxidase (HRP) modified gold nanoparticle nanoprobes for the electrocytosensing with specific recognition and enzymatic signal amplification. The procedures for the fabrication of the nanoprobe are shown in Figure 9.1. They reported 30 cells ml^{-1} detection value in this study [28].

Chen et al. developed a biosensor for the early detection of liver cancer, one of the most common types of cancers in the world. By this biosensor, detection of HepG2 cells with an unlabeled micro-cantilever array aptasensor was investigated. Sensitive micro-cantilevers were functionalized by aptamers specific for HepG2 cells. The reference micro-cantilevers have been modified with 6-mercapto-1-hexanol monolayers to prevent interactions caused by the medium. This aptasensor was designed not only for the normal cells of the human liver, breast, bladder, and cervical tumors but also showed high selectivity over other cancer cells. It is measured that this sensor has a linear relationship and its detection limit is 300 cells ml^{-1}. It can detect cells in the range of $1x10^3$ to $1x10^5$ cells ml^{-1} [29].

9.3.3 Breast Cancer

It is accepted that breast cancer is one of the most common cancers today. It is the second most common form of cancer in women. It occurs as a result of some molecular changes, such as abnormal or irregular receptor expression in epithelial cells. In order to execute efficient treatment, it is important to reach a diagnosis with high sensitivity. Therefore, different methods are developed for diagnosis. Yang et al. obtained an easy, fast, sensitive, and targeted methods for the detection of cancer cells. They developed Au nanocages (Au NCs) and a multi-walled carbon nanotube (MWCNT-NH_2). Au NCs were used for selective and unlabeled detection with the electrode modified via 3D graphene. It was determined that this cytosensor detected MCF-7 cells with low specificity and sensitivity in the

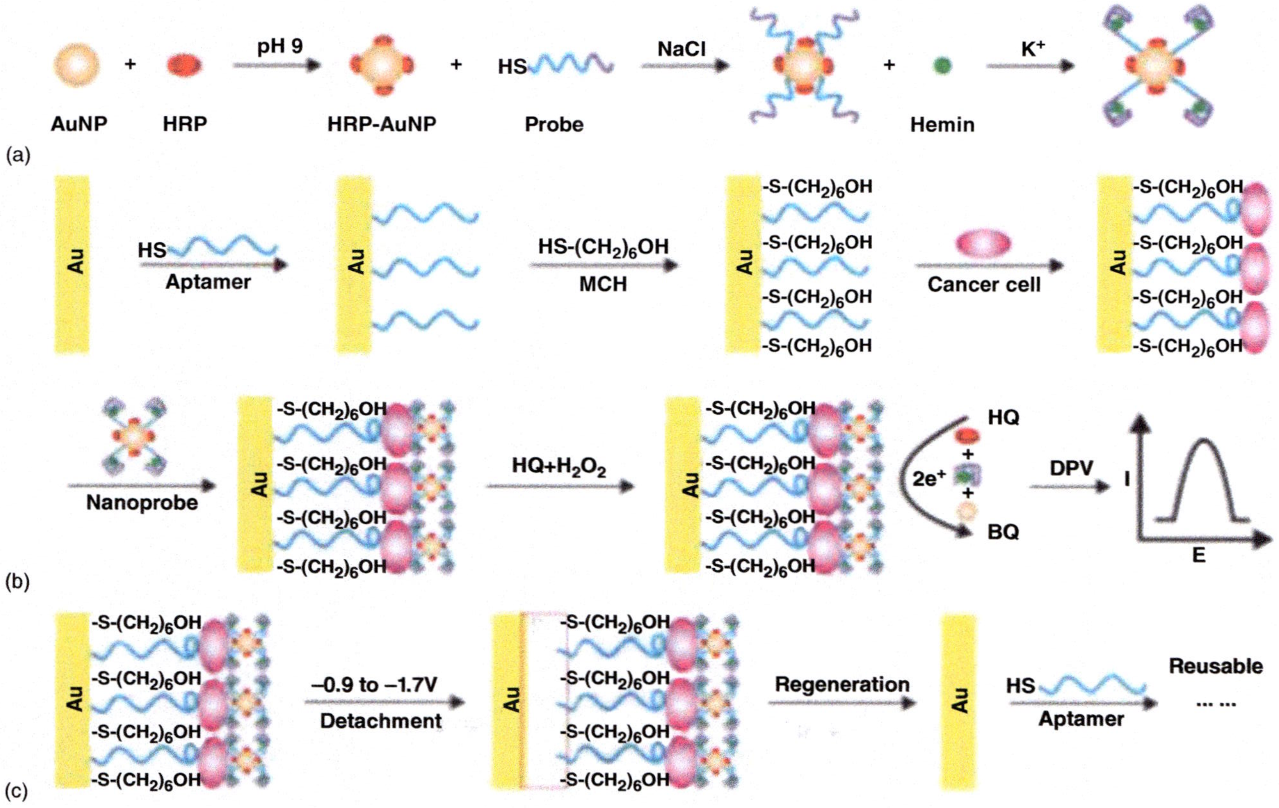

Figure 9.1 The procedures for the fabrication of the nanoprobe. Source: Reprinted with permission from Sun et al. [28]. © 2015, Elsevier.

range of 1.0 to 1.0×10^2 to 1.0×10^6 cells ml^{-1}. They reported 80 cells ml^{-1} as a limit of detection value. DNA labeled antibodies and nanomaterial-based signal amplification strategy were found to be an advantage in detecting cells. In addition, they showed high selectivity for MCF-7 cells via this sensor (Figure 9.2). Therefore, it has significant potential for practical applications [30].

Piezoelectric sensors are one of the common sensor systems that used in detection of cancer cells. Piezoelectric sensors, including quartz crystal microbalance (QCM), are often preferred especially in the detection of biochemical. The working principle of this biosensor is based on the sensitivity of crystal resonance to a disruption that occurs around it. Thanks to this sensitivity response, shifts in resonator frequency can be measured based on chemical bonds and mass losses on the resonator surface. QCM enables the interaction of biomolecules on its surface with crystal release. It is a simple, inexpensive, and high-resolution mass detection technique. In addition to these advantages, it is also preferable as a sensor, as it is a technique that can perform mass determination in nanoscale.

Breast cancer cells can be detected with the help of membrane antigens and specific antibodies. In this context, Bakhshpour et al. aimed to create a QCM biosensor to detect breast cancer cells in real-time sensitively. They used poly-hydroxyethyl methacrylate-based nanoparticles for preparation of QCM chips. They immobilized Notch-4 receptor as a ligand for the detection of cancer cells. They showed a reusable and stable QCM

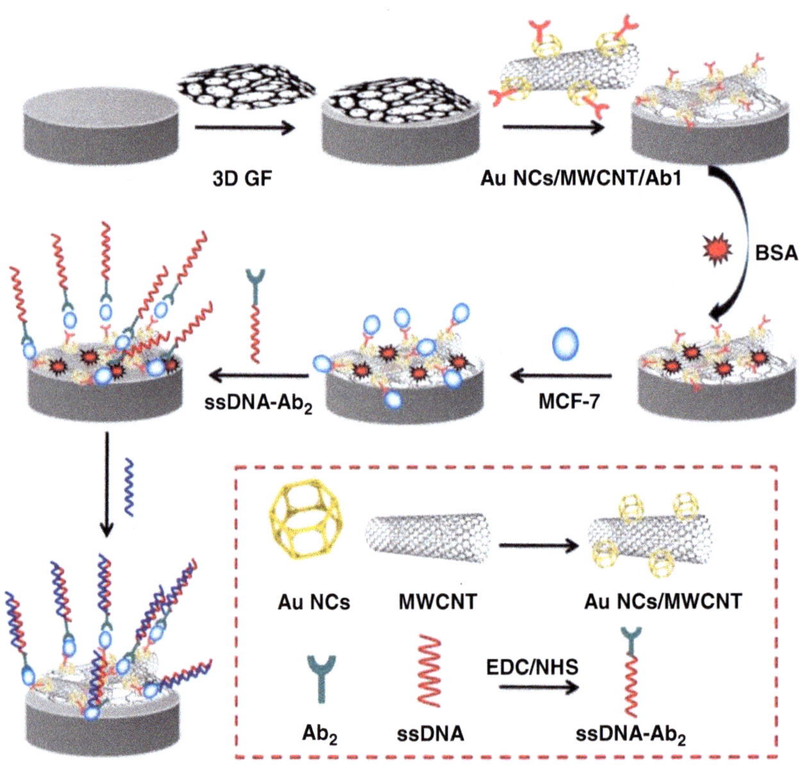

Figure 9.2 The schematic of prepared Au NCs and a multi-walled carbon nanotube. Source: Reprinted with permission from Yang et al. [30]. © 2018, Elsevier.

biosensor over three months. The schematic view of the preparation chip is shown in Figure 9.3 [12].

Zhang et al. aimed to develop a sensitive biosensor with QCM using chitosan (CS) and folic acid (FA). In this biosensor, selective recognizing conjugates were obtained by highly expressing MCF-7 cancer cells with folic acid receptors. The prepared CS-FA conjugate was characterized by UV-vis spectroscopy and Fourier transform infrared spectroscopy. CS-FA conjugation interface morphology was determined by atomic force microscopy (AFM) and scanning electron microscope. Detection of MCF-7 cancer cells caught with FA was investigated with QCM. The low detection limit was found to be 430 cells ml^{-1}. Cell detection range was obtained as 4.5×10^2 to 1.01×10^5 cells ml^{-1}. The specificity and accuracy of the biosensor were confirmed by fluorescence microscopy. In addition, regeneration of the QCM biosensor was examined using lysozyme. The receptor-bound ligand-based QCM biosensor also showed good selectivity and reproducibility in the cell mixture. With the biosensor that they formed, Zhang and his friends succeeded in detecting simple and economically unlabeled chitosan-based QCM [31].

Atay et al. developed a QCM biosensor to detect highly metastatic breast cancer cells. In this sensor, they functionalized the gold sensor surface with transferrin attachment. In this study, MDA-MB-231 breast cancer cells, MCF 7 cells, and transferrin expression were used. MDA-MB-231 cells with serum starvation were used as control cells. Affinity and binding kinetics of cells were investigated with QCM biosensor. As a result, it has been found that nanoparticles and transferrin form a single layer on the QCM surface [32].

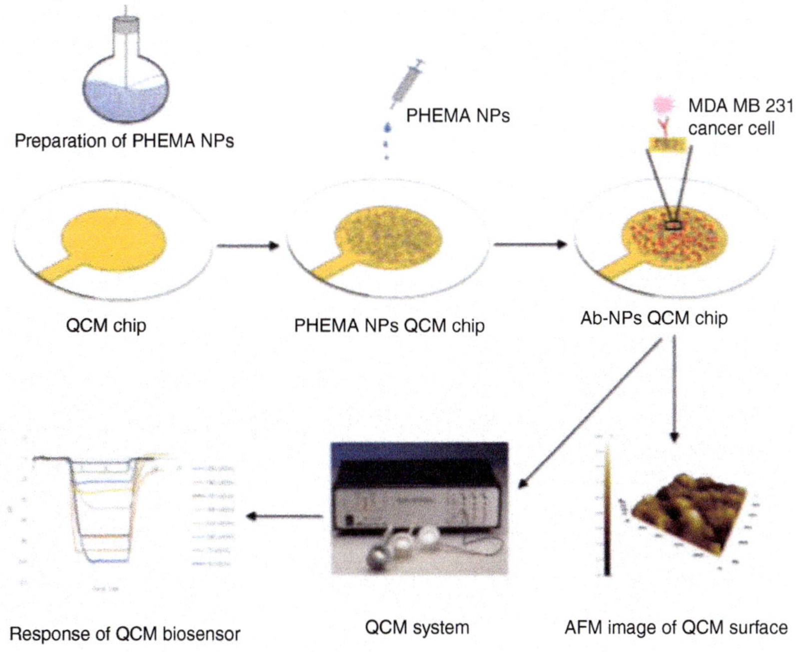

Figure 9.3 Preparation of Notch-4 receptor immobilized sensor. Source: Reprinted with permission from Bakhshpour et al. [12]. © 2019, Elsevier.

Hathaway and her group conducted a study on breast cancer, which has significant mortality, especially in women. Today, mammography makes great progress in the detection of breast cancer, and tumors of very small sizes can be detected. However, mammography could not distinguish about 25% of tumors whether benign or malignant, Hathaway designed antibody-conjugated magnetic nanoparticles with the help of magnetic relaxometry to detect breast cancer cells. For this purpose, the anti-Her2 antibody was conjugated to the developed superparamagnetic iron oxide nanoparticles by using carbodiimide. After the labeled nanoparticles were incubated with breast cancer cell lines, imaging was achieved by various methods. As a result, it was found that the number of conjugated nanoparticles with antibodies bound to the cells increases as a function of time and antigen concentration. In addition, with the use of the early prototype system by measuring the labeled cells with magnetic relaxation, the detection of about one million labeled cells at a distance of 4.5 cm was achieved [33].

9.3.4 Lung Cancer

In a study published by MA et al., human lung carcinoma cells were detected in a different way than that was for breast cancer cells. For this, a new amplification method has been reported by determining the amount of gold accumulated on it, acting as an "insemination" catalyst of an Au-antibody conjugate. QCM was used to measure the amount of catalyzed gold. The gold-coated quartz crystal was physically adsorbed to substrates modified with polystyrene films for antibody immobilization. The morphologies of these polystyrene films were evaluated by AFM. As a result of the evaluations, it was determined that polystyrene films significantly improved the surface smoothness of the electrode. Compared with the Traditional Enzyme-Based Immunosorbent Assay (ELISA), it was found that this method is a less time-consuming method. The detection sensitivity for human lung carcinoma cells was determined as 100 cells ml^{-1} [34].

Chen et al. created a biosensor with SPR-based magnetic nanoparticles. With this biosensor, they aimed to offer an improvement in breast cancer diagnosis by detecting MCF-7 cells. First of all, cancer cells (MCF-7) were caught on the surface by using MUC1 aptamer with the double marker protein recognition system method that used for cell detection. In the double marker protein recognition system, folic acid (FA) tends to be selectively recognized by cancer cells and the folic acid receptor (FR) is highly expressed on the surface of cancer cells. Cancer cells caught by interaction with FA conjugated monodisperse MNP (MNP-FA) with the help of FA and FR. They concluded that SPR signals were increased with these interactions, they performed cell detection at 500 cells ml^{-1} with a low detection limit and high selectivity [35].

9.3.5 Ovarian Cancer

Bayat et al. developed a biosensor to detect SVOK-3 ovarian cells in their study. Single helixes that recognize the CD70 molecule expressed in tumor cell lines were isolated by the systematic evolution of ligands by exponential enrichment (SELEX). Purified CD70 protein and CD70 expressing cells were used to isolate the target cell-specific aptamer. Afterward, aptamer (Apt928) with high affinity and specificity was obtained for CD70.

Specific binding of the aptamer to its target has been noted to prevent CD70 from interacting with the CD27 known as the CD70 receptor. As a result, the dissociation constant of Apt928 was calculated to be 66 $nmol\,l^{-1}$. In addition, the labeled Apt928 was used as a fluorescent aptasensor to detect SKOV-3 cells quickly and precisely. The detection limit of this aptasensor was measured as 14 cells ml^{-1} [36].

Liu et al. identified a mini cell culture with the SPR biosensor. They managed to monitor proteomic biomarker secretion with biomarker secretions from living cells. They used live SKOV-3 ovarian cancer cells as cells and measured the secretion of vascular endothelial growth factor (VEGF) for real-time [37].

Ladd et al. used the carcinoembryonic antigen (CEA) as a biomarker. They developed an SPR biosensor, since an increase in this antigen may cause an increase in autoantibody level in the colon or ovarian cancer diseases. With this biosensor, human serum samples were examined directly based on antibody levels. Using the sandwich assay method with the SPR biosensor, linearity similar to that of the ELISA method was observed. Serum samples were taken from five different healthy individuals and used to compare with a negative serum sample. A threshold value was obtained with the samples taken. When the results were evaluated by ELISA method, it was seen that the CEA antibody level of three positive serum samples was above the critical value [38].

9.3.6 Other Cells

Shan et al. developed a QCM biosensor for detection of leukemia cells. In this study, gold nanoparticles (APBA-AuNPs) modified with aminophenyl boronic acid was used for the selection and precise detection of cells that can be attached to the cell membrane. Other reasons for using these nanoparticles were to enable cell labeling and easier signal acquisition. The selectivity of the sensor was showed with obtained QCM chip. A linear relationship was found between the frequency response and the cell concentration in the 2×10^3–1×10^5 cells ml^{-1} range. The detection limit value was obtained 1160 cells ml^{-1} [21]. Table 9.2 are summarized some studies of the detection of cancer cells with different sensor systems.

Li et al. prepared a dually cross-linked supramolecular hydrogel for detection of Lysophosphatidic acid as the biomarker of early-stage ovarian cancer. They combined the surface plasmon resonance with optical waveguide spectroscopy to high sensitivity and selectivity detect of Lysophosphatidic acid. They developed a new strategy without target spot to detect biomarkers. Figure 9.4 are shown the schematic illustration of optical waveguide spectroscopy SPR. This platform provides a sensitive detection method for small molecules via SPR [43].

Bajaj et al. created gold nanoparticle-green fluorescent protein-based sequences. Thanks to these arrays, they managed to distinguish three different types of cells, such as isogenic normal and a metastatic cell, using 5000 cells. These arrays, which were created to identify mammalian cells based on cell surface properties, were obtained from different cell types. They achieved this cell separation by identifying cells based on their cell surface properties [44]. Bajaj and colleagues created a sequence-based system based on another chemical nose/tongue approach. This approach is in the physicochemical structure of different cell surfaces by special interactions between the target cell and the reporter cell. They used

Table 9.2 The comparison of various cancer cells and biomarkers detection methods.

Detection system	Detection cells	LOD	References
Electrochemical aptamer cytosensor	Hepatocellular carcinoma cells (HepG2)	30 cells ml^{-1}	[28]
QCM biosensor	Leukemia cells	1160 cells ml^{-1}	[21]
Aptamer-based leaky surface acoustic wave biosensor	Human breast cancer cells	32 cells ml^{-1}	[20]
Electrochemical impedance cytosensor	MDA-MB-231 cells	10 cells ml^{-1}	[39]
QCM (with enlarging Au nanoparticles)	Human lung carcinoma cell	100 cells ml^{-1}	[34]
Chitosan-based QCM biosensor	Breast cancer cells	430 cells ml^{-1}	[31]
QCM biosensor	Breast cancer cells	500 cells ml^{-1}	[32]
Electrochemical cytosensor	Cancer cells	79 cells ml^{-1}	[18]
CD70 binding aptamer biosensor	SKOV-3 ovarian cells	14 cells ml^{-1}	[36]
Micro-cantilever biosensor	Liver cancer cells	300 cells ml^{-1}	[29]
QCM biosensor	MDA-MB 231	12 cells ml^{-1}	[12]
DNA-labeled electrochemical biosensor	MCF-7 breast cancer cells	80 cells ml^{-1}	[30]
Sensitive cytosensor (with FA-AuNP)	MCF-7 breast cancer cells	12 cells ml^{-1}	[19]
Au nanoparticle-based SPR sensor	PSA	500 fM	[40]
Au nanoparticle-based SPR sensor	PSA	1 ng ml^{-1}	[41]
Au nanoparticle-based imaging SPR sensor	Nasopharyngeal carcinoma cells	–	[42]

nanoparticles in cell identification and created a sensing strategy by making use of the differential interaction of nanoparticles. They managed to effectively distinguish different cell types. The cell types can be listed as follows (i) normal, cancerous, and metastatic human breast cells; and (ii) isogenic normal, cancerous, and metastatic murine epithelial cell lines [45]. In another study by Bajaj et al., fluorescent conjugated polymers were constructed and these polymers were used to form sequences in cell identification. Thanks to the interactions between the surface of fluorescent conjugated polymers and the cell membranes of normal, cancerous, and metastatic isogenic cell types, a high rate of cell separation is achieved [46].

In a study published by Abdolahad et al., a biosensor was designed to detect cancer cells. This sensor is a biosensor that detects carbon nanotube-based electrical cell impedance (CNT-ECIS). This designed biosensor has been reported to be the first device to make cell detection faster, more precise, and more selective. The impedance of this biosensor, created from vertically aligned carbon nanotube arrays, changes due to mechanical and electrical

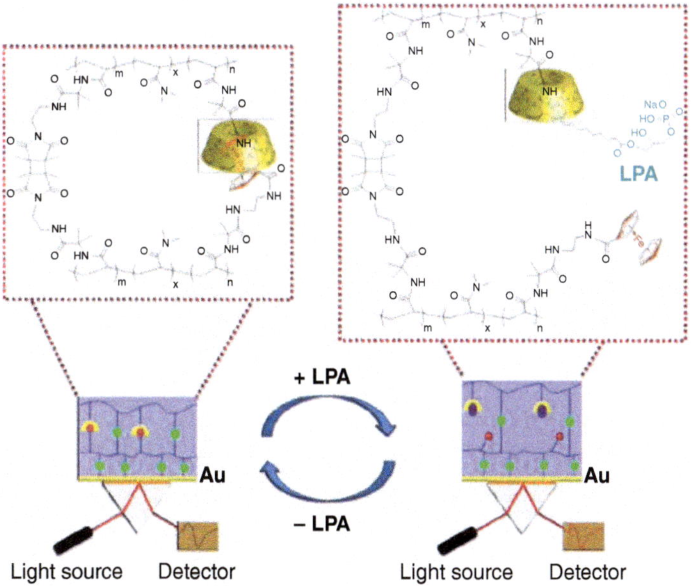

Figure 9.4 Schematic illustration of optical waveguide spectroscopy SPR sensor based on dually cross-linked supramolecular hydrogel toward specific detection of Lysophosphatidic acid in mimic plasm conditions. Source: Reprinted with permission from Li et al. [43]. © 2020, American Chemical Society.

interactions between the CNT terminals and the cell membranes attached to these terminals. CNT arrays adhere to the surface and provide conductivity. Thus, impedance changes take place in a time period of 30 seconds. Cancer cells with a density of $1.7 \times 10^{-3}\,\Omega\,\text{cm}^2$ and a sensitivity of as low as 4000 cells cm^{-2} were detected on the surface of the sensor with this sensor [47].

Ladd et al. conducted experiments on the detection of biomarkers, an important point for cancer diagnosis, with polarization-contrast SPR imaging. SPR imaging is a method that allows examination and evaluation of biomolecular interactions on the basis of SPR. In this study, a protein sequence was created by marking the antibodies by micro staining on self-assembled monolayers (SAM). Then, the detection ranges of activated leukocyte cell adhesion molecule (ALCAM) also knowns as CD166 and transgelin-2 (TAGLN2) used as two probable cancer biomarkers were determined in the experiments. As a result, it is stated that the detection limits of ALCAM and TAGLN2 in buffer solutions were obtained 6 and 3 ng ml^{-1}, respectively. It was observed that the signals received from the separate SPR sensor overlap with the SPR sensor [48].

Jabin et al. developed a bowl-shaped monocore SPR sensor for the detection of cancer cells. This SPR sensor was used for the detection of cancer cell in cervical cancer (HeLa), blood cancer (Jurkat), adrenal glands cancer (PC12), skin cancer (Basal), and breast cancer (MCF-7, MDA-MB-231) etc. They reported a wide spectrum ranging between 0.5 and 2.0 μm with the proposed SPR system. They aimed to show an SPR sensor with low loss, wavelength sensitivity, high amplitude, and high birefringence. They showed a new method for the detection of cancer cell [49].

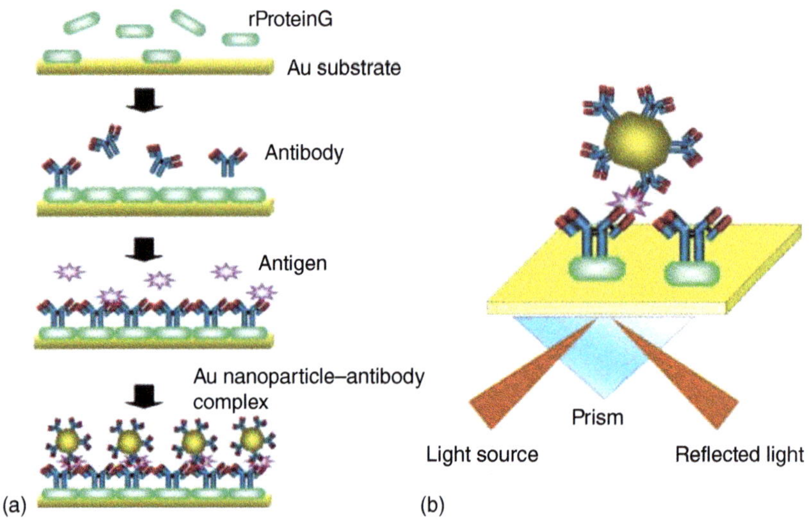

Figure 9.5 (a) Schematic description of sensor surface by self-assemble method and immunoreaction. (b) Schematic description of immunoassay configuration combined with SPR for signal enhancement based on the conjugate of Au nanoparticle and antibody. Source: Reprinted with permission from Choi et al. [40]. © 2008, Elsevier.

Besides, PSA is used as a biomarker antigen for prostate cancer in men, it can be used as a biomarker for the detection of breast cancer in women as well. Therefore, Choi et al. designed a sensitive early diagnostic SPR sensor for the detection of breast cancer in women via conjugation of Au nanoparticles and antibody. Schematic description of sensor surface is shown in Figure 9.5. They showed an ultra-sensitive SPR sensor with 300 fM the detection limit in their study [40].

9.4 Conclusion

Efficient sensor systems with high affinity and selectivity were obtained by the use of sensor techniques. These systems allow rapid detection of cancer cells by recognizing the cancer-related surface molecules, namely current or potential tumor markers. This is of great value for cancer diagnosis not only for cells in the tumor but also for circulating cells. Additional modifications, such as the attachment of nanoparticles, improve the sensing capacity of these systems by providing larger surface of recognition and cell attachment. The research in this area may contribute significantly to the diagnosis of cancer and therefore its treatment when it finds its place in clinical applications.

References

1 Rasooly, J.J. (2006). *Biosens. Bioelectron.* 21 (10): 1851–1858.
2 Visvader, J.E. (2011). *Nature* 469: 314–322.

3 Idil, N., Bakhshpour, M., Perçin, I., and Denizli, A. (2020). *Inamuddin* (ed. A. Asiri), 353–388. Cham: Springer.
4 Mayer, K.M. and Hafner, J.H. (2011). *Chem. Rev.* 111 (6): 3828–3857.
5 Saylan, Y., Erdem, Ö., Cihangir, N., and Denizli, A. (2019). *Chem* 7 (3): 33.
6 Akgönüllü, S., Yavuz, H., and Denizli, A. (2020). *Talanta* 219: 121219.
7 Saylan, Y., Akgönüllü, S., Yavuz, H. et al. (2017). *Sensors (Switzerland).* 17: 898.
8 Saylan, Y. and Denizli, A. (2020). *Nanosensors for Smart Cities*, 501–511. Elsevier.
9 Gür, S.D., Bakhshpour, M., and Denizli, A. (2019). *Mater. Sci. Eng. C* 104: 109869.
10 Bakhshpour, M., Özgür, E., Bereli, N., and Denizli, A. (2017). *Colloids Surf. B. Biointerfaces* 151: 264–270.
11 Akgonullu, S., Cimen, D., Bakhshpour, M. et al. (2020). *Commercial Biosensors and Their Applications: Clinical, Food, and Beyond*, 89. Elsevier.
12 Bakhshpour, M., Piskin, A.K., Yavuz, H., and Denizli, A. (2019). *Talanta* 204: 840–845.
13 Perçin, I., Idil, N., Bakhshpour, M. et al. (2017). *Sensors* 17: 1375.
14 Çalışır, M., Bakhshpour, M., Yavuz, H., and Denizli, A. (2020). *Sens. Actuators B* 306: 127561.
15 Ferrari, M. (2005). *Nat. Rev. Cancer* 5: 161–171.
16 Perfézou, M., Turner, A., and Merkoçi, A. (2012). *Chem. Soc. Rev.* 41: 2606–2622.
17 Schroeder, A., Heller, D.A., Winslow, M.M. et al. (2012). *Nat. Rev. Cancer* 12: 39–50.
18 Han, L., Liu, P., Petrenko, V.A., and Liu, A.H. (2016). *Sci. Rep.* 6: 22199.
19 Cai, H.H., Pi, J., Lin, X. et al. (2015). *Biosens. Bioelectron.* 74: 165–169.
20 Chang, K., Pi, Y., Lu, W. et al. (2014). *Biosens. Bioelectron.* 60: 318–324.
21 Shan, W., Pan, Y., Fang, H. et al. (2014). *Talanta* 126: 130–135.
22 Gown, A.M. (2008). *Mod. Pathol.* 21: S8–S15.
23 Tothill, I.E. (2009). *Semin. Cell Dev. Biol.* 20: 55–62.
24 Smith, D.S., Humphrey, P.A., and Catalona, W.J. (1997). *ACS J* 80: 1852–1856.
25 Ertürk, G., Özen, H., Tümer, M.A. et al. (2016). *Sens. Actuators B* 224: 823–832.
26 Ertürk, G., Hedström, M., Tümer, M.A. et al. (2015). *Anal. Chim. Acta* 891: 120–129.
27 Moscovici, M., Bhimji, A., and Kelley, S.O. (2013). *Lab Chip* 13: 940–946.
28 Sun, D., Lu, J., Chen, Z. et al. (2015). *Anal. Chim. Acta* 885: 166–173.
29 Chen, X., Pan, Y., Liu, H. et al. (2016). *Biosens. Bioelectron.* 79: 353–358.
30 Yang, Y., Fu, Y., Su, H. et al. (2018). *Biosens. Bioelectron.* 122: 175–182.
31 Zhang, S., Bai, H., Luo, J. et al. (2014). *Analyst* 139: 6259–6265.
32 Atay, S., Pişkin, K., Yilmaz, F. et al. (2016). *Anal. Methods* 8: 153–161.
33 Hathaway, H.J., Butler, K.S., Adolphi, N.L. et al. (2011). *Breast Cancer Res.* 13: R108.
34 Ma, Z., Wu, J., Zhou, T. et al. (2002). *New J. Chem.* 26: 1795–1798.
35 Chen, H., Hou, Y., Ye, Z. et al. (2014). *Sens. Actuators B* 201: 433–438.
36 Bayat, P., Taghdisi, S.M., Rafatpanah, H. et al. (2019). *Talanta* 194: 930–940.
37 Liu, C., Lei, T., Ino, K. et al. (2012). *Chem. Commun.* 48: 10389–10391.
38 Ladd, J., Lu, H., Taylor, A.D. et al. (2009). *Colloids Surf. B. Biointerfaces* 70: 1–6.
39 Tang, Y.H., Lin, H.C., Lai, C.L. et al. (2018). *Biosens. Bioelectron.* 116: 100–107.
40 Choi, J.W., Kang, D.Y., Jang, Y.H. et al. (2008). *Colloids Surf A* 655: 313–314.
41 Huang, L., Reekmans, G., Saerens, D. et al. (2005). *Biosens. Bioelectron.* 21: 483.
42 Lee, H., Kang, T., Yoon, K.A. et al. (2010). *Biosens. Bioelectron.* 25: 1669.
43 Li, J., Ji, C., Lü, B. et al. (2020). *ACS Appl. Mater. Interfaces* 12 (33): 36873–36881.

44 Bajaj, A., Rana, S., Miranda, O.R. et al. (2010). *Chem. Sci.* 1: 134–138.
45 Bajaj, A., Miranda, O.R., Kim, I.B. et al. (2009). *Proc. Natl. Acad. Sci. U. S. A.* 106 (27): 10912–10916.
46 Bajaj, A., Miranda, O.R., Phillips, R. et al. (2010). *J. Am. Chem. Soc.* 132: 1018–1102.
47 Abdolahad, M., Taghinejad, M., Taghinejad, H. et al. (2012). *Lab Chip* 12: 1183–1190.
48 Ladd, J., Taylor, A.D., Piliarik, M. et al. (2009). *Anal. Bioanal. Chem.* 393: 1157–1163.
49 Jabin, M.A., Ahmed, K., Rana, M.J. et al. (2019). *IEEE Photonics J* 11 (4): 1–10.

10

Ultrasensitive Sensors Based on Plasmonic Nanoparticles

Ilgım Göktürk[1], Fatma Denizli[2], Erdoğan Özgür[3], and Fatma Yılmaz[4]

[1] *Department of Chemistry, Hacettepe University, Ankara, Turkey*
[2] *Turkish Atomic Energy Authority, Nuclear Research and Training Center, Ankara, Turkey*
[3] *Advanced Technologies Application and Research Center, Hacettepe University, Ankara, Turkey*
[4] *Vocational School of Gerede, Department of Chemistry Technology, Bolu Abant Izzet Baysal University, Bolu, Turkey*

10.1 Introduction

Nanotechnology plays an essential role in current sensor technology. The future of sense depends on the critical factors of simplicity, cost-effectiveness, and rapid response. Nanoparticles (NPs) such as Au, Ag, and Cu are widely used in visual detection thanks to their optical properties known as surface plasmon resonance (SPR). The collective oscillation of free electrons occurs in the visible region due to resonance with the incident light [1]. Plasmon resonance scattering (PRS) feature of Au and Ag nanoparticles has been used in bio-affinity sensing systems. In general, the detection mechanism is based on molecular interaction at the surface of the modified or functionalized substrate with specific functional groups or nanoparticles [2].

Sensor technology has several challenges to detect analytes with high sensitivity and selectivity quickly. Nanoparticle-based sensors have gained wide acceptance due to their highly sensitive and selective response to various analytes despite the obstacles. Developing an effective sensor has many challenges. An ideal sensor should meet analytical properties such as selectivity, sensitivity, robustness, accuracy, precision, minimum error, repeatability, and linearity. A sensor's selectivity refers to its properties to recognize the analyte of interest in many other interfering compounds. The ability of a sensor to detect analytes even at low concentration is called sensitivity. For example, the sensitivity of an immunoassay is determined mainly by the intensity of the output signal [3–5]. The development of novel signal amplification strategies is necessary to maximize signal output. Metal NPs doped with other materials have non-toxicity, chemical stability, sufficient biocompatibility, excellent catalytic activity, and a high surface-to-volume ratio. There have been widely used as essential components of signal amplification strategies to enhance the sensitivity of the immunoassays such as magnetic bead, gold NPs (AuNPs), silver NPs (AgNPs), Fe_3O_4@SiO_2, Ag@ bovine serum albumin (Ag@BSA), zinc oxide nanoflower-bismuth sulfide composites

Plasmonic Sensors and their Applications, First Edition. Edited by Adil Denizli.

($ZNF@Bi_2S_3$), and so on [6–9]. Several chemicals, optical, electrochemical, biological, and pH sensors based on nanoparticles have attracted considerable attention in the literature recently due to their high detection thresholds, low cost, fast response, high surface area, and portability [10–16]. Nanosensors are promising for use in the field of biosensing, such as real-time monitoring of intracellular activities, monitoring of disease biomarkers, and toxic chemicals [17, 18]. Metallic NPs possessing size-, and shape-dependent optical properties are useful in colorimetric and fluorescence sensing applications [19, 20].

Many exciting studies have been reported about the plasmonic nanoparticles' physical and chemical properties, such as good conductivity, easy modification, and light sensitivity [21], which are used in various applications such as molecular recognition, live membrane transfer, and metal ion detection [22]. Notably, some noble metal nanoparticles show low toxicity and biocompatibility in biological research areas such as cell imaging, drug release, biorecognition, and photothermal therapy [23].

Advances in nanoplasmonics cover nanoplasmonic sensor substrates and all aspects of related optical technologies, including SPR, localized surface plasmon resonance (LSPR), and surface-enhanced Raman scattering (SERS). SPR-based unlabeled detection of targeted small molecules poses great challenges and requires significant signal amplification to achieve accurate and precise measurements. The incorporation of noble metal nanoparticles (NPs) like gold AuNPs for the fabrication of SPR biosensor has shown remarkable impact both for anchoring the signal amplification and generate plasmonic resonant coupling between NPs and chip surface [24]. In contrast, Raman spectroscopy can precisely identify chemical molecules with the aid of distinctive molecular vibrational modes. Moreover, surface-enhanced Raman spectroscopy (SERS) extends this capability to detecting ultra-low concentrations [25]. SERS has been performed using various instruments such as nanohole array and semi-3D devices. However, the Raman enhancements of these instruments are moderate (103–106) compared to other SERS substrates, and they struggle to meet the requirements for detecting lower analyte concentrations [26, 27].

Noble metal nanoparticles have found wide applications in bioanalytical and photonic-based studies due to their unique light scattering and absorption properties [28]. These properties are due to the localized surface plasmon resonance caused by the resonance oscillations of the conducting electrons of the nanoparticles.

Many theoretical and experimental studies [29, 30] on metal nanoparticles with different geometries have been conducted to find the best nanoparticle configuration to increase the sensitivity of the plasmon resonance response. Metal nanorod and nanoshell structures have attracted attention in terms of sensitivity. The morphology of metallic nanoparticles also affects spectral properties. For example, longer aspect ratio nanorods are much more sensitive to refractive index changes, but longer surface distortion lengths compromise their surface sensitivity [31]. The customizability features of the LSPR response [32] allow LSPR to offer a versatile, adjustable, and potentially portable platform that is simple, robust, and easily applicable for point of care (POC) diagnosis [33].

Metal nanoshells are a new type of composite spherical nanoparticle consisting of a dielectric core covered with a thin metallic shell-like gold. The optical and chemical properties of nanoshells allow them to be used in biomedical imaging and therapeutic applications. Changing the core and shell's relative dimensions allows for precise and systematic modification of the optical resonance of the nanoparticles over a wide range of regions ranging from near UV to mid-infrared. In addition to their spectrally tunable properties, nanoshells have other advantages over conventional organic dyes. Moreover, the same conjugation protocols used for binding biomolecules to gold colloids can be easily modified for nanoshells [34]. Gold nanoshells have physical properties similar to gold colloid, which has a strong optical absorption caused by the metal's collective electronic response to light. Based on the optical absorption yielding a bright red color, gold colloids can have considerable utility in consumer-related medicinal products such as home pregnancy tests.

An ideal biosensor should detect a specific biomarker in clinically relevant range with a low detection limit and minimum signal-to-noise (S/N) ratio for clinical diagnosis [35] and perform multiple analyses of various biomarkers related to the same disease. It should detect with high specificity and provide sensor-to-sensor reproducibility and have fast testing turnaround times (potentially in minutes). Minimal patient body-fluids (sample) are required with minimal operational complications. Used sensor chips or substrates should have a long shelf-life and stability in a wide working range of temperatures and humidities. Sensing instruments should be fairly compact and portable, as well as user-friendly and cost-effective.

Biomarkers are used to evaluate normal or pathological diagnostic results [36–38]. Biomarkers present in the body fluids like blood, urine, saliva, sputum, and cerebrospinal fluid (CSF) provide a diagnosis of diseases associated. Although biomarkers can be detected with existing analytical techniques, it is still a significant challenge to measure them reproducibly with the required accuracy.

Optical biosensors have attracted significant attention due to their biomarker detection capabilities in real-time with less processing time. When all aspects of nanoplasmonic sensor substrates and related optical technologies were covered, nanoplasmonics can be categorized as SPR, LSPR, and SERS.

10.2 SPR and LSPR

Photons induce a collective oscillation of the free electrons in the conduction band of the metals. When incident light interacts with a noble metal generates surface plasmon polaritons (SPPs), which is a form of electromagnetic waves, propagating at the sensor metal-dielectric interface. The most straightforward geometry that sustains SPPs is a straight interface between a metal and dielectric medium, with dielectric constants of opposite signs and only *p*-polarized electromagnetic or transverse magnetic waves can sustain SPPs [39].

Some metals that can support SPPs are gold, silver, copper, and aluminum, which have the properties of a negative real and a small positive imaginary dielectric constants. Silver (Ag), which is more sensitive to refractive index (RI) changes, has the largest negative real dielectric constant. However, silver that is easily oxidized in the air has poor chemical stability. On the other hand, gold (Au) with lower RI sensitivity than Ag has higher stability and higher chemical affinity. SPR sensor chips are usually made of a thin gold film coated on a dielectric substrate such as glass or silica and can be used in plasmonic biosensors by making them functional to capture the target biomolecule [39]. Figure 10.1a shows a schematic representation of surface plasmon resonance where the free conduction electrons in

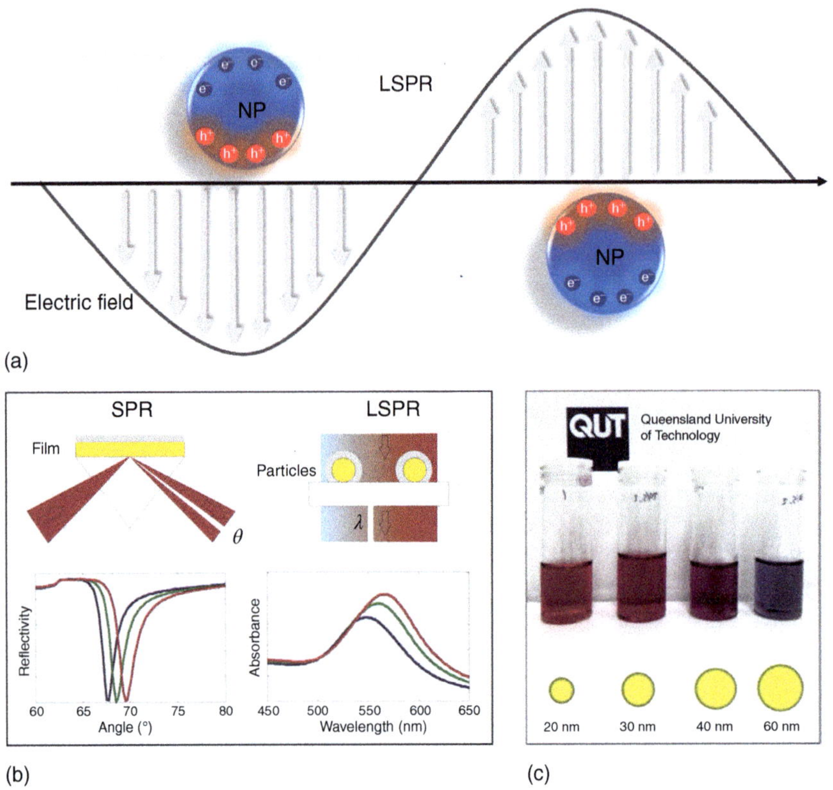

Figure 10.1 (a) Schematic representation of surface plasmon resonance where the free conduction electrons are distributed in the metal nanoparticles. Source: © John Wiley & Sons. (b) Schematic representation of the sensing principle for the studied pSPR (left) and LSPR (right) systems. The graphs below show the calculated shift of dip/peak upon adsorption of 10 nm (green line) and 20 nm (red line) thick molecular film with a refractive index of 1.48 in water ($n = 1.33$). Source: Permission from Ref. [40] © 1969, Elsevier. (c) Au NPs with various sizes (20–60 nm) causes the difference in colors. Source: Permission from Ref. [41], © 2013 Elsevier.

the metal nanoparticles are driven into oscillation due to strong coupling with incident light. Schematics of the sensing principle for the studied propagating surface plasmon resonance (pSPR) and localized surface plasmon resonance (LSPR) systems are shown in Figure 10.1b, and Figure 10.1c represent the solutions of AuNPs of various sizes [40, 41].

The coupling of incoming electromagnetic radiation to the metal surface induces resonant oscillations of conduction electrons at the metal-dielectric interface results in surface plasmons, which gives rise to an evanescent electric field that extends from the metal surface a few hundreds of nanometers into the surrounding medium [42]. Even small changes in the refractive index of the dielectric materials can create large differences in the resonance conditions of these surface plasmons. Traditional SPR sensors using flat metal films generate surface plasmon polaritons, which can propagate tens to hundreds of microns throughout the metal–dielectric interface [43, 44]. El-Sayed and Lee [45] investigated the effect of the size and shape of nanorods and the type of metal (Au vs. Ag) on the sensitivity of the surface plasmon resonance response to changes in their surrounding environment. The first controlling factor is the bulk plasma wavelength, a property dependent on the metal type. The second controlling factor is a geometrical parameter, the aspect ratio of the nanorods.

LSPR caused by the interaction of light and discrete metallic nanoparticles has attracted substantial attention as another optical phenomenon [46, 47]. The LSPR is based on the interaction between the collective oscillation of conduction band electrons and electromagnetic radiation confined to the nanoscale. If the wavelength of incident light is about ten times greater than the nanoparticle's size, the surface electrons will oscillate intensively. The maximum LSPR extinction occurs when the frequencies of the incident light match to the free electrons in the nanoparticles [48, 49]. The active LSPR band leads to unique scattering and absorption spectra as well as excellent catalytic abilities for the nanoparticles [50, 51]. Inducing collective oscillations of free conduction band electrons by incident light leads to the increased electric field intensity and leads to the characteristic extinction peak at the plasmonic resonance frequency. Unlike SPR-based biosensors, the surface plasmon field is not propagating and is localized around the subwavelength sized nanoparticles [52]. When the analyte is attached to the nanoparticles, the refractive index change causes a wavelength shift in the extinction spectrum and makes the LSPR platforms suitable for the biorecognition events [53].

The LSPR of the noble metal nanoparticles intensely is dependent on the particle size, shape, surrounding dielectric environment, and also proximity to other nanoparticles (plasmon coupling) [54]. Plasmonic metal nanoparticles exhibiting synthetically tunable plasmon wavelengths, huge absorption/scattering cross-sections, and high local electromagnetic field enhancements were used for various technological applications, such as biological sensing and imaging, photocatalysis, and optical nanoantennas during the last decade [55–58]. The extremely high sensitivity of LSPR wavelengths to small changes in the dielectric environment around noble metal nanoparticles has proven that the LSPR transduction mechanism can be attractive for chemical and biological sensing, broadly used in detecting proteins, DNAs, vapor molecules, polymers, and metal ions [59–63]. LSPR demanding nanostructured properties onto a substrate requires that thin gold film used in SPR will not be sufficient. As compared to SPR, LSPR is negligibly affected by the bulk sensitivity and bulk temperature fluctuations and also high nanosurface sensitivity features. Unlike SPR, which requires an external light coupling method, light couples directly to the sensor in the LSPR. This reduces instrumentation complexity, and LSPR is considered to have a strong potential for clinical diagnostic applications [64, 65].

Furthermore, LSPR λ_{max} shifts are entirely linear with the number of bound molecules on the nanoparticle surface. Nanoparticle sensors as diagnostic tools for various diseases can be used after functionalizing nanoparticle surfaces with the appropriate receptors. Truong and Sim introduced a simple, label-free, ultrasensitive approach to detect protein biomarkers used in clinical diagnosis. The detection principle based on the resonant Rayleigh scattering response of a single Au nanoparticle causes the LSPR λ_{max} shift [66].

Nanostars, as a new kind of metallic nanoparticle, are composed of a central core from which several protruding tips extend. They typically show LSPR of the core and multiple LSPRs corresponding to the tips and core. Multiple LSPRs corresponding to the tips and core typically enables locally enhanced fields that amplify Raman signals (SERS), allowing molecular detection at zeptomolar levels [67]. Recently, they also facilitated the demonstration of SERS at the single gold nanostar level [68].

AuNPs used to fabricate biosensors can be easily induced to a distinct color by slightly altering their compositions, shapes, sizes, and aggregation states [69, 70]. Therefore, AuNP-based plasmonic ELISA nanosensors have been fabricated by combining with a conventional ELISA platform for colorimetric sensing of DNA, small molecules, bacteria, and other disease biomarkers [71–74]. AuNP aggregation modulating the LSPR [75] changes recognized by the naked eye (color changes from red to blue) yields narrow dynamic linear range for quantitative target detection.

10.3 SERS

Surface-enhanced Raman spectroscopy (SERS) arising from the electric field caused by strong light in the localized nanostructured space (called "hot spots") offers the possibility to develop ultra-sensitive SERS-based biosensors by tremendously amplifying Raman signals by factors up to 10^{14} [76, 77]. SERS requires a minimal sample amount for testing. It is a non-destructive and non-invasive technique. Besides, features such as minimized background signal, peak overlapping, and photobleaching allow SERS to be widely used in biological systems [78]. Due to the amplification of scattering on the surface of nanoscale roughened noble metal substrates, SERS is a promising analytical tool for obtaining molecular fingerprints with good sensitivity [79]. One of the most effective tools in the sensitive and selective detection of biomolecules is SERS, which uses gold and silver-based nanoparticles. Also, photo-excitation of TiO_2 with increased sensitivity beyond the traditional SERS effect by Photo-induced enhanced Raman spectroscopy (PIERS) gives rise to strong Raman enhancement at the metallic nanoparticles sites [80, 81]. SERS-active substrates with metallic nanoscale gaps or edges used in SERS-based chemical sensors have been produced by various approaches such as self-assembly [82], e-beam lithography [83], and nanosphere lithography [84]. Superhydrophobic plasmonic nanostructures have become an attractive tool for ultrasensitive detection in SERS due to their non-wetting and strong plasmonic resonance properties. However, inducing superhydrophobic surfaces on hydrophilic metals (e.g. Au, Ag) provides high plasmonic enhancement but involves complex fabrication processes. Because the noble metals used for plasmonic gaps (e.g. gold, silver) are intrinsically hydrophilic, combining superhydrophobic surfaces with plasmonic nanostructures is not a simple task. Roughening surfaces can make the hydrophobic surface more hydrophobic (or vice versa) by increasing the intrinsic wetting properties [85].

The detection based on advanced Raman spectroscopy (SERS) is provided by enhancing the electromagnetic field near the NP surface by the LSPR of AuNPs and AgNPs [86]. Although the long-range coupling of electromagnetic fields decreases with increasing particle distance, it can extend to a distance of 2.5 times the nanoparticle diameter [87]. Pathogens such as bacteria and viruses that are too large to enter SERS hot spots cause several times lower Raman enhancement factors. The SERS tag used to overcome this problem usually contains a recognition element, a Raman reporter, and a signal transducer [88]. Figure 10.2 shows the monitoring of pathogens in water samples by functionalizing the nanoparticle surface with a pathogen recognition agent [89].

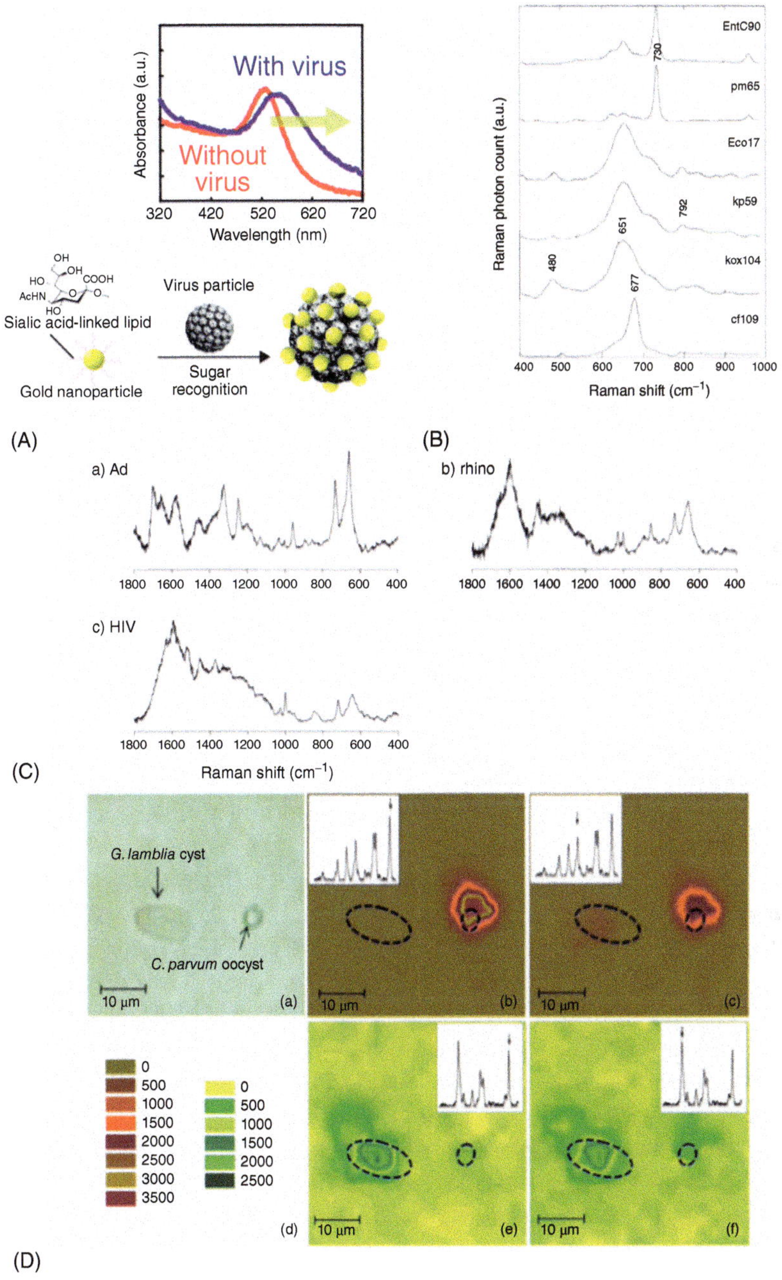

Figure 10.2 Schematic representation of surface plasmon enabling signal transduction: (a) SPR shift for virus detection, (b) intrinsic SERS signals for urinary tract bacteria (EntC90: *Enterococcus* spp.; pm65: *Proteus mirabilis*; Eco17: *Escherichia coli*; kp59: *Klebsiella pneumoniae*; kox108: *Klebsiella oxytoca*; cf109: *Citrobacter freundii*), (c) intrinsic SERS signals for adenovirus (Ad), rhinovirus (rhino), and HIV, (d) FITC (top row) and RBITC (bottom row) label for *Cryptosporidium* and *Giardia*, respectively. Source: Permission from Ref. [89], © 2010 American Chemical Society.

Ammar et al. conducted studies to prepare magneto-plasmonic granular nanostructures and their use as efficient substrates for magnetic-assisted surface-assisted Raman spectroscopy (SERS) [90]. Problems faced in pharmaceutical drug monitoring are solved thanks to innovations in Raman spectroscopic techniques. Plasmonic particles are used in surface-enhanced Raman spectroscopy (SERS) for strong local amplification of Raman signals obtained from pharmaceutical drugs. Frosch et al. summarized the studies about the recent developments in which SERS is increasingly being used for forensic trace detection and therapeutic drug monitoring [91]. Ramon et al. used small aromatic molecules to demonstrate the structural and functional plasticity of DNA. The DNA coating of silver nanoparticles directs the particle assembly into highly efficient SERS clusters by modulating the interaction of the alizarin red S (ARS) chemosensor with the nanomaterial. By this sensing approach, the quantitative determination of Al(III) and Fe(III) ions in tap water at the sub ppb level was successfully performed [92]. In another study, the performance of gold nanoislands for LSPR and Surface SERS applications was investigated by Kökenyesi et al., and the SERS enhancement test was performed by using benzophenone-isopropyl alcohol solution [93].

10.4 Colorimetric Sensing

Colorimetric techniques, due to their low cost, the use of inexpensive equipment, the need for less signal transmission equipment, and above all, easy to understand results, are useful in sensing methods. Observing the color changes with our naked eye would be the most appropriate mechanism for a rapid, field-deployed contaminant detection test. Since the LSPRs of gold and silver colloids fall in the visible spectrum, color changes caused by aggregation are used for sensor production.

Aggregation-dependent sensor designs to detect a range of biomolecules, heavy metal ions, and pathogens have often been applied [94]. When the target is directly bound to a recognition element, it causes aggregation and a color change from red to blue in AuNPs. Considering the results of the reports describing the role of nanoparticles in colorimetric detection, it became clear label-free assay is also possible with nanoparticles [95]. Although most colorimetric assays exploit the LSPRs developed on spherical AuNPs, there are other options, such as nonspherical gold nanorods [96] and titanium oxide clusters [97].

Li and coworkers reported a highly sensitive Bisphenol A (BPA) sensor using BPA-responsive gold nanoparticles (AuNPs) as optical probes. Detection of BPA with fast response (four minutes), a broad linear range (0.1–4 nM), and a low limit of detection (0.02 nM) was provided by the BPA concentration-dependent color and UV absorption spectra changes of AuNPs [98]. Hupp and coworkers described the colorimetric detection of heavy metal ions, including toxic metals such as lead, cadmium, and mercury. Functionalized gold nanoparticles chelate with divalent metal ions and aggregate in solution, causing an easily measurable change in the absorption spectrum [99].

Zhou et al. summarized methods using metal nanoparticles (MeNPs) for signal amplification in immunoassays [100]. Recently, simple and fast colorimetric detection methods that do not require an expensive instrument to achieve high sensitivity have attracted much attention.

Biswas et al. performed the colorimetric detection of Ag^+ ion with Tetrazine-Capped Gold Nanoparticles forming Au–Ag core-shell structure by using the non-aggregation induced system. Au–Ag core–shell NPs formation was fabricated without any external reducing agents to observe a color change from red to brown [101].

Jangde and coworkers developed a plasmonic-based colorimetric sensing strategy to detect lead (Pb) ions using polyvinyl alcohol (PVA) modified silver nanoparticles (AgNPs) and paper-based analytical devices (PADs). The redshift of the localized surface plasmon resonance (LSPR) absorption band of AgNP/PVA using UV–Vis spectrophotometry was measured and the color intensity was recorded with the Smartphone [102].

Demir et al. produced polyacrylamide (PAAm) and AgNPs-based free-standing flexible polymeric films showing intense optical response upon application of mechanical pressure. The pressure could possibly cause plasmonic shifting due to the disassembly of the clusters from blue to individual reddish particles depending on the pressure range [103]. The study performed by Blackburn et al. [104] developed a simple and inexpensive sensor comprising citrate-capped silver nanoparticles (cc-AgNPs) to detect creatinine directly in unprocessed urine rapidly and sensitively.

Detecting of small molecules such as oligonucleotides and metal ions by colorimetric sensing methods based on the aggregation of nanoparticles in a colloidal solution was performed by utilizing ligand-modified nanoparticles with a target specificity [105, 106]. But, loss of a fraction of the nanoparticle occurs due to the self-aggregation of nanoparticles during the ligand modification process that could be overcome by transferring nanoparticles from a colloidal solution to a solid substrate. Jongheop et al. [107] detected Cu^{2+} ions by colorimetric method via the assembly of plasmonic silver nanoparticles on density-controlled plasmonic gold nanoparticles (50 nm cores). They used carboxylate ion (COO^-) moiety surrounded with nanoparticles to induce a Cu^{2+} mediated core–satellite assembly without needing any ligand modification of nanoparticles. The system used for detecting Cu^{2+} ions is illustrated in Figure 10.3.

10.5 Luminescence Applications

Luminescence is any emission of light (electromagnetic waves) from a substance. Kandas controlled the enhancement of fluorescence intensity emissions caused by cerium oxide NPs by investigating the effect of the added different plasmonic nanostructures, including gold nanoparticles (AuNPs) and Cadmium sulfide/selenide quantum dots (CdS/CdSe QDs) [108].

Kato and coworkers used plasmonic nanostructures to control the photoluminescence properties of various emitting materials. Localized surface plasmon resonance of silver nanoprisms (AgNPs) was used to control the luminescence intensity of gold quantum dots (AuQDs) [109]. Ren et al. developed a fluorescent and surface-enhanced Raman spectroscopy (SERS) dual-mode probe to image intracellular Zn^{2+} by *N*-(2-(bis(pyridine-2-ylmethyl)amino)ethyl)-2-mercaptoacetamide (MDPA) modified gold nanoparticles (MDPA-GNPs) [110]. As an optical fluorescent material, ceria nanoparticles with visible emission under UV excitation have been widely used. Kandas and his colleagues introduced studies in which gold nanoparticles were added to enhance both fluorescence emission and lifetime [111].

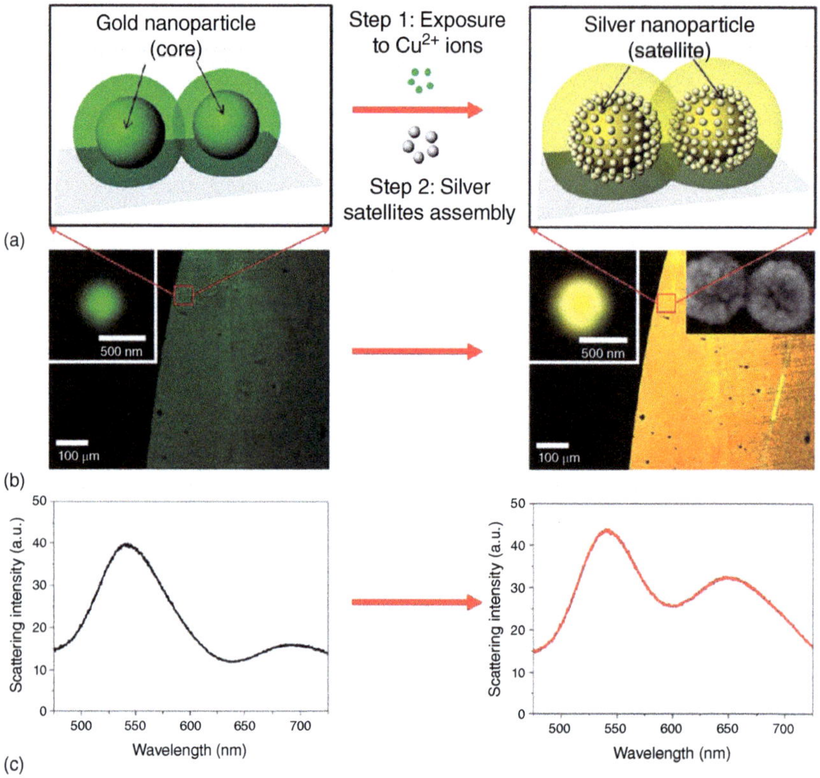

Figure 10.3 Schematic illustration of the colorimetric detection of Cu^{2+} ions via targeted core–satellites nanoassemblies. (a) Illustration of plasmonic coupling between Cu^{2+} ion mediated plasmonic core–satellites structures. (b and c) Representative color and spectral changes after the formation of Cu^{2+} ion medicated core–satellites structures. Scale bars are 500 nm and 100 µm, respectively. Source: Permission from Ref. [107], © 2013 American Chemical Society.

10.6 Conclusion

Nano-sized metallic structures are used as basic materials in the production of nano-sized devices and in the development of new technologies, as they have unique properties compared to macroscopic ones, at certain excitation wavelengths excitation of plasmon polaritons by light results in SPR. With the use of nanostructures, SPR can be induced without a prism or grating. In this case, light stimulates LSPR, representing collective electron oscillations in metallic nanoparticles [112]. A possible great advantage of LSPR as a sensor element an optical transmission setup is used instead of a reflective one. The shape and size of gold, silver, and copper-based nanostructures determine the wavelength or energy of absorption that changes depending on the change in the refractive index of the surrounding medium, so it can also be used for sensor applications [113].

Due to their high sensitivity to dielectric changes in the surrounding environment, plasmonic nanoparticles can allow direct monitoring of the molecular binding occurring on their surface and have been extensively studied for decades because of their extraordinary properties related to localized surface plasmon resonances. The development of the seed-mediated growth method has been a milestone as it provides access to an extraordinary variety of metal nanoparticles of specific size, geometry, and composition. Such a morphological control of plasmonic nanoparticles increased their prospects for implementation in various fields. Surface plasmons having the capability to respond small changes in the surrounding environment or to disturb (increase/quench) optical processes in nearby molecules have been used in a wide range of applications from biomedicine to energy harvesting. Also, localized surface plasmons were used to increase the effectivity of Raman scattering of target molecules adsorbed on the surface of plasmonic nanoparticles in SERS [114].

Most biochemical sensing methods require labels such as fluorescent, radioactive, or enzymes to detect biomolecules. In a label-free sensing system, target molecules can be directly detected without needing any label. In plasmonics, light is coupled to bound charges like electrons in metals. Nanoparticles can be used in sensing applications because the shape of the nanoparticle extinction and scattering spectra and, in particular, the peak wavelength depends on the nanoparticle composition, size, shape, orientation, and dielectric environment for nanoparticles [115].

Metal nanoparticles compatible with bio-functionalization approaches exhibit a large potential for the development of innovative and cost-effective sensing devices having miniaturization potential. LSPR can enable the adjusting of optical properties in the spectral range from UV to IR radiation. Signal amplification, such as fluorescence or Raman spectroscopy and local field enhancement, can be induced by plasmon nanoparticles. In general, plasmonic nanoparticles suit as a label or transducer for different optical detection techniques [116].

In plasmonic nanoparticles, plasmon couplings offer great opportunities to develop colorimetric sensing methods since generated surface plasmon resonance exhibits a powerful and distance-dependent spectral shift. In this regard, ligand-modified nanoparticles with a target specificity were utilized extensively in colorimetric sensing methods based on nanoparticles' aggregation in a colloidal solution.

References

1 Scholl, J.A., Koh, A.L., and Dionne, J.A. (2012). Quantum plasmon resonances of individual metallic nanoparticles. *Nature* 483 (7390): 421–427.

2 Homola, J., Yee, S.S., and Gauglitz, G. (1999). Surface plasmon resonance sensors: review. *Sensors and Actuators B: Chemical* 54 (1–2): 3–15.

3 Qin, X., Xu, A., Liu, L. et al. (2017). Selective staining of CdS on ZnO biolabel for ultrasensitive sandwich-type amperometric immunoassay of human heart-type fatty-acid-binding protein and immunoglobulin G. *Biosensors and Bioelectronics* 91: 321–327.

4 Sun, Q., Luo, J., Zhang, L. et al. (2018). Development of monoclonal antibody-based ultrasensitive enzyme-linked immunosorbent assay and fluorescence-linked immunosorbent assay for 1-aminohydantoin detection in aquatic animals. *Journal of Pharmaceutical and Biomedical Analysis* 147: 417–424.

5 Wang, X., Gao, P., Yan, T. et al. (2018). Ultrasensitive photoelectrochemical immunosensor for insulin detection based on dual inhibition effect of CuS-SiO_2 composite on CdS sensitized C-TiO 2. *Sensors and Actuators B: Chemical* 258: 1–9.

6 Zhang, Y., Sun, G., Yang, H. et al. (2016). Multifunctional reduced graphene oxide trigged chemiluminescence resonance energy transfer: novel signal amplification strategy for photoelectrochemical immunoassay of squamous cell carcinoma antigen. *Biosensors and Bioelectronics* 79: 55–62.

7 Liu, L., Chang, Y., Xia, N. et al. (2017). Simple, sensitive and label–free electrochemical detection of microRNAs based on the in situ formation of silver nanoparticles aggregates for signal amplification. *Biosensors & Bioelectronics* 94: 235–242.

8 Cui, L., Li, Y., Lu, M. et al. (2018). An ultrasensitive electrochemical biosensor for polynucleotide kinase assay based on gold nanoparticle-mediated lambda exonuclease cleavage-induced signal amplification. *Biosensors and Bioelectronics* 99: 1–7.

9 Jalal, U.M., Jin, G.J., Eom, K.S. et al. (2018). On-chip signal amplification of magnetic bead-based immunoassay by aviating magnetic bead chains. *Bioelectrochemistry* 122: 221–226.

10 Jeong, H.-H., Mark, A.G., Alarcón-Correa, M. et al. (2016). Dispersion and shape engineered plasmonic nanosensors. *Nature Communications* 7 (1): 11331.

11 Kong, D., Liu, L., Song, S. et al. (2016). A gold nanoparticle-based semi-quantitative and quantitative ultrasensitive paper sensor for the detection of twenty mycotoxins. *Nanoscale* 8 (9): 5245–5253.

12 Lei, J., Wang, L., and Zhang, J. (2010). Ratiometric pH sensor based on mesoporous silica nanoparticles and Förster resonance energy transfer. *Chemical Communications* 46 (44): 8445.

13 Perfézou, M., Turner, A., and Merkoçi, A. (2012). Cancer detection using nanoparticle-based sensors. *Chemical Society Reviews* 41 (7): 2606–2622.

14 Jiang, Z., Le, N.D.B., Gupta, A., and Rotello, V.M. (2015). Cell surface-based sensing with metallic nanoparticles. *Chemical Society Reviews* 44 (13): 4264–4274.

15 Kumar, A., Kim, S., and Nam, J.-M. (2016). Plasmonically engineered nanoprobes for biomedical applications. *Journal of the American Chemical Society* 138 (44): 14509–14525.

16 Wei, X., Chen, Z., Tan, L. et al. (2017). DNA-catalytically active gold nanoparticle conjugates-based colorimetric multidimensional sensor array for protein discrimination. *Analytical Chemistry* 89 (1): 556–559.

17 Chen, G., Roy, I., Yang, C., and Prasad, P.N. (2016). Nanochemistry and nanomedicine for nanoparticle-based diagnostics and therapy. *Chemical Reviews* 116 (5): 2826–2885.

18 Wang, M., Hou, X., Wiraja, C. et al. (2016). Smart magnetic nanosensors synthesized through layer-by-layer deposition of molecular beacons for noninvasive and longitudinal monitoring of cellular mRNA. *ACS Applied Materials & Interfaces* 8 (9): 5877–5886.

19 Bardhan, R., Grady, N.K., Cole, J.R. et al. (2009). Fluorescence enhancement by Au nanostructures: nanoshells and nanorods. *ACS Nano* 3 (3): 744–752.

20 Pollinger, K., Hennig, R., Ohlmann, A. et al. (2013). Ligand-functionalized nanoparticles target endothelial cells in retinal capillaries after systemic application. *Proceedings of the National Academy of Sciences* 110 (15): 6115–6120.

21 Jain, P.K., Huang, X., El-Sayed, I.H., and El-Sayed, M.A. (2007). Review of some interesting surface plasmon resonance-enhanced properties of noble metal nanoparticles and their applications to biosystems. *Plasmonics* 2 (3): 107–118.

22 Nusz, G.J., Marinakos, S.M., Curry, A.C. et al. (2008). Label-free plasmonic detection of biomolecular binding by a single gold nanorod. *Analytical Chemistry* 80 (4): 984–989.

23 Alkilany, A.M., Lohse, S.E., and Murphy, C.J. (2013). The gold standard: gold nanoparticle libraries to understand the nano–bio interface. *Accounts of Chemical Research* 46 (3): 650–661.

24 Bhardwaj, H., Sumana, G., and Marquette, C.A. (2020). A label-free ultrasensitive microfluidic surface plasmon resonance biosensor for Aflatoxin B1 detection using nanoparticles integrated gold chip. *Food Chemistry* 307: 125530.

25 Willets, K.A. and Van Duyne, R.P. (2007). Localized surface plasmon resonance spectroscopy and sensing. *Annual Review of Physical Chemistry* 58 (1): 267–297.

26 Yu, Q., Braswell, S., Christin, B. et al. (2010). Surface-enhanced Raman scattering on gold quasi-3D nanostructure and 2D nanohole arrays. *Nanotechnology* 21: 355301.

27 Wang, D., Yu, X., and Yu, Q. (2012). X-shaped quasi-3D plasmonic nanostructure arrays for enhancing electric field and Raman scattering. *Nanotechnology* 23: 405201.

28 Murray, W.A. and Barnes, W.L. (2007). Plasmonic materials. *Advanced Materials* 19 (22): 3771–3782.

29 Mcfarland, A.D. and Van Duyne, R.P. (2003). Single silver nanoparticles as real-time optical sensors with zeptomole sensitivity. *Nano Letters* 3 (8): 1057–1062.

30 Gole, A. and Murphy, C.J. (2004). Seed-mediated synthesis of gold nanorods: role of the size and nature of the seed. *Chemistry of Materials* 16 (19): 3633–3640.

31 Kelly, K.L., Coronado, E., Zhao, L.L., and Schatz, G.C. (2003). The optical properties of metal nanoparticles: the influence of size, shape, and dielectric environment. *The Journal of Physical Chemistry B* 107 (3): 668–677.

32 Petryayeva, E. and Krull, U.J. (2011). Localized surface plasmon resonance: nanostructures, bioassays and biosensing—a review. *Analytica Chimica Acta* 706 (1): 8–24.

33 Li, Z., Leustean, L., Inci, F. et al. (2019). Plasmonic-based platforms for diagnosis of infectious diseases at the point-of-care. *Biotechnology Advances* 37 (8): 107440.

34 Loo, C., Lin, A., Hirsch, L. et al. (2004). Nanoshell-enabled photonics-based imaging and therapy of cancer. *Technology in Cancer Research & Treatment* 3 (1): 33–40.

35 Justino, C.I.L., Duarte, A.C., and Rocha-Santos, T.A.P. (2016). Critical overview on the application of sensors and biosensors for clinical analysis. *TrAC Trends in Analytical Chemistry* 85: 36–60.

36 Wu, L. and Qu, X. (2015). Cancer biomarker detection: recent achievements and challenges. *Chemical Society Reviews* 44 (10): 2963–2997.

37 Çalışır, M., Bakhshpour, M., Yavuz, H., and Denizli, A. (2020). HbA1c detection via high-sensitive boronate based surface plasmon resonance sensor. *Sensors and Actuators B: Chemical* 306: 127561.

38 Erdem, Ö., Saylan, Y., Cihangir, N., and Denizli, A. (2019). Molecularly imprinted nanoparticles based plasmonic sensors for real-time Enterococcus faecalis detection. *Biosensors and Bioelectronics* 126: 608–614.

39 Kravets, V.G., Kabashin, A.V., Barnes, W.L., and Grigorenko, A.N. (2018). Plasmonic surface lattice resonances: a review of properties and applications. *Chemical Reviews* 118 (12): 5912–5951.

40 Jatschka, J., Dathe, A., Csáki, A. et al. (2016). Propagating and localized surface plasmon resonance sensing — a critical comparison based on measurements and theory. *Sensing and Bio-Sensing Research* 7: 62–70.

41 Notarianni, M., Vernon, K., Chou, A. et al. (2014). Plasmonic effect of gold nanoparticles in organic solar cells. *Solar Energy* 106: 23–37.

42 Ozbay, E. (2006). Plasmonics: merging photonics and electronics at nanoscale dimensions. *Science* 311 (5758): 189–193.

43 Homola, J. (2008). Surface plasmon resonance sensors for detection of chemical and biological species. *Chemical Reviews* 108 (2): 462–493.

44 Homola, J. (2003). Present and future of surface plasmon resonance biosensors. *Analytical and Bioanalytical Chemistry* 377 (3): 528–539.

45 Jain, P.K., Lee, K.S., El-Sayed, I.H., and El-Sayed, M.A. (2006). Calculated absorption and scattering properties of gold nanoparticles of different size, shape, and composition: applications in biological imaging and biomedicine. *The Journal of Physical Chemistry B* 110 (14): 7238–7248.

46 Xie, T., Jing, C., and Long, Y.-T. (2017). Single plasmonic nanoparticles as ultrasensitive sensors. *The Analyst* 142 (3): 409–420.

47 Jackman, J.A., Rahim Ferhan, A., and Cho, N.-J. (2017). Nanoplasmonic sensors for biointerfacial science. *Chemical Society Reviews* 46 (12): 3615–3660.

48 Nehl, C.L. and Hafner, J.H. (2008). Shape-dependent plasmon resonances of gold nanoparticles. *Journal of Materials Chemistry* 18 (21): 2415.

49 Stewart, M.E., Anderton, C.R., Thompson, L.B. et al. (2008). Nanostructured Plasmonic Sensors. *Chemical Reviews* 108 (2): 494–521.

50 Daniel, M.-C. and Astruc, D. (2004). Gold nanoparticles: assembly, supramolecular chemistry, quantum-size-related properties, and applications toward biology, catalysis, and nanotechnology. *Chemical Reviews* 104 (1): 293–346.

51 Eustis, S. and El-Sayed, M.A. (2006). Why gold nanoparticles are more precious than pretty gold: noble metal surface plasmon resonance and its enhancement of the radiative and nonradiative properties of nanocrystals of different shapes. *Chemical Society Reviews* 35 (3): 209–217.

52 Brolo, A.G. (2012). Plasmonics for future biosensors. *Nature Photonics* 6 (11): 709–713.

53 Chen, P., Chung, M.T., Mchugh, W. et al. (2015). Multiplex serum cytokine immunoassay using nanoplasmonic biosensor microarrays. *ACS Nano* 9 (4): 4173–4181.

54 Rycenga, M., Cobley, C.M., Zeng, J. et al. (2011). Controlling the synthesis and assembly of silver nanostructures for plasmonic applications. *Chemical Reviews* 111 (6): 3669–3712.

55 Giannini, V., FernáNdez-DomíNguez, A.I., Heck, S.C., and Maier, S.A. (2011). Plasmonic nanoantennas: fundamentals and their use in controlling the radiative properties of nanoemitters. *Chemical Reviews* 111 (6): 3888–3912.

56 Saha, K., Agasti, S.S., Kim, C. et al. (2012). Gold nanoparticles in chemical and biological sensing. *Chemical Reviews* 112 (5): 2739–2779.

57 Neumann, O., Urban, A.S., Day, J. et al. (2013). Solar vapor generation enabled by nanoparticles. *ACS Nano* 7 (1): 42–49.

58 Linic, S., Aslam, U., Boerigter, C., and Morabito, M. (2015). Photochemical transformations on plasmonic metal nanoparticles. *Nature Materials* 14 (6): 567–576.

59 Karakouz, T., Vaskevich, A., and Rubinstein, I. (2008). Polymer-coated gold island films as localized plasmon transducers for gas sensing. *The Journal of Physical Chemistry B* 112 (46): 14530–14538.

60 Kim, S., Cheng, N., Jeong, J.-R. et al. (2008). Localized surface plasmon resonance (LSPR) sensitivity of Au nanodot patterns to probe solvation effects in polyelectrolyte brushes. *Chemical Communications* 31: 3666–3668.

61 Sannomiya, T., Hafner, C., and Voros, J. (2008). In situ sensing of single binding events by localized surface plasmon resonance. *Nano Letters* 8 (10): 3450–3455.

62 Dondapati, S.K., Sau, T.K., Hrelescu, C. et al. (2010). Label-free biosensing based on single gold nanostars as plasmonic transducers. *ACS Nano* 4 (11): 6318–6322.

63 Wu, Y., Zhan, S., Wang, F. et al. (2012). Cationic polymers and aptamers mediated aggregation of gold nanoparticles for the colorimetric detection of arsenic(iii) in aqueous solution. *Chemical Communications* 48 (37): 4459.

64 Svedendahl, M., Chen, S., Dmitriev, A., and KäLl, M. (2009). Refractometric sensing using propagating versus localized surface plasmons: a direct comparison. *Nano Letters* 9 (12): 4428–4433.

65 Estevez, M.-C., Otte, M.A., Sepulveda, B., and Lechuga, L.M. (2014). Trends and challenges of refractometric nanoplasmonic biosensors: a review. *Analytica Chimica Acta* 806: 55–73.

66 Truong, P.L. and Sim, S.J. (2015). Development of individual plasmonic nanosensors for clinical diagnosis. In: *5th International Conference on Biomedical Engineering in Vietnam. IFMBE Proceedings*, vol. 46 (eds. V. Toi and T. Lien Phuong). Cham: Springer https://doi.org/10.1007/978-3-319-11776-8_1.

67 RodríGuez-Lorenzo, L., ÁLvarez-Puebla, R.A., Pastoriza-Santos, I. et al. (2009). Zeptomol detection through controlled ultrasensitive surface-enhanced raman scattering. *Journal of the American Chemical Society* 131 (13): 4616–4618.

68 Hrelescu, C., Sau, T.K., Rogach, A.L. et al. (2009). Single gold nanostars enhance Raman scattering. *Applied Physics Letters* 94 (15): 153113.

69 Du, J., Shao, Q., Yin, S. et al. (2012). Colorimetric chemodosimeter based on diazonium-gold-nanoparticle complexes for sulfite ion detection in solution. *Small* 8 (22): 3412–3416.

70 Klajn, R., Stoddart, J.F., and Grzybowski, B.A. (2010). Nanoparticles functionalised with reversible molecular and supramolecular switches. *Chemical Society Reviews* 39 (6): 2203.

71 Yuan, J., Wu, S., Duan, N. et al. (2014). A sensitive gold nanoparticle-based colorimetric aptasensor for *Staphylococcus aureus*. *Talanta* 127: 163–168.

72 Guo, Y., Wu, J., Li, J., and Ju, H. (2016). A plasmonic colorimetric strategy for biosensing through enzyme guided growth of silver nanoparticles on gold nanostars. *Biosensors and Bioelectronics* 78: 267–273.

73 Huang, X., Chen, R., Xu, H. et al. (2016). Nanospherical brush as catalase container for enhancing the detection sensitivity of competitive plasmonic ELISA. *Analytical Chemistry* 88 (3): 1951–1958.

74 Zhang, Z., Chen, Z., Cheng, F. et al. (2016). Iodine-mediated etching of gold nanorods for plasmonic sensing of dissolved oxygen and salt iodine. *The Analyst* 141 (10): 2955–2961.

75 Satija, J., Punjabi, N., Mishra, D., and Mukherji, S. (2016). Plasmonic-ELISA: expanding horizons. *RSC Advances* 6 (88): 85440–85456.

76 He, Y., Su, S., Xu, T. et al. (2011). Silicon nanowires-based highly-efficient SERS-active platform for ultrasensitive DNA detection. *Nano Today* 6 (2): 122–130.

77 Li, P., Long, F., Chen, W. et al. (2020). Fundamentals and applications of surface-enhanced Raman spectroscopy–based biosensors. *Current Opinion in Biomedical Engineering* 13: 51–59.

78 Jimenez De Aberasturi, D., Henriksen-Lacey, M., Litti, L. et al. (2020). Using SERS tags to image the three-dimensional structure of complex cell models. *Advanced Functional Materials* 30 (14): 1909655.

79 Zhao, B., Shen, J., Chen, S. et al. (2014). Gold nanostructures encoded by non-fluorescent small molecules in polyA-mediated nanogaps as universal SERS nanotags for recognizing various bioactive molecules. *Chemical Science* 5 (11): 4460–4466.

80 Man, T., Lai, W., Xiao, M. et al. (2020). A versatile biomolecular detection platform based on photo-induced enhanced Raman spectroscopy. *Biosensors and Bioelectronics* 147: 111742.

81 Wang, X., Liu, B., Xiao, M. et al. (2020). Fractal SERS nanoprobes for multiplexed quantitative gene profiling. *Biosensors and Bioelectronics* 156: 112130.

82 Ko, H., Singamaneni, S., and Tsukruk, V.V. (2008). Nanostructured surfaces and assemblies as SERS media. *Small* 4 (10): 1576–1599.

83 Carlson, A., Bowen, A.M., Huang, Y. et al. (2012). Transfer printing techniques for materials assembly and micro/nanodevice fabrication. *Advanced Materials* 24 (39): 5284–5318.

84 Fang, X., Zheng, C., Yin, Z. et al. (2020). Hierarchically ordered silicon metastructures from improved self-assembly-based nanosphere lithography. *ACS Applied Materials & Interfaces* 12 (10): 12345–12352.

85 Yoon, D., Chae, S., Kim, W. et al. (2018). Superhydrophobic plasmonic nanoarchitectures based on aluminum hydroxide nanotemplates. *Nanoscale* 10 (36): 17125–17130.

86 Wei, H., Hossein Abtahi, S.M., and Vikesland, P.J. (2015). Plasmonic colorimetric and SERS sensors for environmental analysis. *Environmental Science: Nano* 2 (2): 120–135.

87 Schatz, G.C., Young, M.A., and Duyne, R.P. (2006). *Electromagnetic Mechanism of SERS* 103: 19–46.

88 Lin, H.-Y., Huang, C.-H., Hsieh, W.-H. et al. (2014). On-line SERS detection of single bacterium using novel SERS nanoprobes and a microfluidic dielectrophoresis device. *Small* 10 (22): 4700–4710.

89 Vikesland, P.J. and Wigginton, K.R. (2010). Nanomaterial enabled biosensors for pathogen monitoring – a review. *Environmental Science & Technology* 44 (10): 3656–3669.

90 Nguyen, T.T., Lau-Truong, S., Mammeri, F., and Ammar, S. (2020). Star-shaped $Fe_{3\text{-}x}O_4$-Au core-shell nanoparticles: from synthesis to SERS application. *Nanomaterials* 10 (2): 294.

91 Frosch, T., Knebl, A., and Frosch, T. (2019). Recent advances in nano-photonic techniques for pharmaceutical drug monitoring with emphasis on Raman spectroscopy. *Nanophotonics* 9 (1): 19–37.

92 Guerrini, L. and Alvarez-Puebla, R.A. (2019). Multiplex SERS chemosensing of metal ions via DNA-mediated recognition. *Analytical Chemistry* 91 (18): 11778–11784.

93 Bonyár, A., Csarnovics, I., Veres, M. et al. (2018). Investigation of the performance of thermally generated gold nanoislands for LSPR and SERS applications. *Sensors and Actuators B: Chemical* 255: 433–439.

94 Liu, L., Li, S., Liu, L. et al. (2012). Simple, sensitive and selective detection of dopamine using dithiobis(succinimidylpropionate)-modified gold nanoparticles as colorimetric probes. *The Analyst* 137 (16): 3794.

95 Nath, N. and Chilkoti, A. (2004). Label free colorimetric biosensing using nanoparticles. *Journal of Fluorescence* 14 (4): 377–389.

96 Zhai, M., Chen, Q., Yuan, W. et al. (2019). Directional self-assembly of gold nanorods into 1D and 2D arrays by quadruple hydrogen bonding. *Materials and Materials Chemistry* 3 (9): 1888–1891.

97 Liu, C., Hu, J., Biswas, S. et al. (2020). Surface-enhanced Raman scattering of phenols and catechols by a molecular analogue of titanium dioxide. *Analytical Chemistry* 92 (8): 5929–5936.

98 Ma, Y., You, F., Promthaveepong, K., and Li, N. (2017). Ultrasensitive Bisphenol A sensing based on responsive plasmonic nanoparticles. *Sensors and Actuators B: Chemical* 245: 369–374.

99 Kim, Y., Johnson, R.C., and Hupp, J.T. (2001). Gold nanoparticle-based sensing of "spectroscopically silent" heavy metal ions. *Nano Letters* 1 (4): 165–167.

100 Yang, H., Xu, W., and Zhou, Y. (2019). Signal amplification in immunoassays by using noble metal nanoparticles: a review. *Microchimica Acta* 186: 859.

101 Ray, S., Biswas, R., Banerjee, R. et al. (2019). Non-aggregation-induced colorimetric detection of Ag + by Tetrazine-capped gold nanoparticles based on the formation of Au-Ag core-shell nanoparticles. *ChemistrySelect* 4 (42): 12409–12417.

102 Shrivas, K., Sahu, B., Deb, M.K. et al. (2019). Colorimetric and paper-based detection of lead using PVA capped silver nanoparticles: experimental and theoretical approach. *Microchemical Journal* 150: 104156.

103 Topcu, G., Guner, T., Inci, E., and Demir, M.M. (2019). Colorimetric and plasmonic pressure sensors based on polyacrylamide/Au nanoparticles. *Sensors and Actuators, A: Physical* 295: 503–511.

104 Alula, M.T., Karamchand, L., Hendricks, N.R., and Blackburn, J.M. (2018). Citrate-capped silver nanoparticles as a probe for sensitive and selective colorimetric and spectrophotometric sensing of creatinine in human urine. *Analytica Chimica Acta* 1007: 40–49.

105 Marinakos, S.M., Chen, S., and Chilkoti, A. (2007). Plasmonic detection of a model analyte in serum by a gold nanorod sensor. *Analytical Chemistry* 79 (14): 5278–5283.

106 Sener, G., Uzun, L., and Denizli, A. (2014). Colorimetric sensor array based on gold nanoparticles and amino acids for identification of toxic metal ions in water. *ACS Applied Materials & Interfaces* 6: 18395–18400.

107 Song, H.D., Choi, I., Lee, S. et al. (2013). On-chip colorimetric detection of Cu^{2+} ions via density-controlled plasmonic core–satellites nanoassembly. *Analytical Chemistry* 85 (16): 7980–7986.

108 Shehata, N., Samir, E., and Kandas, I. (2020). Gold/QDs-embedded-ceria nanoparticles: optical fluorescence enhancement as a quenching sensor. *Applied Sciences* 10 (4): 1236.

109 Lertvachirapaiboon, C., Kiyokawa, I., Baba, A. et al. (2020). Controlling the luminescence of gold quantum dots by the plasmonic effect of silver nanoprisms. *Plasmonics* 15: 3–9.

110 Li, D., Ma, Y., Duan, H. et al. (2018). Fluorescent/SERS dual-sensing and imaging of intracellular Zn^{2+}. *Analytica Chimica Acta* 1038: 148–156.

111 Shehata, N., Samir, E., and Kandas, I. (2018). Plasmonic-ceria nanoparticles as fluorescence intensity and lifetime quenching optical sensor. *Sensors* 18 (9): 2818.

112 Du, G.X., Mori, T., Suzuki, M. et al. (2010). Evidence of localized surface plasmon enhanced magneto-optical effect in nanodisk array. *Applied Physics Letters* 96 (8): 081915.

113 de Mello Donegá, C. (2014). The nanoscience paradigm: "size matters!". In: *Nanoparticles* (ed. C. de Mello Donegá). Berlin, Heidelberg: Springer https://doi.org/10.1007/978-3-662-44823-6_1.

114 Taylor, J., Huefner, A., Li, L. et al. (2016). Nanoparticles and intracellular applications of surface-enhanced Raman spectroscopy. *The Analyst* 141 (17): 5037–5055.

115 Ameen, A., Gartia, M.R., Hsiao, A. et al. (2015). Ultra-sensitive colorimetric plasmonic sensing and microfluidics for biofluid diagnostics using nanohole array. *Journal of Nanomaterials* 2015: 1–21.

116 Csáki, A., Thiele, M., Jatschka, J. et al. (2015). Plasmonic nanoparticle synthesis and bioconjugation for bioanalytical sensing. *Engineering in Life Sciences* 15 (3): 266–275.

11

Surface-Enhanced Raman Scattering Sensors for Chemical/Biological Sensing

Huma Shaikh, Zaib un Nisa Mughal, Saeed Memon, and Shahabuddin Memon

National Center of Excellence in Analytical Chemistry, University of Sindh, Jamshoro, Pakistan

11.1 Introduction

In this chapter, we will discuss in detail sensors based on Surface-Enhanced Raman Scattering (SERS). We will also highlight methods for preparation of SERS sensors and their ability to become efficient chemical or biological sensor. We will start chapter with basic principle and brief history of SERS so that new reader can easily understand the working mechanism of SERS sensors.

The molecules can scatter light by *Rayleigh scattering* or *Raman scattering*. The Rayleigh scattering is elastic therefore photon's energy and state of the molecule remain unchanged. Consequently, Rayleigh scattering does not provide any information about structure of molecular states [1]. On the other hand, the Raman scattering is inelastic in which photons of monochromatic light change their frequency when interact with vibrational modes or states of molecules. Smekal et al. hypothesized this effect theoretically in 1923 and C. V. Raman discovered this effect first time experimentally during experimentation based on sunlight [2].

There can be two inelastic procedures of Raman scattering. They are (i) Strokes process and (ii) Anti-stokes process. In strokes process, red shift occurs as an incident photon ($h\nu_0$) stimulates molecular vibration and correspondingly scatters with respect to change in energy [h ($\nu_0 - \nu_{vib}$)]. However, anti-strokes process follows blue shift as the incident photon gains vibrational energy and scatters with higher energy [h ($\nu_0 + \nu_{vib}$)]. Thus, information about low frequency, rotational, and vibrational transitions is acquired with these shifts. Not only solids and liquids but also the gaseous samples can be studied using Raman spectroscopy (Figure 11.1).

However, the Raman Effect has limitations of producing very weak signals due to the fact that one in 107 photons undergoes Raman Scattering. Therefore, detection of analytes at trace level concentration and analytes with poor Raman scatterers is not reliable with Raman Effect. Occasionally, Raman signals are also incomprehensible due to high fluorescence produced from molecules. Ultimately, enhancement of signals produced by Raman Effect was

Plasmonic Sensors and their Applications, First Edition. Edited by Adil Denizli.

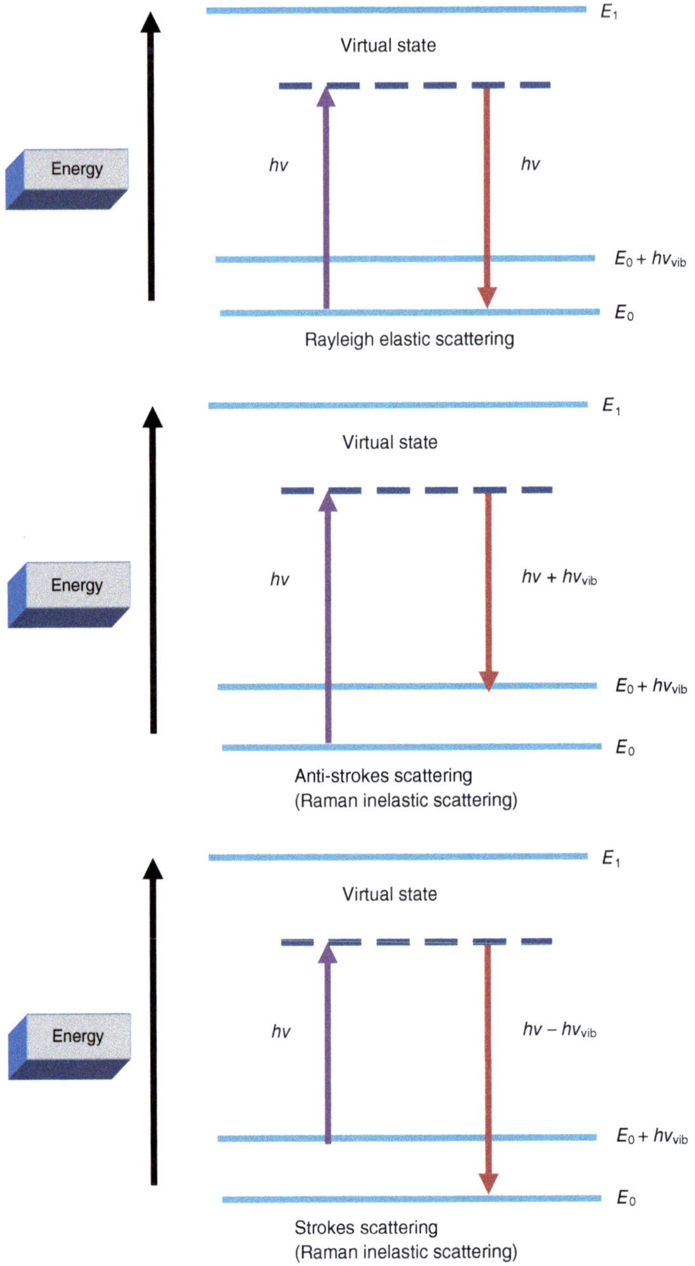

Figure 11.1 Energy level diagram for the representation of energy changes during Rayleigh and Raman Scattering processes. Source: © John Wiley & Sons.

required. This need brought Surface-Enhanced Raman Spectroscopy (SERS) into being. In due course, the utmost target of SERS was to amplify Raman signals produced by molecules, by several orders of magnitude [3]. Thus, in SERS metal, colloidal nanoparticles or rough metal surfaces are used to enhance the Raman Effect; when molecules are adsorbed on their surfaces, they produce much higher Raman scattering. Fleischmann et al. discovered SERS effect for the first time during an experiment in which pyridine produced many folds enhanced Raman signal in the presence of roughened silver electrode [4]. Initially, it was thought that the enhancement of signal is due to increased surface area but later it was shown in different reports that the unknown intensity could not be accounted for increased surface area and was due to the novel phenomenon known as SERS [5, 6]. From the time of its discovery, numerous enhancement mechanisms are proposed; however, electromagnetic (EM) theory and chemical enhancement (CE) theory are accepted broadly [3, 7]. In electromagnetic models, the molecules are treated as a point dipole that responds to the enhanced local field at or near the metal surface. These enhanced fields are due to the roughness of metal surfaces that actually couple the incident fields to surface plasmons [8]. Conversely, in chemical models, the characteristic SERS intensity is due to the modified molecular polarizability that occurs due to the interaction with metal and results in molecular resonance. This molecular resonance leads to the enhanced resonance Raman scattering [9]. The CE theory is based on chemical interaction between metal and probe molecules and enhances the magnitude of Raman signal by two to three orders [10]. In SERS, both enhancements work simultaneously and it is difficult to investigate them separately (Figure 11.2).

Plasmonics is precisely defined as the study of the optical properties of noble metals, in particular gold and silver. However, recent research and advancement in nanotechnology have broadened its scope as it defines the properties and characteristics of newly designed nanomaterials and occasionally referred to as nano-plasmonics. SERS and plasmonics are mutually benefiting each other as SERS phenomenon is originated from plasmonics. So, SERS can help to understand the optical properties of newly prepared plasmonic substrates and, at the same time some plasmonic substrates can be used as good SERS substrates. In fact, the electromagnetic enhancements of SERS are totally dependent on plasmons.

The sensitivity and specificity of SERS due to higher signal intensity and molecule specific information provided by Raman spectrum, respectively, make it an eligible candidate in the field of sensing. Depending on the plasmonic properties of SERS substrates, SERS-based sensors are capable of sensing very low concentration of analyte molecules. These properties of

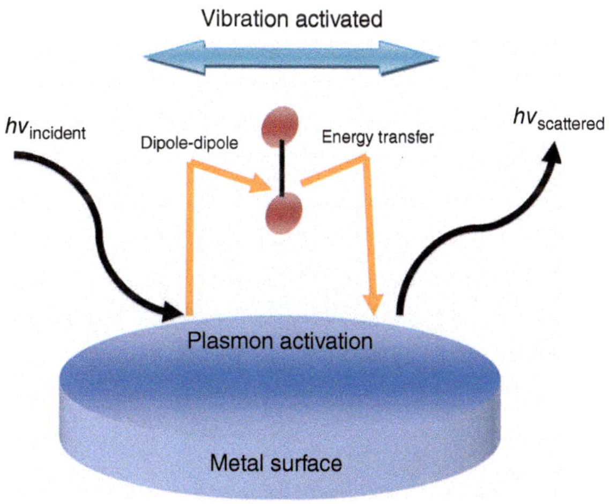

Figure 11.2 Schematic diagram to show phenomenon of SERS. Source: © John Wiley & Sons.

SERS-based sensors highlight their ability for practical applications [11]. Consequently, SERS sensors have already been developed for the detection of trace amount of chemicals [11, 12], such as food additives [13, 14], bioanalysis [15–17], explosives detection [18], etc.

For the sensing applications of SERS, the substrates are required to be extraordinary reproducible, specific, and reliable. Therefore, the choice of substrate in terms of material morphology, its manufacture, and modification is purely based on sensor's applications. Consequently, the conception of SERS substrate is not limited to solid substrates only and colloidal nanoparticles (NPs) dispersions are also applicable. Furthermore, the colloidal NPs dispersions are more reliable, reproducible, and cost effective with adapted optical properties because their shape, size, and surface functionalization can always be programmed according to the requirements [19, 20].

As we discussed earlier that the Raman signals amplification is dependent on interactions between plasmonic waves and electromagnetic waves on the surface of metal surface, the nanoparticles with different shape, size, and morphology produce different Raman enhancement characteristics [21]. Thus, nano-structures with diverse morphology and excellent ratio of surface area to volume cross-section such as nanoparticles [22], nanocavities [23], nanorods/nanowires, [21, 24], and nanoshells [25, 26] have been recommended for SERS sensors.

Analytical approaches dealing with SERS are either responsible for the detection or quantification of substances, or both. In general, it is possible to classify a molecule/component by means of direct or indirect techniques.

11.2 Direct Method

Direct measurements are distinguished by the recording and examination of the analyte own fingerprint (Figure 11.3a), whereas indirect techniques use a secondary molecule's SERS reaction, triggered by the analyte presence (Figure 11.3b). Usually, direct techniques

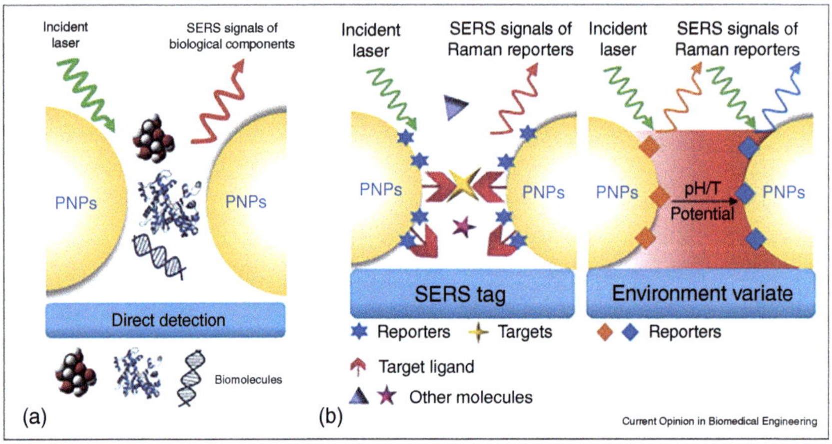

Figure 11.3 (a) Pictorial of direct detection through SERS. (b) Indirect detection through SERS. Source: Reprinted with permission from Li et al. [27] © 2020, Elsevier.

give the benefit of expressing the molecule's instantaneous situation (binding position, direction, and molecular conformation, surface and surrounding molecules interaction) but suggest label-free SERS measurements, which are much more difficult for most molecules because of very low Raman cross sections. Using a bright molecule as a messenger label, featuring an extremely high SERS cross section, will solve this problem. Indirect strategies belong to analytical identification systems focused on mark molecules or SERS marks.

11.3 Indirect Method

In the classical sense, indirect mechanisms are also possible without marks, such as when the SERS reaction of a secondary molecule varies or relies on the analyte interaction. These molecules can be of a different nature and are selective for a particular analyte in general.

SERS's broad signal improvements provide stronger and more reliable signals than those from chemical fluorophores; however, the lack of specific control about the degree of amplification usually results in SERS signals that are partially reproducible and nonquantifiable. In the narrowest gap distance (sub-nanometer size), the fastest SERS improvement will be achieved before quantum quenching effects dominate [28–30]. Many facets of excitation and detection matching, such as plasmon resonances in wavelength, often need further optimization of SERS detection [31, 32], polarization[33, 34], and direction of emission [35].

11.4 SERS-based Chemical Sensors (Chemosensors)

A chemical sensor is basically made up of a molecular recognition unit, a receptor, and a transducer that can transform a measurable signal into a binding recognition case. As an analytical tool, SERS is exciting, as it provides molecular-specific data with adequate sensitivity to study either a single analyte or different species simultaneously; because molecular vibrations can be easily separated (as SERS peaks are normally narrow peaks) [36]. SERS can be distinguished from many other techniques for analytical applications because of the rich vibrational spectroscopic information it provides. This has applications in several different directions, including electrochemistry, catalysis, biology, medicine, conservation of art, materials science, and others.

It is possible to consider chemosensors (also molecular sensors) as an analytical concept (or device) which is used for analyte sensing. Usually, they consist of a signaling moiety and a recognition moiety that produces a visible signal or a signal shift in the presence of the analyte. Surface-enhanced Raman dispersion-based chemosensors include plasmatic NPs or nanostructures functionalized as receptors with directly binding legends (chemosensors), which can selectively classify the analyte of interest. The interaction of the analyte-ligand removes the initial SERS signal of the ligand or gives rise to a new SERS signal. The advantages of this approach include the production of highly selective sensors and the use of the chemo-sensor for quantitative purposes as an analytical standard. The high sensitivity and narrow bandwidth of the molecular fingerprint that enables molecules with identical structures to be differentiated is an advantage of SERS chemosensors.

Chemical contaminants include certain organic herbicides and insecticides, other molecules produced for use in different fields are phthalates, polychlorinated byphenyls (PCBs), by-products of natural or industrial operations, polycyclic aromatic hydrocarbons (PAHs), dioxin, etc. The nonspecific coadsorption into metallic nanostructures of other organisms in the matrix solution also impairs direct SERS analysis in normal and polluted waters. This greatly increases the vibrational assignment difficulty or even entirely eliminates the contact with the target analyte, thus reducing the detection assay's sensitivity. A SERS substrate based on a microporous silica capsule with gold nanoparticles (NPs) has been produced to circumvent these problems. The microporous structure functions as a molecular sieve, preventing the plasmonic portion from accessing large biomolecules and cells while imparting colloidal stability. Using microporous silica capsule with gold NPs river water spiked with dichlorodiphenyl trichloroethane (DDT), a pesticide listed as a chronic organic pollutant was detected [37]. A bimetallic Au/Ag ternary film-packaged chip was manufactured as a robust SERS sensor to measure thiabendazole fungicide in drinking water. Interestingly, as a proof-of-concept for creating more robust and functional sensors for on-site tracking, the plasmonic substrate was sealed with polymer films. Polydopamine spheres coated with a gold shell carrying holes and voids (hotspots) were used to detect the pesticide thiram [38]. The LOD of $2.4\,\mu g\,l^{-1}$ in spiked river water was obtained by the nanowaxberry substrate. Although the chemical pollutant content in aquatic waters is usually within the range of $ng\,l^{-1}$ to $\mu g\,l^{-1}$ [39], the active versatile membrane of this SERS stuck and condense chemical and water contaminants showing concentration of crystal violet dye spiked up to $4.1\,pg\,l^{-1}$ in estuary water samples [40]. Super hydrophobic platform was used to concentrate Rhodamine 6 G in an evaporating liquid droplet along with plasmonic nanoparticles, allowing this environmentally dangerous dye to be detected down to $35.9\,fg\,l^{-1}$ [41]. In this context, core–shell HKUST-1@AgNP composites demonstrated strong sensing capabilities in environmental samples for polycyclic aromatic hydrocarbons (PAHs), while retaining the cyclability and selectivity needed for effective quantitative study [42]. This plasmonic composite was found as efficient a gas chromatography mass spectrometry (GC-MS) when compared. This revealed comparable detection capabilities, indicating its ability for detecting these contaminants on location. For their quantitative, label-free and multiplex SERS identification, host–guest approaches focusing on PAHs have been developed [43]. MIP's function was to trap the PAH close to the surface of Au. Pyrene identification has been shown in creek water and seawater [44].

Toxic polyatomic anions (e.g., perchlorate, nitrate, nitrite ions) and heavy transition metal (copper, arsenic, chromium, lead, mercury, cadmium, and copper) cations are chief environmental pollutants. The main objectives of environmental screening are to detect trace quantities of such possible pollutants. Thus, the innovative and enhanced methods with extended sensitivity and selectivity are required for the detection of trace level contaminants.

The sensing procedures of SERS based sensors are as follows

1) Absolute intensity-based method,
2) Wavenumber shift-based method and
3) Ratiometric method [45–48].

11.5 Absolute Intensity-based Method

An absolute intensity-based sensor was given by Li et al. for the detection of fluorine anions (F^{-1}) (Figure 11.4) [46]; 1,4-diketo-3,6-diphenylpyrrolo[3,4-*c*]pyrrole (DPP) was applied as Raman reporter and a selective unit in this method. The sorption approach of diketopyrrolopyrrole (DPP) molecules is equivalent to the substrate surface, and hence, only a weak Raman signal of DPP has been detected in the Surface-Enhanced Raman Spectroscopy spectrum due to the surface chosen method. The identification of DPP to fluorine anions (F^{-1}) modifies the molecular orientation from a parallel adsorption assembly to a vertical one. As a consequence, the Surface-Enhanced Raman Spectroscopy spectrum intensity rises steadily with the increasing strength of fluorine anions (F^{-1}).

11.6 Wavenumber Shift-based Method

In this method, an example has been reported by Chen et al. In this study, DASS (dimethyldi-thiocarbamic acid sodium salt) was used as a molecular probe for the wavenumber shift-dependent measurement of mercury substances, which contain organic mercury and Hg^{2+} ions. DASS attached to the silver nanoparticles (Ag NPs) surface in the shape of bidentate chelation through the carbodithioate ($-CS_2$) group. When the sensing procedure was exposed to mercury ions, a transition by the bidentate chelation to the monodentate chelation occurred due to the definite binding of mercury ions to sulfur atoms cleaving the S–Ag bond. The significant reduction in the Raman intensity was observed due to orientational modification of CSS bending modes. In the intervening time, a band at 1374 cm^{-1} linked with Carbon–Hydrogen deforming modes, shifts to a maximum wavenumber due to the rise in the degrees of independence for the methylene moiety. The wavenumber shifts data could be easily achieve through Surface-Enhanced Raman Spectroscopy spectrum [48].

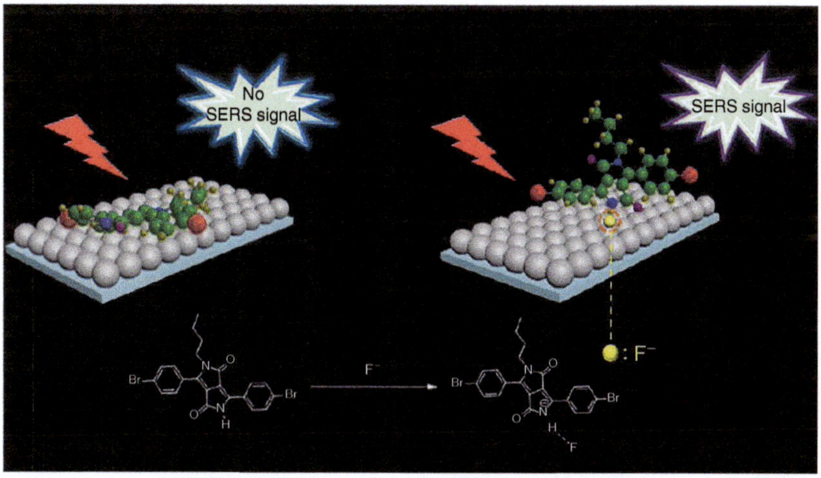

Figure 11.4 A schematic diagram representing the sensing of F^- ions using DPP as a reporter. Source: Reprinted with permission from Li et al. [46]© 2018, Elsevier.

11.7 Ratiometric Method

Ratiometric SERS has been developed as the new methodology to improve the accuracy of SERS measurement via built-in signal correction [49]. The rapid charge transfer between different molecules and substrate varies with the nature of binding force between inorganic ion and molecule. The specific concentrations of various inorganic ions are proportional to total number of symmetric vibrations as stated by selection rules. The ultrasensitive detection of F^- ions (Figure 11.5) with Ag NPs-decorated silicon wafer by derivatising reagent 2-mercaptoboric acid (MPBA) was determined by a ratiometric SERS [47]. In this highly sensitive system, the force between F^- ions as ligand and boronic acid involves changes in the symmetrical orientations as well as charge rearrangement of phenylboronic acid. Resultantly, the IR mode of C–C stretching from 1589 to 1576 cm^{-1} has been shifted due to the bonding between fluoride and MPBA. A ratiometric signal area ratio of this value (A_{1576}/A_{1589}) between MPBA molecules and the fluoride ion and MPBA molecules without binding of fluoride ion can be used for the quantitation of F^- ions. This sensor signifies specific selectivity for F^- ions which is due to dative bonding. The proposed SERS sensor has a quick response to F^- within 120 seconds under the suitable conditions. Li et al. described a Zn^{2+} ions sensor by using *N*,*N*′-bis (2 hydroxybenzylidene)-4-aminophenyl disulfide (HBA) as both the Raman reporter and Zn^{2+}receptor [48]. Zn^{2+} could form coordinated complex with HBA through the oxygen of phenol and Nitrogen of the formazyl molecule both acting as ligands. A monolayer of HBA probes was first adjusted onto the Ag NPs surface by the covalent bond between S–Ag. The complex formation between HBA and Zn^{2+} changes the three-dimensional conformation and structure of HBA, which brings changes in SERS spectrum. The shifts in wavenumber of Raman spectra were observed in this system. This could be interrelated to the difference in molecular or electronic structure of HBA after its complexion with Zn^{2+} ions. The large Raman spectra in Zn^{2+} and chelation formation of HBA probes made this sensor possible to accomplish an ultrasensitive Zn^{2+} examination with a considerable LOD of 10^{-14} M. Likewise, Teng et al. established a new Zn^{2+} sensor for the primary identification of prostate tumor with a reasonable detection limit of 0.1 μM [50].

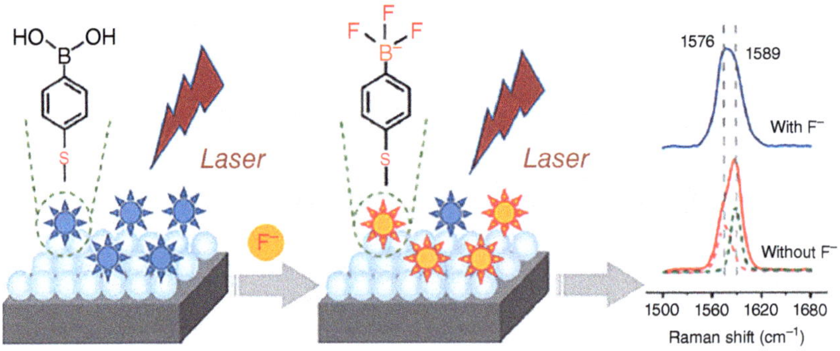

Figure 11.5 A schematic diagram representing the sensing mechanism for the detection of F^- ions. Source: Reprinted with permission from Yue et al. [47] © 2019, American Chemical Society.

Generally, oxyanions, particularly those with Raman cross-sections (e.g. perchlorate) can be detected by their vibrational modes. In contrast, direct SERS detection of monatomic metal ions is more difficult due to insignificant cross-section. Sensitive detection of perchlorate anions by SERS depends on the functionality of surface of the plasmonic substances with cystamine, 2-dimethylaminoethanethiol, or poly(ethyleneimine). The SERS substrates based on gold ellipse dimers functionalized with 2-(dimethylamino) ethanethiol were used to detect and quantify ClO^{-4} contamination at the $\mu g\,l^{-1}$ level within ground waters, thereby demonstrating the applications of this method for field measurements [51]. Arsenic species, including arsenate (As^{5+}) and arsenite (As^{3+}), which usually exist in the environment as $AsO_4{}^{3-}$ or $AsO_3{}^{3-}$, respectively, can be distinguished by SERS based on the typical vibration of As–O stretch mode [52]. The combined interactions of arsenic species with the Ag NPs, solvent, and sodium dodecyl sulfate surfactant made possible arsenic speciation and SERS detection at $0.1\,\mu g\,l^{-1}$, indicating the potential of this method for separation and qualitative SERS analysis [53].

Metal organic frameworks (MOF)-based SERS substrates have been effectively employed to identify environmental pollutants, especially organic dyes contaminants, pesticide residues [54, 55] VOCs, and PAHs [43, 56]. Guselnikova et al. developed MOF-5 coated SERS active gold gratings as SERS platform for detection of organic contaminants naturally present in soil such as organophosphorus pesticides. The LOD was found as 10^{-12} M ($2.8\times10^{-7}\,mg\,l^{-1}$) not larger than residue levels of OPPs ($10^{-2}\,mg\,l^{-1}$) in plant and food published by Regulation of the European Parliament. Besides, it showed $r = 0.972$ for Paraoxon and 0.976 for Fenitrothion and recognition of several relevant organic contaminants (azo-dye, mycotoxin, and pesticide) from the simulated soil [57]. Koh et al. created a ZIF-encapsulated silver nanocube array surface-enhanced Raman spectroscopy substrate for the identification of different VOCs and PAHs. The LOD of naphthalenethiol was found as 50 part per billion (ppb), which is 15-times less than the exposure limit of $0.005\,g\,m^{-3}$ (0.7 part per million) [43]. MOF incorporated SERS-active substrates can be used to discriminate pesticide residues [58, 59], illegal additives [60], or other threatening ingredients to ensure food safety, which is a hot topic closely related to our daily life and has really great importance to human health. A MOF shell isolated Au nanoparticles substrate, Au@MIL-100, was developed for sensitive detection and quantitative analysis of malachite green in aquaculture water with LOD down to 8×10^{-9} M [61]. By using identical core-shell Au@MIL-101 substrates, Cai et al. discovered direct detection of trace methenamine, an illegal additive can result in harm to human organs and viscera, in vermicelli with a strong linear relationship from 3.16×10^{-6} to 1.0×10^{-8} M that complies with the standard exposure limit [60]. Zhou et al. prepared ZIF-8 wrapped urchin-like Au-Ag particles (UAANs@ZIF-8) substrate for SERS detection of trace volatile hexachlorocyclohexane (HCH) pesticides with LOD below 1.5 ppb through ZIF-8 wrapping-enhanced SERS effect [62].

11.8 SERS-based Biological Sensors (Biosensors)

Biomolecule is the term that refers to the molecule originated from living organism, comprises of the substances of low molecular weight such as amino acids, nucleotides, fatty acids, monosaccharide's, vitamins, etc. which are the building blocks of life. These compounds are mainly composed of carbon, oxygen, hydrogen, sulfur, nitrogen, and phosphorus. Furthermore, the macromolecules such as nucleic acid, proteins, lipids, and carbohydrates are formed by these biopolymers monomers having the high biological relevance [63].

Biosensors are used for qualitative and quantitative analyzing of biomolecules. The main task of biosensor is to monitor the biological process going on in the body and to diagnose the diseases in the body. Biochemical reaction takes time to proceed. Mostly, the biological reactions are performed in complex water environment. For analyzing purpose in-vivo and in-vitro methods are widely used. Study of temporal evolution of biochemical reactions takes place by in-vitro methods and detailed mechanisms and dynamic reactions are analyzed by in-vivo methods [64]. SERS has also taken great attention as powerful analytical technique for studying the biological samples. It provides the detailed information regarding living cells and biological reactions. It helps to correlate the interaction with drug or toxic agents, diseases, and cell death. Moreover, instead of traditional nonresonant Raman spectroscopy, the resonant Raman spectroscopy, and coherent anti strokes Raman spectroscopy are taking much more contribution in studying the living cells. SERS has been taking the tremendous attention in the biomedical and bioengineering applications. Because SERS provides the high sensitivity and selectivity due to the optical property of plasmonic nanostructure [65]. Therefore, SERS has become the best analytical instrument for examining the reaction phenomenon at molecular level, especially the enzyme-catalyzed reactions. Although, the problem occurred commonly is retaining of biomolecular structure such as protein denaturation on the undecorated metal surface, that can reduce the bioactivity and function [66]. SERS has been enhancing the million folds signals in traditional Raman signal by using the suitable SERS substrates [27]. That's why SERS has long been used in biological related research like any other analytical tool.

In biological area, the SERS has been divided in two main parts, mainly fundamental in which SERS is used to examine biomolecules of structures, conformation, and charge transfer. In the second case, it further splits into two applications i.e. direction biomedical diagnostic and indirection biomedical diagnostic [67].

These are the advantages of SERS-based biosensors:

a) It has the power of reflecting basic fingerprint molecular information of biomolecules and has a high sensitivity to examine the molecule at singular level.
b) If compared it with fluorescence spectroscopy, it produces the sharp peaks with good resistance to photo-degradation and photo-bleaching.
c) There are multiple choices of signal enhancement substrates having multiple sizes and shapes that make it appropriate for different applications.
d) The large depth laser penetration makes it applicable for in-vivo and in-vitro diagnosis and imaging.

There are numerous factors to be considered, while analyzing the biological samples. Such as, preparation methods of biological samples to be analyzed, suitable SERS probes, an appropriate instrumental configuration, optimal data processing method, and analysis.

Sample preparation depends upon the type of sample. The sample can be analyzed by (i) In-vivo, (ii) In-vitro and (iii) Ex-vivo.

In-vivo method, there is no need of sample preparation because in this method the measurements are performed in alive patients. Therefore, the SERS probes have to be designed to integrate into the living environment. In-vitro method, the sample should be live or fixed measurements will often only be physiologically relevant if obtained using live cells, for example, using SERS probes for pH sensing [68]. In ex-vivo method blood, serum, plasma, or tissues are analyzed [69]. Probe selection depends upon number of factors. For biological samples, most commonly probe used in SERS is either gold or silver because of their plasmon resonances that lie in visible and near-infrared range and most favorable optical properties. Moreover, gold and silver show minimum toxicity in biological system. In particular, gold nanoparticles are inert and have already been approved for use in live humans for particular applications. For this reason, in vitro and in vivo applications will often use gold nanoparticles as the substrate of choice. Silver, however, tends to exhibit superior scattering properties and a larger enhancement of Raman signals. Therefore, in ex vivo applications, where nanoparticles are not being applied directly to living systems, silver may be preferable [68]. These metals probes are functionalized with Raman reporter label, a biomolecule which is either an antibody or an oligonucleotide. And in some cases, label-free analyzing has been performed. Instrumentation selection for analyzing biological sample depends upon which type of information regarding the samples is required. Once a probe is selected, the first decision to make is often laser wavelength. Raman spectroscopy has ability to use more than one laser excitation wavelengths for measurement. For biological samples, the problem is occurred in autofluorescence of sample and poor tissue penetration. Selection of visible wavelength provides higher signal intensity and autofluorescence, while in NIR, the signal intensity is low but with less autofluorescence. Therefore, NIR is selected mostly [70].

The unique, but weak, Raman scattering patterns from the excitation or relaxation of molecules' vibrational modes can be used to identify and characterize molecular systems. As roughened/nanoscale surfaces provide large orders of enhancement, this spectroscopic technique is no longer confined to strong scattering targets or high concentration systems [9].

Analyzing DNA is one of the major fields in the biosensing, because it is used to identify the different diseases state and bacteria, virus in the blood sample. For this purpose, the fluorescence technique is considered as the best because of its sensitivity and specificity. However, SERS has been also used because of its sensitivity and specificity. As it has ability to provide the specific spectra of molecule in the mixtures of products. Due to this capability, it has more advantages than the fluorescence technique, and it has ability to detect simultaneously the multiple substances without using the separation steps. Moreover, it gains more attention over the fluorescence techniques due to very effective surface quenching fluorescence. It makes SERS availability to sense the fluorophores currently used in the industry and reduce the barrier in commercialization [71]. Fragments of DNA and RNA are considered to be the most important analytes targeted for SERS sensing. DNA analysis of real samples mostly includes separation, denaturation, fragmentation, and sensing [72].

In the field of rapid detection and analyzing, the single-molecule SERS are the great advantage. They can analyze the single DNA and bases in DNA fragment without using fluorescent or radioactive labels [71]. The specific DNA targets identification is done by hybridization of a nucleic acid to its complementary sequence. SERS signal of hybridization reaction can be observed by keeping the Raman active molecules near to the metal surface. It can be seen by Figure 11.6 which shows detection of DNA hybridization by functionalized metal nanoparticles approach [71].

Protein identification and detection have great importance in different fields including biotechnology, medicine, and pharmacology. Usually, the techniques commonly used are mass spectroscopy and immunoassay-based techniques. They are considered as powerful techniques but they have certain disadvantages for example, mass spectroscopy is expensive technique; however, the immunoassay-based sensors have low sensitivity. On the basis of remarkable sensitivity, SERS is considered to be the best one. And nowadays, it has been taking a place of different techniques for the protein detection and identification. In the class of biomacromolecules, proteins are the most important class and their detection methods are required. Therefore, the use of SERS as label-free detection and identification of protein is pursued [72]. By using SERS, protein interaction with other materials, its structure, chemical environment of protein, and orientation toward the metallic surface can be analyzed [63]. In several indirect protein detection analyses, fluorescence and staining steps are replaced by SERS. There are numerous methods or techniques that are replaced by SERS for protein detection. And several different methods of SERS are also used to characterize the protein [73]. SERS has been used to analyze cytochromes. Cytochromes are redox-active protein. For the detection of cytochromes, SERS worked on bases in which the silver electrode adsorbed the redox reaction of cytochromes that helped to give information regarding the kinetics and structural information. SERS-based biosensors are also used for the detection of myoglobin, one of the types of protein, Raman spectra of myoglobin can be recorded at individual molecular level [74].

Cellular microenvironment includes the ionic concentration, pH values, and redox potential charges. The physiological and pathological diseases depend upon the intra and extracellular pH values of living cell. Therefore, sensitive and reliable detection of these values that help to diagnose many diseases at early stage is required. For pH sensing, 4-mercaptobenzoic acid (4MBA) and 4-mercaptopyridine (4MPy) pH sensing molecules are mostly used for SERS-based detection. These pH sensing molecules are used because they provide the large range of pH (pH 4–9) and possess the good chemical affinity towards the surface of gold NPs [75]. Figure 11.7 shows the intracellular pH sensing with targeted gold nanoparticles [78].

Mostly, the indirect SERS sensor has been used in cellular environmental changes because the direct or label-free SERS provide the less satisfactory Raman spectra [76]. Like pH, the redox potential also plays a vital role in detecting the pathological and physiological diseases. For understanding the cellular functions i.e. apoptosis, cell cycle, and signaling pathways, the redox potential changes play a major role. Quantitative measurement of intracellular redox potential of cells has been detected by SERS substrates. Researchers recently developed the single cell-based chip to investigate the intracellular and extracellular redox potential of neural cells (PC12). This chip is combined with linear sweep voltammetry (LSV) to record the biochemical changes in the cell [74].

Nowadays, many methods for SERS-based sensing are introduced that can directly analyze the cell. The main challenge of this era is to establish the method, which is compatible and stable toward the plasmonic material exposed to biological environment. The surface modification and bioconjugation are essential steps to impart site selectivity for SERS

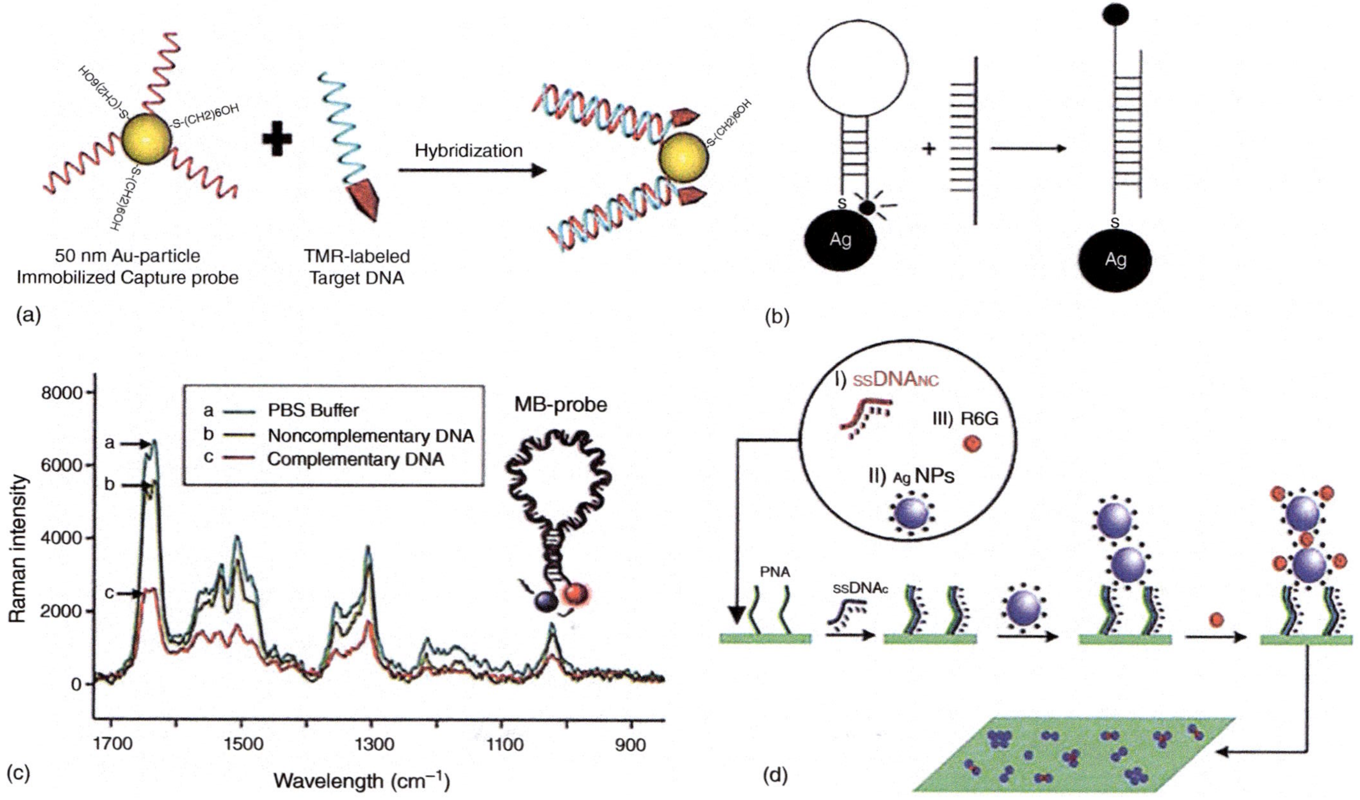

Figure 11.6 Detection of DNA hybridization by functionalized nanoparticles approach. Source: Reprinted with permission from Huh et al. [75] © 2009, Springer Nature.

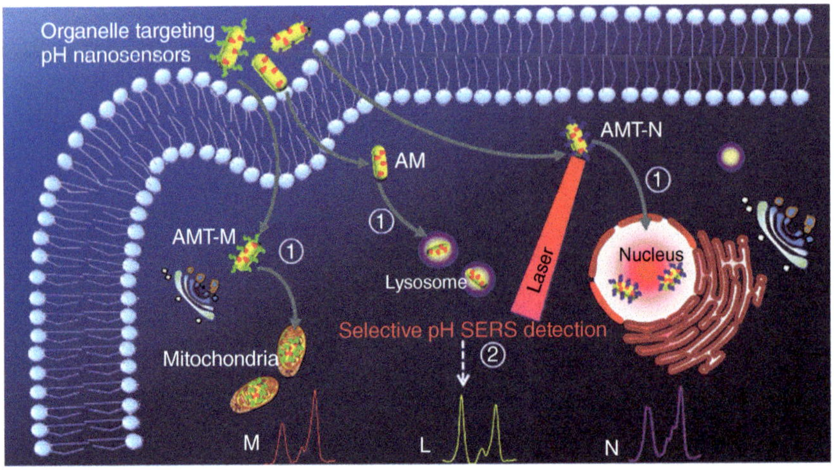

Figure 11.7 Intracellular pH sensing with targeted Au Nps. Source: Reprinted with permission from Shen et al. [78] © 2021, Royal Society of Chemistry. Licensed under CC BY 4.0.

diagnostics and imaging. The use of bioconjugated plasmonic vesicles from amphiphilic polymer brush coated with gold nanoparticles SERS-based cancer targeting and drug delivery was one of the recent advances [77]. Recently, the most important approach of early diagnosis of different diseases particularly cancer is introduced. It works on the basis of analyzing biomarkers in the body fluids, example are blood, saliva, and urine. As a matter of fact, Au NPs specially labeled by 5,50-dithiobis (succinimidyl-2-nitrobenzoate) (DSNB) followed by antibody immobilization and surface blocking by BSA, were developed for the SERS detection of two pancreatic adenocarcinoma biomarkers in sandwich bioassays [77].

11.9 Conclusion

In new era, the flexible sensors have taken place of the conventional sensors because of their ease to operate, rapid diagnosis, and cost-effective properties. Various flexible sensors have been introduced, but SERS has gained much attention because of its label-free and fingerprint detection. Moreover, it has ability to analyze at trace level. Furthermore, the development of handheld portable Raman spectrometer attached with flexible SERS nanotechnologies provides the promising result in the field of identification of biochemical and other molecules for point of care diagnostics and on-spot detection, respectively. However, there are numerous challenges which we have to overcome to commercialize it such as macromolecules identification, less sensitivity, reliability, and cost of portable SERS sensors.

References

1 Willes, H.W. and Roberto, M. (eds.) (2000). *Raman Scattering in Materials Science. Raman Scattering in Materials Science*, vol. 42, 493. Springer, Berlin: Berlin Heidelberg.

2 Smekal, A. (1923). Zur Quantentheorie der Dispersion. *Naturwissenschaften* 11 (43): 873–875.

3 Moskovits, M. (1985). Surface-enhanced spectroscopy. *Reviews of Modern Physics* 57 (3): 783–826.

4 Fleischmann, M., Hendra, P.J., and McQuillan, A.J. (1974). Raman spectra of pyridine adsorbed at a silver electrode. *Chemical Physics Letters* 26 (2): 163–166.

5 Jeanmaire, D.L. and Van Duyne, R.P. (1977). Surface raman spectroelectrochemistry: part I. Heterocyclic, aromatic, and aliphatic amines adsorbed on the anodized silver electrode. *Journal of Electroanalytical Chemistry and Interfacial Electrochemistry* 84 (1): 1–20.

6 Albrecht, M.G. and Creighton, J.A. (1978). Intense Raman spectra at a roughened silver electrode. *Electrochimica Acta* 23 (10): 1103–1105.

7 Ru, E.C.L., Etchegoin, P.G., and Meyer, M. (2006). Enhancement factor distribution around a single surface-enhanced Raman scattering hot spot and its relation to single molecule detection. *Journal of Chemical Physics* 125 (20): 204701.

8 Zhao, Y., Liu, X., Lei, D.Y., and Chai, Y. (2014). Effects of surface roughness of Ag thin films on surface-enhanced Raman spectroscopy of graphene: spatial nonlocality and physisorption strain. *Nanoscale* 6 (3): 1311–1317.

9 McNay, G., Eustace, D., Smith, W.E. et al. (2011). Surface-Enhanced Raman Scattering (SERS) and Surface-Enhanced Resonance Raman Scattering (SERRS): a review of applications. *Applied Spectroscopy* 65 (8): 825–837.

10 Su, J.-P., Lee, Y.-T., Lu, S.-Y., and Lin, J.S. (2013). Chemical mechanism of surface-enhanced raman scattering spectrum of pyridine adsorbed on Ag cluster: Ab initio molecular dynamics approach. *Journal of Computational Chemistry* 34 (32): 2806–2815.

11 Schlücker, S. (2014). Surface-enhanced Raman spectroscopy: concepts and chemical applications. *Journal of Computational Chemistry* 53 (19): 4756–4795.

12 Zhang, L., Xu, J., Mi, L. et al. (2012). Multifunctional magnetic-plasmonic nanoparticles for fast concentration and sensitive detection of bacteria using SERS. *Biosensors & Bioelectronics* 31 (1): 130–136.

13 Granger, J.H., Schlotter, N.E., Crawford, A.C., and Porter, M.D. (2016). Prospects for point-of-care pathogen diagnostics using surface-enhanced Raman scattering (SERS). *Chemical Society Reviews* 45 (14): 3865–3882.

14 Craig, A.P., Franca, A.S., and Irudayaraj, J. (2013). Surface-enhanced Raman spectroscopy applied to food safety. *Annual Review of Food Science and Technology* 4 (1): 369–380.

15 Jarvis, R.M. and Goodacre, R. (2004). Discrimination of bacteria using surface-enhanced Raman spectroscopy. *Analytical Chemistry* 76 (1): 40–47.

16 Wu, X., Huang, Y.-W., Park, B. et al. (2015). Differentiation and classification of bacteria using vancomycin functionalized silver nanorods array based surface-enhanced Raman spectroscopy and chemometric analysis. *Talanta* 139: 96–103.

17 Xu, K., Zhou, R., Takei, K., and Hong, M. (2019). Toward flexible surface-enhanced Raman scattering (SERS) sensors for point-of-care diagnostics. *Advanced Science* 6 (16): 1900925.

18 Hakonen, A., Andersson, P.O., Stenbæk Schmidt, M. et al. (2015). Explosive and chemical threat detection by surface-enhanced Raman scattering: A review. *Analytica Chimica Acta* 893: 1–13.

19 Rodríguez-Lorenzo, L., Álvarez-Puebla, R.A., Pastoriza-Santos, I. et al. (2009). Zeptomol detection through controlled ultrasensitive surface-enhanced Raman scattering. *Journal of the American Chemical Society* 131 (13): 4616–4618.
20 Hrelescu, C., Sau, T.K., Rogach, A.L. et al. (2009). Single gold nanostars enhance Raman scattering. *Applied Physics Letters* 94 (15): 153113.
21 Tian, Z.-Q., Ren, B., and Wu, D.-Y. (2002). Surface-enhanced Raman scattering: from noble to transition metals and from rough surfaces to ordered nanostructures. *The Journal of Physical Chemistry B* 106 (37): 9463–9483.
22 Ringler, M., Schwemer, A., Wunderlich, M. et al. (2008). Shaping emission spectra of fluorescent molecules with single plasmonic nanoresonators. *Physical Review Letters* 100 (20): 203002.
23 Kahraman, M., Tokman, N., and Culha, M. (2008). Silver nanoparticle thin films with nanocavities for surface-enhanced Raman scattering. *Chemphyschem: A European Journal of Chemical Physics and Physical Chemistry* 9 (6): 902–910.
24 Gunawidjaja, R., Peleshanko, S., Ko, H., and Tsukruk, V.V. (2008). Bimetallic nanocobs: decorating silver nanowires with gold nanoparticles. *Advanced Materials* 20 (8): 1544–1549.
25 Lal, S., Grady, N.K., Kundu, J. et al. (2008). Tailoring plasmonic substrates for surface enhanced spectroscopies. *Chemical Society Reviews* 37 (5): 898–911.
26 Oldenburg, S.J., Jackson, J.B., Westcott, S.L., and Halas, N.J. (1999). Infrared extinction properties of gold nanoshells. *Applied Physics Letters* 75 (19): 2897–2899.
27 Li, P., Long, F., Chen, W. et al. (2020). Fundamentals and applications of surface-enhanced Raman spectroscopy–based biosensors. *Current Opinion in Biomedical Engineering* 13: 51–59.
28 Zhu, W. and Crozier, K.B. (2014). Quantum mechanical limit to plasmonic enhancement as observed by surface-enhanced Raman scattering. *Nature Communications* 5 (1): 1–8.
29 Mao, L., Li, Z., Wu, B., and Xu, H. (2009). Effects of quantum tunneling in metal nanogap on surface-enhanced Raman scattering. *Applied Physics Letters* 94 (24): 243102.
30 Chen, W., Zhang, S., Kang, M. et al. (2018). Probing the limits of plasmonic enhancement using a two-dimensional atomic crystal probe. *Light: Science & Applications* 7 (1): 1–11.
31 Dieringer, J.A., Wustholz, K.L., Masiello, D.J. et al. (2009). Surface-enhanced Raman excitation spectroscopy of a single rhodamine 6G molecule. *Journal of the American Chemical Society* 131 (2): 849–854.
32 Chu, Y., Banaee, M.G., and Crozier, K.B. (2010). Double-resonance plasmon substrates for surface-enhanced Raman scattering with enhancement at excitation and stokes frequencies. *ACS Nano* 4 (5): 2804–2810.
33 Shegai, T., Li, Z., Dadosh, T. et al. (2008). Managing light polarization via plasmon–molecule interactions within an asymmetric metal nanoparticle trimer. *Proceedings of the National Academy of Sciences* 105 (43): 16448.
34 Wei, H., Hao, F., Huang, Y. et al. (2008). Polarization Dependence of Surface-Enhanced Raman Scattering in Gold Nanoparticle–Nanowire Systems. *Nano letters* 8 (8): 2497–2502.
35 Li, Z., Shegai, T., Haran, G., and Xu, H. (2009). Multiple-particle nanoantennas for enormous enhancement and polarization control of light emission. *ACS Nano* 3 (3): 637–642.
36 Hamon, C. and Liz-Marzán, L.M. (2018). Colloidal design of plasmonic sensors based on surface enhanced Raman scattering. *Journal of Colloid and Interface Science* 512: 834–843.

37 Mariño-Lopez, A., Sousa-Castillo, A., Blanco-Formoso, M. et al. (2019). Microporous plasmonic capsules as stable molecular sieves for direct SERS quantification of small pollutants in natural waters. *ChemNanoMat* 5 (1): 46–50.

38 Wang, K., Sun, D.-W., Pu, H. et al. (2019). Stable, flexible, and high-performance SERS chip enabled by a ternary film-packaged plasmonic nanoparticle array. *ACS applied materials & interfaces* 11 (32): 29177–29186.

39 Chen, D., Zhu, X., Huang, J. et al. (2018). Polydopamine@ gold nanowaxberry enabling improved SERS sensing of pesticides, pollutants, and explosives in complex samples. *Analytical Chemistry* 90 (15): 9048–9054.

40 Fateixa, S., Nogueira, H.I., and Trindade, T. (2018). Surface-enhanced Raman scattering spectral imaging for the attomolar range detection of crystal violet in contaminated water. *ACS omega* 3 (4): 4331–4341.

41 Lee, H.K., Lee, Y.H., Koh, C.S.L. et al. (2019). Designing surface-enhanced Raman scattering (SERS) platforms beyond hotspot engineering: emerging opportunities in analyte manipulations and hybrid materials. *Chemical Society Reviews* 48 (3): 731–756.

42 Choi, J., Kim, J.-H., Oh, J.-W., and Nam, J.-M. (2019). Surface-enhanced Raman scattering-based detection of hazardous chemicals in various phases and matrices with plasmonic nanostructures. *Nanoscale* 11 (43): 20379–20391.

43 Li, D., Cao, X., Zhang, Q. et al. (2019). Facile in situ synthesis of core–shell MOF@Ag nanoparticle composites on screen-printed electrodes for ultrasensitive SERS detection of polycyclic aromatic hydrocarbons. *Journal of Materials Chemistry A* 7 (23): 14108–14117.

44 Castro-Grijalba, A. V. Montes-García, M.J. Cordero-Ferradás, , et al., SERS-based molecularly imprinted plasmonic sensor for highly sensitive PAH detection. *ACS Sensors*, 2020. 5(3): p. 693-702.

45 Chen, L., Zhao, Y., Wang, Y. et al. (2016). Mercury species induced frequency-shift of molecular orientational transformation based on SERS. *Analyst* 141 (15): 4782–4788.

46 Li, X., Zhang, M., Wang, Y. et al. (2018). Direct detection of fluoride ions in aquatic samples by surface-enhanced Raman scattering. *Talanta* 178: 9–14.

47 Yue, X., Su, Y., Wang, X. et al. (2019). Reusable Silicon-Based SERS Chip for Ratiometric Analysis of Fluoride Ion in Aqueous Solutions. *ACS Sensors* 4 (9): 2336–2342.

48 Zhuang, H., Wang, Z., Zhang, X. et al. (2017). A highly sensitive SERS-based platform for Zn (II) detection in cellular media. *Chemical Communications* 53 (11): 1797–1800.

49 Teng, Y., Ren, Z., Zhang, Y. et al. (2020). Determination of prostate cancer marker Zn2+ with a highly selective surface-enhanced Raman scattering probe on liquid–liquid self-assembled Au nanoarrays. *Talanta* 209: 120569.

50 Jubb, A.M., Hatzinger, P.B., and Gu, B. (2017). Trace-level perchlorate analysis of impacted groundwater by elevated gold ellipse dimer nanoantenna surface-enhanced Raman scattering. *Journal of Raman Spectroscopy* 48 (4): 518–524.

51 Hao, J., Han, M.-J., Han, S. et al. (2015). SERS detection of arsenic in water: a review. *Journal of Environmental Sciences* 36: 152–162.

52 Yang, M., Liamtsau, V., Fan, C. et al. (2019). Arsenic speciation on silver nanofilms by surface-enhanced raman spectroscopy. *Analytical chemistry* 91 (13): 8280–8288.

53 Zhang, Y., Hu, Y., Li, G., and Zhang, R. (2019). A composite prepared from gold nanoparticles and a metal organic framework (type MOF-74) for determination of 4-nitrothiophenol by surface-enhanced Raman spectroscopy. *Microchimica Acta* 186 (7): 477.

54 Xuan, T., Gao, Y., Cai, Y. et al. (2019). Fabrication and characterization of the stable Ag-Au-metal-organic-frameworks: An application for sensitive detection of thiabendazole. *Sensors and Actuators B: Chemical* 293: 289–295.
55 Koh, C.S.L., Lee, H.K., Han, X. et al. (2018). Plasmonic nose: integrating the MOF-enabled molecular preconcentration effect with a plasmonic array for recognition of molecular-level volatile organic compounds. *Chemical Communications* 54 (20): 2546–2549.
56 Guselnikova, O., Postnikov, P., Elashnikov, R. et al. (2019). Metal-organic framework (MOF-5) coated SERS active gold gratings: A platform for the selective detection of organic contaminants in soil. *Analytica Chimica Acta* 1068: 70–79.
57 Zhou, X., Liu, G., Zhang, H. et al. (2019). Porous zeolite imidazole framework-wrapped urchin-like Au-Ag nanocrystals for SERS detection of trace hexachlorocyclohexane pesticides via efficient enrichment. *Journal of Hazardous Materials* 368: 429–435.
58 Li, Q., Gong, S., Zhang, H. et al. (2019). Tailored necklace-like Ag@ ZIF-8 core/shell heterostructure nanowires for high-performance plasmonic SERS detection. *Chemical Engineering Journal* 371: 26–33.
59 Lai, H., Li, G., Xu, F., and Zhang, Z. (2020). Metal–organic frameworks: opportunities and challenges for surface-enhanced Raman scattering–a review. *Journal of Materials Chemistry C* 8 (9): 2952–2963.
60 Liao, J., Wang, D., Liu, A. et al. (2015). Controlled stepwise-synthesis of core–shell Au@ MIL-100 (Fe) nanoparticles for sensitive surface-enhanced Raman scattering detection. *Analyst* 140 (24): 8165–8171.
61 Qiao, X., Su, B., Liu, C. et al. (2018). Selective surface enhanced Raman scattering for quantitative detection of lung cancer biomarkers in superparticle@ MOF structure. *Advanced materials* 30 (5): 1702275.
62 Zheng, X.-S., Jahn, I.J., Weber, K. et al. (2018). Label-free SERS in biological and biomedical applications: Recent progress, current challenges and opportunities. *Spectrochimica Acta Part A: Molecular and Biomolecular Spectroscopy* 197: 56–77.
63 Notingher, I. (2007). Raman spectroscopy cell-based biosensors. *Sensors* 7 (8): 1343–1358.
64 Cialla-May, D., Zheng, X.S., Weber, K., and Popp, J. (2017). Recent progress in surface-enhanced Raman spectroscopy for biological and biomedical applications: from cells to clinics. *Chemical Society Reviews* 46 (13): 3945–3961.
65 Ji, W., Zhao, B., and Ozaki, Y. (2016). Semiconductor materials in analytical applications of surface-enhanced Raman scattering. *Journal of Raman Spectroscopy* 47 (1): 51–58.
66 Fan, M., Andrade, G.F., and Brolo, A.G. (2011). A review on the fabrication of substrates for surface enhanced Raman spectroscopy and their applications in analytical chemistry. *Analytica Chimica Acta* 693 (1-2): 7–25.
67 Jamieson, L.E., Jaworska, A., Jiang, J. et al. (2015). Simultaneous intracellular redox potential and pH measurements in live cells using SERS nanosensors. *Analyst* 140 (7): 2330–2335.
68 Premasiri, W., Lee, J., and Ziegler, L. (2012). Surface-enhanced Raman scattering of whole human blood, blood plasma, and red blood cells: cellular processes and bioanalytical sensing. *The Journal of Physical Chemistry B* 116 (31): 9376–9386.
69 Kearns, H., Bedics, M.A., Shand, N.C. et al. (2016). Sensitive SERS nanotags for use with 1550 nm (retina-safe) laser excitation. *Analyst* 141 (17): 5062–5065.

70 Szlag, V.M., Rodriguez, R.S., He, J. et al. (2018). Molecular affinity agents for intrinsic surface-enhanced Raman scattering (SERS) sensors. *ACS applied materials & interfaces* 10 (38): 31825–31844.

71 Culha, M., Cullum, B., Lavrik, N., and Klutse, C.K. (2012). Surface-Enhanced Raman Scattering as an Emerging Characterization and Detection Technique. *Journal of Nanotechnology* 2012: 971380.

72 Huh, Y.S., Chung, A.J., and Erickson, D. (2009). Surface enhanced Raman spectroscopy and its application to molecular and cellular analysis. *Microfluidics and nanofluidics* 6 (3): 285.

73 Pilot, R., Signorini, R., Durante, C. et al. (2019). A review on surface-enhanced Raman scattering. *Biosensors* 9 (2): 57.

74 Jahn, I.J., Mühlig, A., and Cialla-May, D. (2020). Application of molecular SERS nanosensors: where we stand and where we are headed towards? *Analytical and Bioanalytical Chemistry* 412 (24): 5999–6007.

75 Wang, J., Geng, Y., Shen, Y. et al. (2019). SERS-active fiber tip for intracellular and extracellular pH sensing in living single cells. *Sensors and Actuators B: Chemical* 290: 527–534.

76 Song, J., Zhou, J., and Duan, H. (2012). Self-assembled plasmonic vesicles of SERS-encoded amphiphilic gold nanoparticles for cancer cell targeting and traceable intracellular drug delivery. *Journal of the American Chemical Society* 134 (32): 13458–13469.

77 Granger, J.H., Granger, M.C., Firpo, M.A. et al. (2013). Toward development of a surface-enhanced Raman scattering (SERS)-based cancer diagnostic immunoassay panel. *Analyst* 138 (2): 410–416.

78 Shen, Y., Liang, L., Zhang, S. et al. (2018). Organelle-targeting surface-enhanced Raman scattering (SERS) nanosensors for subcellular pH sensing. *Nanoscale* 10 (4): 1622–1630.

12

Carbon Nanomaterials as Plasmonic Sensors in Biotechnological and Biomedical Applications

Tahira Qureshi[1,2], Kemal Çetin[3], and Adil Denizli[1]

[1]*Department of Chemistry, Hacettepe University, Ankara, Turkey*
[2]*National Centre of Excellence in Analytical Chemistry, Sindh University, Jamshoro, Pakistan*
[3]*Department of Biomedical Engineering, Necmettin Erbakan University, Konya, Turkey*

12.1 Introduction

Primarily, in plasmonic sensing moreover the surface plasmon process engaged directly for analyte sensing or it facilitates the other spectroscopic techniques such as surface enhanced Raman spectroscopy (SERS) and surface plasmon-enhanced fluorescence. In brief, analytes penetrate its near-field system and alter the local refractive index, which results the shift in resonance frequency of the whole system [1–4]. The advancement of analytical instruments in the last decade has facilitated the nanofabrication and characterization techniques for plasmonic nanosensors. Plasmonic nanosensors have been fabricated with variety of nanostructures, such as nanorods, nanotriangles, nanocrescent, etc. these fabricated materials directly influence the cost, analysis time, and feasibility for portable sensing device. The fabrication of nanosensors plays key role to achieve specific outcome of sensing analysis [5–8]. Not long ago, carbon nanomaterials were observed the plasmonic effect induced by π electrons. Afterward, carbon-based nanomaterials have attracted the attention of the scientific community for the fabrication of plasmonic-based sensors. This is because of their unique electrical, mechanical, chemical, thermal, and optical properties. In case of carbon, nanomaterials have shown quite feasible nature to tune their plasmonic properties such as controlling their shape, etc. to exhibiting desired optical properties [9–12]. Among all the carbon nanomaterials (CNMs), the carbon nanotubes and graphene are promising candidates for various applications in plasmonic sensing. Till now, number of reports have been observed by using CNMs as sensing materials for biomedical and biotechnological applications [13–16].

The CNMs in comparison to conventional plasmonic sensors such as gold (Au), silver (Ag), copper (Cu), chromium (Cr), and aluminum (Al), were well used for a long time have been viewed as the finest plasmonic materials in the field of sensing. Nevertheless, these materials have some limitations, such as large energy losses (e.g. Ohmic loss and radiative loss) and limited tunability. To overcome these limitations, there is plenty room available

Plasmonic Sensors and their Applications, First Edition. Edited by Adil Denizli.

for new materials with plasmonic properties having minimal loss, decent tunability, and offer upturn in sensitivity for sensing [17–24]. The CNMs have appeared as the better option for plasmonic materials, with the considerable sensing applications in the fields of biomedical and biotechnological areas [25–27]. In this chapter, the focus is on graphene and carbon nanotubes for their last five year applications in biomedical and biotechnological plasmonic sensors. Following given brief description of graphene and carbon nanotubes.

12.1.1 Graphene

Novoselov and Geim (2004) prepared a profound material graphene from graphite, which possesses hexagonal honeycomb structure [28]. Lately, graphene has surfaced as an alternating substance for plasmonic sensing in the terahertz (THz) to the mid-infrared scale area [29–31]. The graphene plasmonics are quite attractive, even though there is disparate flanked by the graphene plasmons wavevector and free space light wavevector, which interfere in the graphene plasmons outcome [32, 33]. In the graphene structure, the electron acts as massless "dirac fermions" which also lead to draw remarkable properties, such as ultra-high-mobility carriers, gate-tunable carrier densities, fine structure constant defined optical transmission, and long mean free path, etc. As compared to conventional noble metals, the captivity of surface plasmons in graphene is impregnable. The graphene plasmons are beneficial over conventional noble metals plasmons in a manner as they have relatively low loss, impregnable, tunable, flexible nature, and so on. These attainments are evident of advantages in the control of EM wave compared to the conventional noble metal materials [19, 22, 34].

12.1.2 Carbon Nanotubes

Carbon nanotubes (CNTs) were synthesized in 1991. They are categorized on the number of their graphitic layers [35]. The single-walled carbon nanotubes (SWCNTs) are single graphitic sheets, while multi-walled carbon nanotubes (MWCNTs) are multiple parallel layers [36, 37]. CNTs plasmons behave in a longitudinal charge oscillations coupled to infrared or terahertz optical regions [38, 39]. The extreme strength of electromagnetic grounds stemming from the nanotubes' one dimensionality allows the plasmons both to confine light to the nanometer scale and to boost light-matter interactions by Purcell factors that are expected to be ≤106. CNTs plasmonics at infrared frequencies behave extremely responsive via absorption spectroscopy through surface-enhanced infrared absorption (SEIRA). CNTs have elevated absorptions in the NIR spectral range and they are often employed as photonic materials for biological imaging [40–42].

12.2 Biomedical and Biotechnological Applications of Carbon Nanomaterials as Plasmonic Sensors

The plasmonic sensing has noteworthy place for label-free, intensifying signal, and real-time sensing means for bioparticles and bioprocesses at the molecular level. The frequent use of these sensors has been observed in automated industry like robotics, aeronautics and aerospace, biomedical devices, and the manufacturing industry. CNMs based plasmonic sensors have been gradually achieving status in the fields of biomedical analysis (therapeutic applications, drug and gene delivery, and diagnostics) and biotechnological biosensing (biomolecular imaging, various advance industrial manufacturing and product evaluation via Terahertz plasmonics, plasmonic photothermal, agricultural advancement, food processing, etc.) to improve health care and investigating chronic infectious viral diseases (AIDS, ebola, etc.) or current pandemic of corona virus. We discuss several recent research studies and previous five years using the CN-PS applications.

12.2.1 Graphene-based Plasmonic Sensors

Performing graphene on SPR sensors exhibits some advantages. For instance, π–π stacking of the graphene layer lets an increment in the adsorption of organic molecules. Graphene provides an anticorrosive coating for the surfaces. Detrimental effects of plasmonic features are also minimized by coating with graphene. In literature, there are many examples of coating metals with graphene in the field of SPR technology [18, 43, 44]. For instance, a graphene-metal surface was prepared and integrated with a microfluidic device as a potential biosensor [43]. They coated Au and Ag films (each is 50 nm in thickness) with graphene and investigated their surface plasmon ability. Nonspecific physical interaction between graphene layer and proteins was also studied with various concentrations (40–500 nM) of bovine serum albumin. Graphene and carbon nanotubes are the most popular materials used as carbon-based nanomaterials in plasmonic biosensors. Chiu and Haung fabricated graphene oxide sheet (GOS)-based surface plasmon resonance (SPR) sensors for protein immobilization detection [45]. The GOS- bovine serumalbumin(BSA) conjugates were prepared using activated functional groups present on treated GOS surfaces. The carboxyl end groups on the surfaces of the GOSs were activated for covalent bond formation to immobilize hydrocarbon chains using a 4 : 1 mixture of 400 M 1-ethyl-3-(3-dimethylaminopropyl)carbodiimide (EDC)/100 M *N*-hydroxysuccinimide (NHS) for 10 minutes at room temperature. Treatment with EDC/NHS produces a water-stable ester that is able to react with primary amines (Figure 12.1). In another study, Xue et al. binded GO with single-stranded DNA (ssDNA) as biosensor [46]. The hydrogen bonding plays a key role to the interaction between GO and ssDNA. The biosensor, based on the principle of indirect competitive inhibition, exhibits a linear dynamic range of 10^{-14} to 10^{-6} M for the detection of target complementary single strand (csDNA) (HIV-1 U5 long terminal repeat sequence). Zhu et al. prepared reduced GO (rGO) sheets wrapped in Ag nanocubes (AgNCs) sponge surface-enhanced Raman scattering (SERS) substrate for detection of Dithiocarbamate (DTC) pesticides such as, thiram and ferbam. The AgNC–rGO sponge can achieve LODs of 10 and 16 ppb for thiram and ferbam, respectively [47]. Yu et al. improved a glucose biosensor by using both graphene and carbon nanotubes (CNTs) as a hybrid material. They prepared poly(diallyldimethylammonium chloride) (PDDA)-capped AuNPs functionalized graphene (G)/multi-walled CNTs. Glucose oxidase was utilized to monitor the glucose levels. Sensitivity and LOD of the biosensor were found to be 29.72 mA M^{-1} cm^{-2} and 4.8 μM, respectively [48]. Localized surface plasmon resonance (LSPR)-based optic

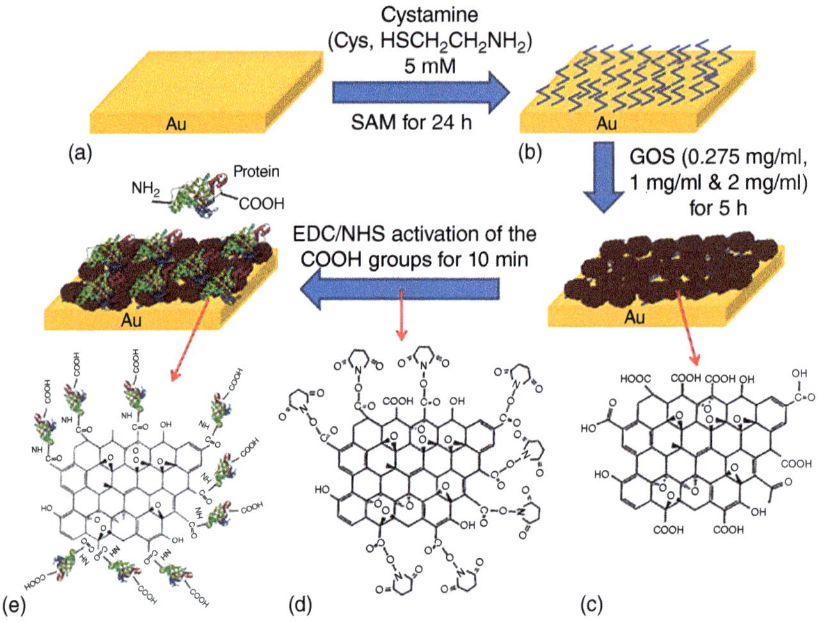

Figure 12.1 Schematic representation of protein immobilization on the modified surface of a GOS. (a) A layer of Cr (2 nm) and one of Au (47 nm) were deposited on a glass substrate.(b) Then, an SAM of the Cys linker was deposited. (c) Next, GOSs were allowed to be adsorbed onto the Au–Cys films using GOS solutions with concentrations of 0.275 $mg\,ml^{-1}$, 1 $mg\,ml^{-1}$, and 2 $mg\,ml^{-1}$. (d) The chips were treated with EDC/NHS to couple the adsorbed GOSs to the surfaces of the sensors. (e) The surfaces of the GOSs were used to sense proteins via binding interactions. Source: Reprinted with permission from Chiu and Huang [45]. © 2014, Elsevier.

sensors exhibit some advantages including label-free detection, real-time monitoring, shorter detection time, ease of production, biocompatibility, portability, and so on [49]. GO nanosheets have been widely used for fabrication of LSPR-based optic sensors because of its high specific surface area, high optical transmittance, high intrinsic mobility, and also satisfactory electrical and mechanical properties [9]. Semwal and Gupta fabricated a SPR-based fiber optic sensor using silver NPs coated with GO nanosheets to detect cholesterol [50]. For this aim, three different probes containing enzyme, i.e. cholesterol oxidase, were fabricated: While the enzyme was entrapped in a hydrogel in probe I, it was immobilized on the GO nanosheets in probe II. The probe III composed of PVA entrapped AgNPs over GO nanosheets and before immobilization of the enzyme. Among the three probes, the last one exhibited the best sensitivity and LOD in which are 5.14 nm/mM and 1.131 mM, respectively. Effect of pH on probes was also examined in the range of 5–8 and the best performance of the sensor was observed at ph 7. Another SPR-based fiber optic sensor was reported for the detection of uric acid in human serum [51]. Two different probes were improved: a micro-ball fiber sensor probe was only coated with AuNPs in probe I while AuNPs and GO in probe II. The authors reported that combination of localized surface plasmon resonance (LSPR) property of AuNPs and surface plasma property of GO increases the sensitivity of sensor. LOD and sensitivity of the probe containing both AuNPs and GO were found to be 65.60 μM and 2.1%/mM, respectively. Kim et al. reported graphene as a replacement for conventional metal films in their SPR sensor [52]. This SPR sensor is composed of an optical fiber as light supporting material and graphene as plasmon supporting material. Performance of the SPR sensor was investigated with Biotinylated Double Crossover DNA (DXB) lattice and protein Streptavidin (SA). The red-shift of 7.276 nm was observed for the combination of DXB and SA. Tabassum et al. developed a gas sensor based on plasmonic crystal combining a thin layer of GO [53]. The plasmonic crystal consists of a periodic array of polymeric nanoposts with gold (Au) disks at the top and nanoholes in an Au thin film at the bottom. The gas adsorbed on GO which alters the refractive index of the plasmonic structure and returns a shift in the resonance wavelength of the surface plasmon polariton excited at the GO coated Au surface. They used principal component analysis (PCA)-based pattern recognition algorithm for optical response of a gaseous mixture. The acclaimed sensor demonstrates a refractive index sensitivity of 449.63 nm/RIU. In addition, volatile organic compounds, such as ethylene, methanol that serve as plant health indicators, and ammonia that plays a key role in the ecosystem, were detected by the sensor. The GO coated sensor exhibits sensitivities of 0.6 pm/ppm to gaseous ethylene, 3.2 pm/ppm to methanol, and 12.84 pm/ppm to ammonia. The study about SPR signal enhancement through graphene was reported by Singh et al. The purpose of using graphene in this study was not only to amplify the signal but also to control immobilization of biotinylated cholera toxin antigen on copper coordinated nitrilotriacetic acid (NTA). LOD of this SPR immunosensor was reported as 4 pg mL^{-1} for the specific antibody anticholera toxin [54].

Dopamine was detected by fiber optic surface plasmon resonance-based sensor. The dopamine sensor was developed by using molecular imprinted graphene nanoaplatelets/tin oxide (SnO2) nanocomposite. The linear range of concentration was obtained from 0 to 100 μM. The limit of detection (LOD) of the sensor was observed 0.031 μM [55].

Zhang et al. developed a spectrometer-free sensing design, which has characterized via optical numerical simulations [56]. The sensor comprised of a GSPR structure with gratings for coupling to incident EM waves and a metallic reflector to boost the performance to have a higher absorption peak. Dual sensor modes are operable for sensing such as optical mode and electrical mode of operation. In the optical mode of operation, a sensitivity and figures of merits 1566.03 nm/RIU and 250.6 RIU^{-1}, respectively were obtained. On the other hand, for the electrical mode of operation, these were 720.75 meV/RIU and 287.2 RIU^{-1}. Plasmonics behaves as suitable contender for integrated circuits, as one of its characteristic is confinement of electromagnetic field at metal-dielectric interface. Whereas in comparison to noble metals such as Au, Ag, etc. not effective SPs at the THz and far-infrared (FIR) regions. Rezaei and Zarifkar presented a graphene-based plasmonic electro-optical multi-logic gate (MLG) operating at THz frequency. The designed MLG supports, logic gates with two different structures based on graphene plasmonic switch, exclusive digital logic gate XNOR, and digital logic gates NOR in a graphene-dielectric-metal structure, simultaneously. The propagation length of the SPs, stimulated by a 27.35 THz incident TM wave, is about 62 times larger in ON state compared to OFF state. Simulation results by the finite difference time domain (FDTD) method show the minimum extinction ratio (ER) of 29.41,

97.38, and 29.40 dB for AND, XNOR, and NOR logic gates, respectively. Also, the minimum modulation depth (MD) is obtained 99.99% for XNOR and 99.88% for AND and NOR logic gates [57].

Nanohybrid systems comprising CNMs possess distinctive characteristics of hybrid nanomaterials. They are beneficial to many fields, such as optical and electronic materials, biomaterials, catalysis, sensing, coating, and energy storage. Nanohybrid materials contain organic and inorganic components that are linked together by noncovalent bonds or covalent bonds at the nanometer scale. Senevirathne et al. developed the tumor permittivity profile reconstruction on an artificially simulated skin tumor by using the proposed nanohybrid assembly [58]. Recently, Omar et al. developed a SPR sensor for dengue virus (DENV) 2 E-proteins. They prepared dithiobis (succinimidyl undecanoate, DSU)/amine-functionalized reduced graphene oxide–polyamidoamine dendrimer (DSU/amine-functionalized rGO–PAMAM) thin film-based SPR. Different concentrations of DENV 2 E-proteins were successfully tested by the developed SPR sensor-based system. The figure of merit and signal-to-noise ratio, values were observed in the range of 0.08–0.5 pM ($S = 0.2576°/pM$, $R_2 = 0.92$), and a high equilibrium association constant (KA) of 7.6452 TM^{-1}. The developed sensor also showed a sensitive and selective response toward DENV 2 E-proteins compared to DENV 1 E-proteins and ZIKV (Zika virus) E-proteins [59].

Xu et al. developed a monolayer graphene-based terahertz planar metamaterial. The material works on plasmon-induced transparency (PIT) event. The destructive interference between the bright and dark modes in this structure demonstrated that the PIT window can be carefully tuned by varying the Fermi energy of the planar graphene array. The sharp transmission dips and absorption peaks with narrow line widths were observed by finite difference time domain simulation. In addition, they theoretically studied its resonance mechanism by coupled mode theory [60].

12.2.2 Carbon Nanotube-based Plasmonic Sensors

Carbon nanotubes (CNTs) have been reported quite often in literature for their applications in biotechnological and biomedical fields because of their distinctive shape and properties [61–64].

In many studies, carbon-based nanomaterials, are used as sensitivity enhancement materials for plasmonic sensors. For example, Lee et al. built a SPR-based immunoassay for the detection of human erythropoietin (EPO) and human granulocyte macrophage colony-stimulating factor [63]. They conjugated the CNT with a polyclonal antibody to amplify the SPR signal. Antibody conjugated CNT enhanced the SPR signals for both EPO and GM–CSF more than 30 times compared with the direct label-free immunoassay. The detection sensitivity for both analytes was found in the range of 0.1–1000 $ng\,ml^{-1}$. Zhang et al. prepared SWCNT dispersive for pooled chemo/thermal therapy platform. They reported 1/5th the SWCNT weight of Evans Blue (EB) was able to disperse SWCNTs with more than 60% recovery efficiency. They acclaimed, the SWCNT complexes were loaded with paclitaxel with efficient therapeutic effect over chemotherapy or photothermal therapy to treat tumors [65].

CNTs are reported for signal enhancement of various spectroscopic techniques such as plasmon-enhanced Raman scattering by carbon nanotubes partially suspended in a dimer cavity. Polarization-dependent and spatially resolved Raman measurements obtained due to dimer cavity enhanced the Raman signal of a metallic tube inside a small bundle by 10^3. The signal has arisen from a 65 nm long nanotube segments, oriented at 45° with respect to the cavity axis [66]. In another report, Huang et al. proposed efficient CNT infrared (IR) detectors based on a barrier-free bipolar diode. The use of an ax-like plasmonic electrode suggests a considerably stronger field near the electrode with efficient collection of carriers, simultaneously enabling an enhancement from 1400 to 2100 nm. The enhancement observed has referred to the position of CNTs corresponding to the ax-like structure. Here, the plasmonic electrode significantly improves the utilization of incident light in the IR detection application [67]. Zhang et al. proposed SERS substrate fabricated by multi-walled carbon nanotubes (MWNTs) arrays with silver nanoparticles (AgNPs) on silicon, via magnetron sputtering and annealing method. The MWNTs-AgNPs hybrids with 3D structure increase the coverage of plasmonic nanostructures. The effect of inter-particle and inter-tube coupling creates a multitude of hot spots, with a strong SERS effect. They used finite-difference time-domain (FDTD) method to simulate E-field distribution of the hybrid structures with multilayer AgNPs decorated on MWNTs. The Raman scattering enhancement factor (EF) $10^7$was obtained [68]. Onac et al. detected melamine (MEL) from milk by using a carbon nanotube attached with polymeric membrane (CNT/PIM). The linear range was observed from 0.1–2.0 nM for MEL [69]. Liu et al. proposed graphene-encapsulated gold nanoparticles (GNPs) functionalized with covalently attached SWCNTs onto surface (GNP-CNT) as substrate for SERS sensing. The simulated outcome confirmed by the heterostructures as an effective SERS substrate [70].

Zhang proposed the plasmonic gold nanorods encircled by carbon nitride nanotubes (Au NRs/ CNNTs) photocatalysts were aimed and prepared by impregnation–annealing approach. The LSPR peaks of Au NRs can be adjusted simply by changing the aspect ratios, the long-wavelength visible even near-infrared light can be utilized, which extends the visible absorption range of CNNTs. The photocatalytic H_2 progress rate was 207.0 $\mu mol\,h^{-1}$ obtained [71].

Kamil et al. developed a nanocomposite of $Fe_2H_2O_4$-MWCNT for SPR analysis of As(III) and As(V). The As(III) and As(V) linear range of 0.2 and 1 0.0 ppb yielded sensitivity values of 1.756 °ppb^{-1} and 0.575 °ppb^{-1}, respectively. For both As ions, the sensor detection limit was 0.2 ppb. The AFM images shown in Figure 12.2 the interaction between As(III) and Au/$Fe_2H_2O_4$-MWCNT sensing layer changed surface morphology. The thickness and roughness increased from 8.2 to 23 nm and 1.323 to 8.16 nm, respectively [59].

12.3 Final Statement and Further Outlook

Plasmonic sensing has developed as the sensitive label-free technique to detect various molecular species in solutions and has already proved crucial in drug discovery, food safety, and studies of bioreactions. Extensive research work has been observed on CNMs as one of the most widely used classes of nanomaterials. Their distinctive characteristics such as mechanical, optical, electrochemical, and electrical facilitate the used in multidisciplinary fields. Furthermore, due to their adaptable surface properties, size, and shape over the past decade, CNMs have attracted biomedical and biotechnological researchers. Remarkably,

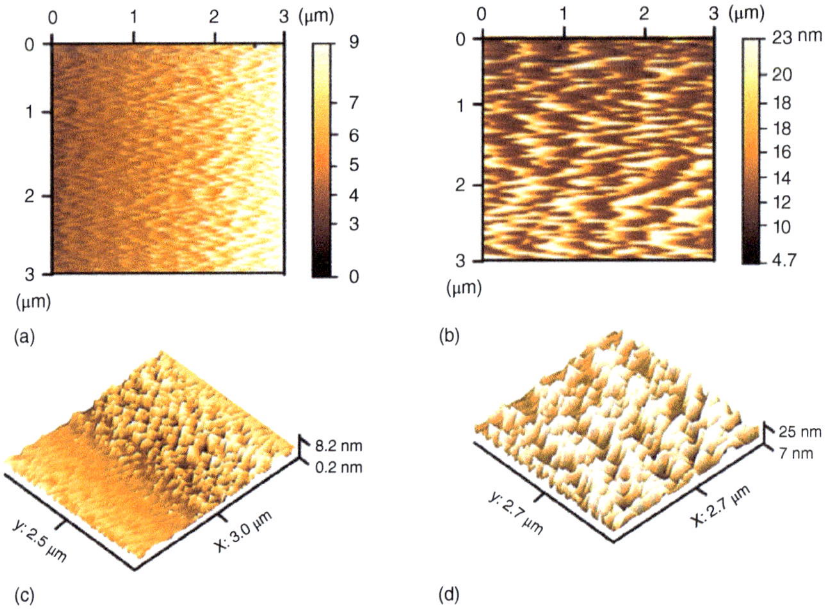

Figure 12.2 2D and 3D AFM images the sensing layer surface before (a) and (c) and after (b) and (d) adsorption of the As(III) ion. Source: Reprinted with permission from kamil et al. [59]. © IEEE.

CNMs are becoming promising materials due to the existence of both inorganic semiconducting properties and organic π–π stacking characteristics. Hence, it could effectively interact with biomolecules and response to the light simultaneously.

CN-PSs have been developed as the sensitive label-free technique to detect various molecular species in solutions and have already proved crucial in drug discovery, food safety, and studies of bioreactions. This method operates on the surface plasmon resonances (~50 nm) of metallic films and the selectivity of functionalities on the surface to acquire targeted analysis. The increased number of research reports in surface plasmons (SPs) sensor related to sensitivity improvement and innovative target materials for specificity. Nanotechnological progress has improved the SPR sensor research vastly by using nanomaterials in the design of SPs based sensors, due to their manifold properties. Carbon-based nanomaterials, like graphene and its derivatives GO, reduced graphene oxide (rGO), carbon nanotubes (CNTs), their hybrids, and nanocomposites have transformed the field of sensing due to their unique characteristics, such as large surface area, easy synthesis, tunable optical properties, and strong compatible adsorption of biomolecules. Make the most of such materials, carbon nanomaterials plasmonic sensors (CN-PS) possess tendency of biomedical and biotechnological applications. Concerning their toxic effect in the biological system, several chemical modification strategies have been developed and successfully used in biomolecular applications including drug delivery, tissue engineering, detection of biomolecules, advanced biomolecule immobilized sensing, and cancer therapy.

The current ongoing pandemic (COVID-19) has drawn attention toward inexpensive, rapid, and authentic sensing for virulence of virus strains [72, 73]. The plasmonic sensing is quite active strategy for clinical diagnosis of various cancerous diseases (Carcinoma,

Lukemia, etc.) and viral or infectious diseases (Zika, Ebola, etc.). The SPR and LSPR as label-free detection systems have often reported for biomolecules.

Typical plasmonic biosensing strategies rely on the versatility of SPR and LSPR as label-free sensing binding interactions in real time. Even so, the innovation in analytical instrumentation has improved the optical sensing fields in sensitivity, selectivity, and alignments. In addition, the SERS coupled with fluorescence, or luminescence enhancement for variety of analysis. The CN-PS research is upgrading day by day, these surface plasmonic sensing facilitating the rapid and sensitive analysis for biomedical and biotechnological fields.

Graphene possesses better biocompatibility and higher surface area, which has facilitate better functioning as PS than CNTs. Either for biomedical applications or biotechnological applications, graphene, its derivatives, composites, or hybrids are versatile in their performances. Graphene compounds are biocompatible with different cell types (such as mammalian and bacterial cells) both *in vitro* and *in vivo* and produce antibacterial effects.

In spite with all the improvements discussed above, there is still plenty of room for approaching conventional laboratory-based sequencing and immunoanalytical techniques. Furthermore, the automation and integration of microfluidics besides the prevention of nonspecific molecules linkage in complex media still remain a challenge in many cases of CN-PS.

References

1 Arsalani, S., Ghodselahi, T., Neishaboorynejad, T., and Baffa, O. (2019). *Plasmonics* 14: 1419–1426.

2 Sari, E., Üzek, R., Duman, M., and Denizli, A. (2016). *Talanta* 150: 607–614.

3 Hu, X., Tian, J., Xue, Y. et al. (2017). *ChemCatChem* 9: 1511–1516.

4 Qu, L.L., Liu, Y.-Y., Liu, M.-K. et al. (2016). *ACS Applied Materials & Interfaces* 8: 28180–28186.

5 Yang, A., Hoang, T.B., Dridi, M. et al. (2015). *Nature Communications* 6: 6939.

6 Vecchi, G., Giannini, V., and Rivas, J.G. (2009). *Physical Review B* 80: 201401.

7 Kim, J., Liu, G., Lu, Y., and Lee, L. (2006). *IEE Proceedings-Nanobiotechnology* 153: 42–46, IET Digital Library.

8 Le Ru, E., Etchegoin, P., Grand, J. et al. (2008). *Current Applied Physics* 8: 467–470.

9 Bursill, L., Stadelmann, P.A., Peng, J., and Prawer, S. (1994). *Physical Review B* 49: 2882.

10 Eberlein, T., Bangert, U., Nair, R.R. et al. (2008). *Physical Review B* 77: 233406.

11 Khanna, V.K. (2008). *Sensor Review* 28: 39–45.

12 Qi, H. and Hegmann, T. (2008). *Journal of Materials Chemistry* 18: 3288–3294.

13 Zeng, S., Baillargeat, D., Ho, H.-P., and Yong, K.-T. (2014). *Chemical Society Reviews* 43: 3426–3452.

14 Hartmann, R.R., Kono, J., and Portnoi, M.E. (2014). *Nanotechnology* 25: 322001.

15 Zeng, S., Hu, S., Xia, J. et al. (2015). *Sensors and Actuators B: Chemical* 207: 801–810.

16 Venkatesh, S., Badiya, P.K., and Ramamurthy, S.S. (2016). *Physical Chemistry Chemical Physics* 18: 681–684.

17 Peng, L.-M., Zhang, Z., and Wang, S. (2014). *Materials Today* 17: 433–442.

18 Novoselov, K.S., Geim, A.K., Morozov, S.V. et al. (2005). *Nature* 438: 197–200.

19 Novoselov, K.S., Geim, A.K., Morozov, S.V. et al. (2004). *Science* 306: 666–669.
20 Laurentius, L.B., Owens, N.A., Park, J. et al. (2016). *Expert Review of Molecular Diagnostics* 16: 883–895.
21 Bezzon, V.D.N., Montanheiro, T.L.A., de Menezes, B.R.C. et al. (2019). *Advances in Materials Science and Engineering* 2019: 4293073.
22 Wang, Y., Li, Z., Wang, J. et al. (2011). *Trends in Biotechnology* 29: 205–212.
23 Singh, V., Joung, D., Zhai, L. et al. (2011). *Progress in Materials Science* 56: 1178–1271.
24 Marcano, D.C., Kosynkin, D.V., Berlin, J.M. et al. (2010). *ACS Nano* 4: 4806–4814.
25 Mishra, S.K., Tripathi, S.N., Choudhary, V., and Gupta, B.D. (2015). *Plasmonics* 10: 1147–1157.
26 Lai, X.-F., Zou, Y.-X., Wang, S.-S. et al. (2016). *Analytical Chemistry* 88: 5385–5391.
27 Wang, J.T., Fabbro, C., Venturelli, E. et al. (2014). *Biomaterials* 35: 9517–9528.
28 Novoselov, K.S., Geim, A.K., Morozov, S.V. et al. (2004). *Science* 306: 666.
29 Gu, X., Lin, I.-T., and Liu, J.-M. (2013). *Applied Physics Letters* 103: 071103.
30 Gan, C.H., Chu, H.S., and Li, E.P. (2012). *Physical Review B* 85: 125431.
31 Low, T. and Avouris, P. (2014). *ACS Nano* 8: 1086–1101.
32 Farhat, M., Guenneau, S., and Bağcı, H. (2013). *Physical Review Letters* 111: 237404.
33 Constant, T.J., Hornett, S.M., Chang, D.E., and Hendry, E. (2016). *Nature Physics* 12: 124–127.
34 Feng, L., Wu, L., and Qu, X. (2013). *Advanced Materials* 25: 168–186.
35 Iijima, S. (1991). *Nature* 354: 56–58.
36 Smart, S., Cassady, A., Lu, G., and Martin, D. (2006). *Carbon* 44: 1034–1047.
37 Tasis, D., Tagmatarchis, N., Bianco, A., and Prato, M. (2006). *Chemical Reviews* 106: 1105–1136.
38 Abouelsayed, A., Anis, B., Hassaballa, S. et al. (2017). *Materials Chemistry and Physics* 189: 127–135.
39 Ji, Y., Huang, Y.Y., Rungsawang, R., and Terentjev, E.M. (2010). *Advanced Materials* 22: 3436–3440.
40 Hümmer, T., Noe, J., Hofmann, M.S. et al. (2016). *Nature Communications* 7: 1–7.
41 Descombin, A., Poncharal, P., Pascale-Hamri, A. et al. (2019). *Nano Letters* 19: 1534–1538.
42 Tao, H., Yang, K., Ma, Z. et al. (2012). *Small* 8: 281–290.
43 Song, B., Li, D., Qi, W. et al. (2010). *ChemPhysChem* 11: 585–589.
44 Szunerits, S., Maalouli, N., Wijaya, E. et al. (2013). *Analytical and Bioanalytical Chemistry* 405: 1435–1443.
45 Chiu, N.-F. and Huang, T.-Y. (2014). *Sensors and Actuators B: Chemical* 197: 35–42.
46 Xue, T., Cui, X., Guan, W. et al. (2014). *Biosensors and Bioelectronics* 58: 374–379.
47 Zhu, C., Wang, X., Shi, X. et al. (2017). *ACS Applied Materials & Interfaces* 9: 39618–39625.
48 Yu, Y., Chen, Z., He, S. et al. (2014). *Biosensors and Bioelectronics* 52: 147–152.
49 Agrawal, N., Saha, C., Kumar, C. et al. (2020). *IEEE Sensors Journal.*
50 Semwal, V. and Gupta, B.D. (2017). *IEEE Sensors Journal* 18: 1039–1046.
51 Kumar, S., Singh, R., Zhu, G. et al. (2019). *IEEE Transactions on Nanobioscience* 19: 173–182.
52 Kim, J.A., Hwang, T., Dugasani, S.R. et al. (2013). *Sensors and Actuators B: Chemical* 187: 426–433.
53 Tabassum, S., Kumar, R., and Dong, L. (2017). *IEEE Sensors Journal* 17: 6210–6223.

54 Singh, M., Holzinger, M., Tabrizian, M. et al. (2015). *Journal of the American Chemical Society* 137: 2800–2803.

55 Sharma, S. and Gupta, B.D. (2018). *Journal of Lightwave Technology* 36: 5956–5962.

56 Zhang, L., Farhat, M., and Salama, K.N. (2020). *Spectrometer-Free Graphene Plasmonics Based Refractive Index Sensor* 20: 2347.

57 Rezaei, M.H. and Zarifkar, A. (2018). *Applied Optics* 57: 10109–10116.

58 Senevirathne, V., Hapuarachchi, H., Mallawaarachchi, S. et al. (2019). *Journal of Physics: Condensed Matter* 31: 085302.

59 Kamil, Y.M., Al-Rekabi, S.H., Mohamed, H.A. et al. (2019). *IEEE Photonics Journal* 11: 1–9.

60 Xu, H., Xiong, C., Chen, Z. et al. (2018). *Journal of the Optical Society of America B: Optical Physics* 35: 1463–1468.

61 Hartmann, N., Piredda, G., Berthelot, J. et al. (2012). *Nano Letters* 12: 177–181.

62 Kadochkin, A., Moiseev, S., Dadoenkova, Y. et al. (2017). *Optics Express* 25: 27165–27171.

63 Lee, E.G., Park, K.M., Jeong, J.Y. et al. (2011). *Analytical Biochemistry* 408: 206–211.

64 Kruss, S., Hilmer, A.J., Zhang, J. et al. (2013). *Advanced Drug Delivery Reviews* 65: 1933–1950.

65 Zhang, L., Rong, P., Chen, M. et al. (2015). *Nanoscale* 7: 16204–16213.

66 Heeg, S., Clark, N., Oikonomou, A. et al. (2014). *Physica Status Solidi RRL: Rapid Research Letters* 08: 785–789.

67 Huang, H., Zhang, D., Wei, N. et al. (2017). *Advanced Optical Materials* 5: 1600865.

68 Zhang, J., Zhang, X., Chen, S. et al. (2016). *Carbon* 100: 395–407.

69 Onac, C., Alpoguz, H.K., Yola, M.L., and Kaya, A. (2018). *Innovative Food Science & Emerging Technologies* 45: 467–470.

70 Liu, H., Li, Y., Dykes, J. et al. (2018). *Carbon* 140: 306–313.

71 Zhang, L., Ding, N., Lou, L. et al. (2019). *Advanced Functional Materials* 29: 1806774.

72 Moulahoum, H., Ghorbanizamani, F., Zihnioglu, F. et al. (2020). *Talanta*: 121534.

73 Jeong, H., Rogers, J.A., and Xu, S. (2020). *Science Advances* 6: eabd4794.

13

Surface Plasmon Resonance Sensors Based on Molecularly Imprinted Polymers

Cem Esen[1] *and Sergey A. Piletsky*[2]

[1]*Department of Chemistry, Faculty of Arts and Sciences, Aydın Adnan Menderes University, Aydın, Turkey*
[2]*School of Chemistry, College of Science and Engineering, University of Leicester, Leicester, UK*

13.1 Introduction

It is well established that molecularly imprinted polymers (MIPs) can serve many applications ranging from separation, chromatography, drug delivery, bioimaging, catalysis, and sensing. Among them, sensing applications of MIPs where MIPs act as recognition elements capable of selective binding target analytes have attracted increasing interest. The useful examples of such devices are MIP based surface plasmon resonance sensors (MIP-SPR sensors). Surface plasmon resonance (SPR) has been used for monitoring processes at metal interfaces [1] and characterization of thin films [2] since late 70s. Nylander and Liedberg were first to use SPR for biosensing and gas detection in the last decades of twentieth century [3–5]. Since then, SPR sensors have been used for measuring physical, chemical, and biological analytes [6]. SPR sensors offer the advantages of label-free and real-time monitoring, low cost, and easy sample preparation [7–10].

13.1.1 Surface Plasmon Resonance

SPR was first observed in 1968 using a prism in contact with a pair of quartz slides which had a 100 nm thick silver film deposited between them as the plasmonic material [11]. Since this initial observation and the development of the associated theory [11–12], SPR sensing has been used in many fields due to its relatively non-invasive nature and localized probing of the sample. In short, SPR sensing is a technique that utilizes the evanescent field of a surface plasmon propagated across a metallic surface to detect changes in the dielectric constant of the sample that is within of 100 nm of the plasmonic material. As this evanescent field interacts with the analyte, the resulting change in light intensity, or shift in transmitted excitation light wavelength or angle, is directly related to the changes in the refractive index of the sample [13–15]. This change in transmitted light properties (i.e. intensity, wavelength, or angle) allows for the non-invasive detection of chemical or biochemical species of interest. In addition, by modifying the plasmonic surface with receptors, specific

Plasmonic Sensors and their Applications, First Edition. Edited by Adil Denizli.

and sensitive SPR sensors can be fabricated. In this context, MIPs and other artificial receptors have been combined with SPR in order to prepare sensors. This chapter describes the latest advances in SPR sensors integrated with MIPs.

13.1.2 Molecularly Imprinted Polymers

MIPs are tailor-made recognition materials which have been prepared by a template-assisted synthesis with high affinity and selectivity for their targets [16–22]. The synthesis, illustrated in Figure 13.1, is based on the polymerization of the functional monomers and the crosslinker in the presence of a target molecule serving as the template. Driven by thermodynamics, the template interacts with the functional monomers to form a pre-polymerization complex stabilized by molecular interactions, and that is later "frozen" by polymerization. Upon polymerization, the template is removed from the three-dimensional polymeric matrix by simple extraction thereby creating specific binding sites capable of recognizing the template and its analogs [21].

As compared to antibodies, MIP synthesis is relatively simple, cost-effective, and does not require use of animals. In addition, MIPs are physically and chemically stable and are not degraded by proteases and nucleases [22]. Owing to remarkable selectivity and affinity toward their targets, MIPs have been extensively investigated for different applications, such as separation/chromatography [23–26], sensing [27–32], drug delivery [33–37], binding assay [38–43], bioimaging/theranostics [44–55], and catalysis [56–60]. In these applications, various types of molecules were used as template in order to capture and recognize the desired analyte.

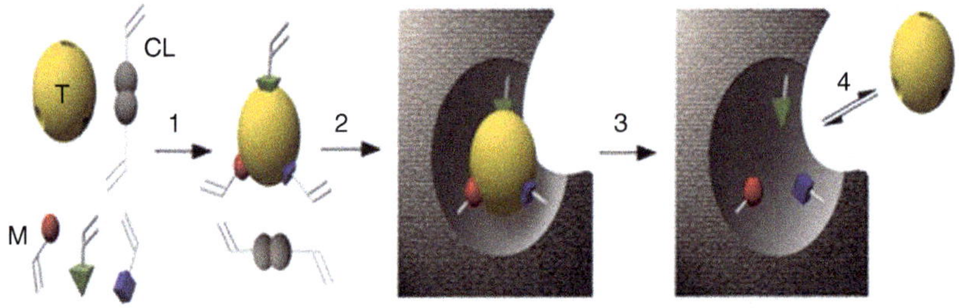

Figure 13.1 The concept of molecular imprinting process. T: template molecule, M: functional monomers, CL: crosslinker, 1: pre-polymerization self-assembly step, 2: polymerization step, 3: removal of the template, 4: rebinding of the target molecule. Source: Reprinted with permission from reference [22]. Copyright (2020) American Chemical Society.

13.2 MIP Based SPR Sensors

In this type of optical MIP sensing considered as MIP-affinity sensor, the changes in refractive index caused by the binding of templates to MIPs immobilized on the surface of an SPR are monitored [61]. With this regard, first MIP-based SPR sensor was developed by Lai et al. for sorbent assay of theophylline, caffeine, and xanthine [62]. Since then, the use of MIPs as recognition elements for realizing SPR sensors with improved sensitivity and selectivity has become an attractive strategy for monitoring various type of analytes such as metal ions [63–65], biomarkers [66–69], pharmaceuticals [70–72], bacteria [73, 74], pesticides [75] and biomolecules [76]. MIPs can be used either in film or nanostructured format as recognition units of SPR sensors. Therefore, SPR sensors based on these MIP formats are exemplified and discussed in the following two sections.

13.2.1 MIP Film Based SPR Sensors

In this type, it is very important to obtain a surface imprinting and a control of the thickness of the polymer layer to increase the reproducibility of the measurements [77]. One of the simplest approaches to prepare MIP films of controlled thickness consists of depositing the pre-polymerization mixture on a substrate and spreading it homogeneously in a step prior to its polymerization. In this context, the spin-coating technique allows a nanosized control of the thickness of the obtained layers [77, 78].

An SPR sensor combined with MIP nanofilm as recognition element was prepared for selective detection of antibiotic ciprofloxacin [79]. In this study, MIP film was prepared on SPR sensor chip by in situ photo-initiated polymerization as monolayer. MIP-SPR sensor exhibited high selectivity, sensitivity, and good stability for ciprofloxacin compared to ofloxacin measured over the range 10^{-11}–$10^{-7}\,mol\,l^{-1}$.

In another study, kanamycin, a widely used aminoglycoside antibiotic to treat some Gram-negative and Gram-positive infectious diseases, was employed as the template for the preparation of MIP film on the surface of the SPR chip using 4-vinylbenzeneboronic acid, polyethylene glycol acrylate, and 2,2′-azo isobutyl amidine hydrochloride respectively, as functional monomer, crosslinker, and initiator [80]. MIP film-based SPR sensor demonstrated a linear response to kanamycin concentration in the range of 1.0×10^{-7} to $1.0\times10^{-5}\,mol\,l^{-1}$ with LOD values $4.33\times10^{-8}\,mol\,l^{-1}$ and $1.20\times10^{-8}\,mol\,l^{-1}$ respectively, for milk powder and honey.

The thickness of the polymer layer can also be controlled by micro-contact imprinting method. The method involves confining the pre-polymerization mixture between two flat materials and applying continuous pressure during the polymerization process. One of these flat materials usually acts as a substrate or support for MIP while the other can be functionalized with the template molecule [81]. This method is particularly useful for the imprinting of high molecular mass species, such as biomolecules [82] or microorganisms [73]. In this case, the polymer is immobilized on the gold surface using monomers or molecules that allow the direct bonding of the MIP during polymerization process.

For example, MIP-based SPR sensor was developed for the detection and quantification of a bacterial factor (RoxP) secreted from skin using this approach [82]. Five different MIPs were prepared and evaluated for sensitivity, selectivity, affinity, and kinetic measurements. 2-Hydroxyethyl methacrylate and poly(ethylene glycol) dimethacrylate monomer-crosslinker composition was found to produce best MIP sensors with LOD 0.23 nM. As seen from the sensorgrams in Figure 13.2, this study offers a very efficient tool for the detection and quantification of RoxP as an early indicator for some oxidative skin diseases.

The control of the thickness and composition of MIP films in SPR sensors could be achieved efficiently using an iniferter, i.e. surface initiator [83].

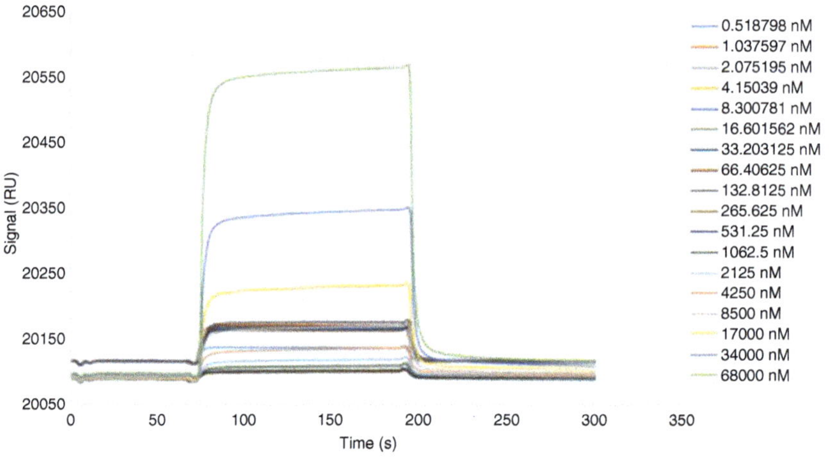

Figure 13.2 Real-time MIP sensorgrams for RoxP binding versus time. Source: Reprinted with permission from reference [82]. Copyright (2019) American Chemical Society.

It is possible to prepare and integrate uniform thin films on SPR sensors by means of hybrid materials consisting of organic and inorganic constituents. Integration of sol–gel-based hybrid MIP films with sensor platforms enhances the analytical sensitivity and selectivity due to rapid mass transport, easy access of analyte molecules to the binding sites, and faster recognition [84].

In this sense, Ayankojo et al. reported the successful synthesis and integration of a selective hybrid organic–inorganic MIP film (AMO-MIP) for amoxicillin (AMO) with an SPR sensor by sol–gel and spin-coating techniques [85]. The hybrid thin film was prepared using methacrylamide as organic functional monomer, tetraethoxysilane as inorganic precursor, and vinyltrimethoxysilane as coupling agent by sol–gel on the SPR gold sensor surface. The AMO-MIP film showed about 16 times higher binding capacity to AMO than non-imprinted polymer (NIP) with SPR characterization. AMO-MIP-modified SPR sensor accomplished the detection of AMO down to 73 pM distinguishing AMO among its analogs in aqueous media with high stability.

Nanostructured metals generally gold (Au) and silver (Ag) promote the SPR phenomenon due to enhancement of the surface-area-to-volume ratio increasing the sensitivity compared with SPR studies in films [61]. In this sense, Yao et al. developed a ractopamine (RAC) SPR sensor consisting of a sensing nano-hybrid film that is made up of MIPs coated with gold nanoparticles (AuNPs) and reduced graphene oxide (GO) [86]. This sensor demonstrated class-specific selectivity toward RAC and its analogs and had a wide linear range over RAC concentration from 20.0 to 1000.0 $ng\,ml^{-1}$ with a 5.0 $ng\,ml^{-1}$ LOD.

Another example to this type of study was carried out by Brule et al. reporting a portable SPR sensor for 1,3,5-trinitroperhydro-1,3,5-triazine (RDX) quantification in environmental water samples based on the RDX selective MIP [87]. In this study, SPR chips were functionalized with RDX selective MIP which was crosslinked with AuNPs and templated with Kemp's acid during electropolymerization. The developed SPR sensor was first validated under optimized conditions with uncontaminated natural water samples spiked with known concentrations of RDX near the EPA limit of 2 ppb and finally used on the field obtaining good agreement in accuracy test with a standard HPLC method. In this study, to overcome the limitation with mass transfer and difficulty in controlling film thickness, a thin electropolymerized polymer film was produced to enhance the reproducibility, simplicity, and rapidity of SPR sensor.

Silver nanoparticles (AgNPs) produce a much stronger and sharper plasmon resonance in comparison with AuNPs [88]. The structural properties such as poor chemical belonging to AgNPs prevent their broad use in SPR sensing. Because the plasmonic properties of AgNPs are subject to changes when exposed to water [89]. High susceptibility to oxidation and consequent deterioration of the silver layer coating are other obstacles about silver-based SPR [90]. To overcome this problem, the most common solution is the addition of protective layer over silver. In this context, recently an effective SPR nanosensor based on core-shell nanoparticles (Ag@AuNPs) incorporated hexagonal boron nitride (HBN) nanosheets and MIP was presented for etoposide (ETO) detection by Özkan et al. [91]. ETO imprinted SPR nanosensor based on Ag@AuNPs-HBN nanocomposite was constructed using poly(2-hydroxyethyl methacrylate-methacryloylamidoglutamic acid) and showed linearity in the range 1.70×10^{-12} M–1.70×10^{-9} M with 4.25×10^{-13} M LOD. Furthermore, the sensor was successfully applied to urine samples with high selectivity and sensitivity.

The first example of epitope-imprinted sensor for cardiac biomarker troponin T (TnT), which is released from the cardiac tissue in the blood stream upon the damage of myocytes, was developed by Scarano's group [92]. The SPR chip imprinted with four peptides (MIP-1234) was evaluated for TnT and human serum albumin (HSA) binding efficiency. In spite of the reduced size and complexity of the epitopes in comparison to whole protein, MIP-1234 sensor retained the linear response to the troponin concentrations observed for MIP-TnT, with even larger sensitivity of 3.15 RU nM^{-1}, lower LOD as 14.8 nM, and less non-specific interaction with HSA.

Recently, Scarano's group developed a polynorepinephrine (PNE) – based imprinted sensor for the early detection of Troponin I (TnI), a crucial biomarker for heart failure, by coupling the MIP with SPR detection [93]. Actually, this is the first example on the use of norepinephrine as functional monomer for imprinted optical biosensors. The imprinting of two short TnI peptides on PNE was performed on gold SPR sensor chips and the sensing efficiency and selectivity of the optical sensor was investigated in buffer and in human plasma by SPR. The developed sensor demonstrated high selectivity and kinetic binding constant for the target protein in very low nanomolar range.

13.2.2 Molecularly Imprinted Polymer Nanoparticles Based SPR Sensors

Until so far, mostly MIPs have been integrated SPR sensing via film format either by generation of films or the use of planar gold chips. However, recently there have been attempts on using molecularly imprinted polymer nanoparticles (nanoMIPs) in SPR sensors through the immobilization of nanoMIPs on the gold chips [94–100]. NanoMIPs are mainly prepared using solid-phase synthesis approach and mini-emulsion polymerization method.

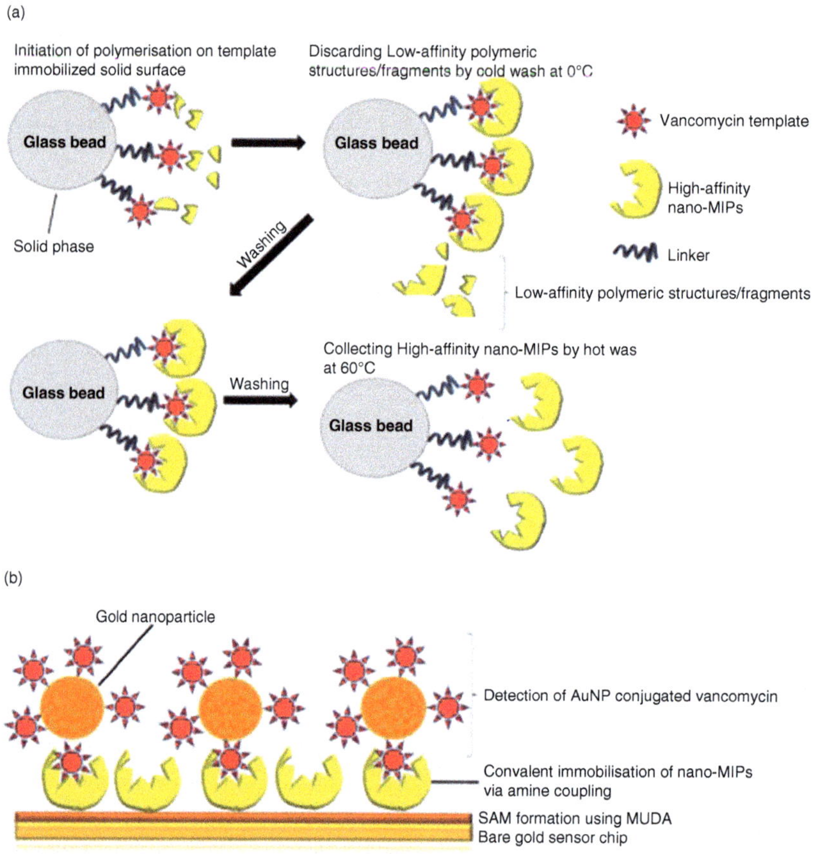

Figure 13.3 (a) Scheme for solid-phase preparation of vancomycin nanoMIPs (b) Schematic illustration of sensor development using nanoMIPs for vancomycin detection. Source: Reprinted with permission from reference [96]. Copyright (2018) Springer Nature.

As schematically illustrated in Figure 13.3 nanoMIPs were rationally designed using solid-phase synthesis to produce synthetic receptors to be coupled with an SPR sensor for the analysis of vancomycin in milk samples by Altintas [96].

In this study, the nanoMIP-SPR sensor enabled vancomycin quantification with the LODs of 4.1 and 17.7 ng ml^{-1} using direct and competitive assays, respectively. The affinity between the nanoMIPs and vancomycin ($K_D = 1.8$ nM) was very high. Moreover, as seen in Figure 13.4 significant difference was obtained for the binding of vancomycin on NIP and nanoMIP surfaces in the concentration range of 125–1000 ng ml^{-1} which proved the high selectivity of nanoMIPs toward vancomycin.

Using the same approach, Ashley et al. reported the synthesis of nanoMIPs for α-casein detection using SPR as a milk allergen sensor [97]. For this, nanoMIPs with high affinity toward bovine α-casein were synthesized and incorporated into label-free SPR sensor. Owing to the high binding affinity and selectivity of the nanoMIPs toward α-casein ($K_D \sim 10$ nM), the resultant SPR sensor enabled the quantitative detection of α-casein with an LOD of 127 ± 97.6 ng ml^{-1} which is superior to existing commercially available ELISA kits.

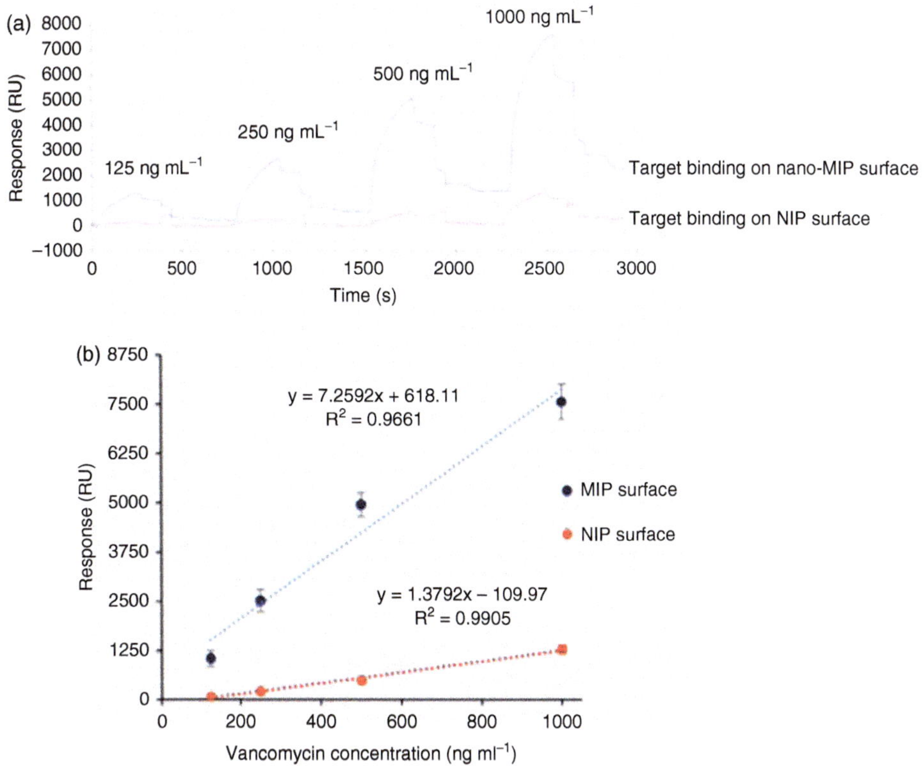

Figure 13.4 (a) Vancomycin binding on nanoMIP and NIP immobilized surfaces in real-time SPR sensorgram (b) NanoMIP and NIP assays for comparison. Source: Reprinted with permission from reference [96]. Copyright (2018) Springer Nature.

In another study, Erdem et al. developed a label-free molecular fingerprinting strategy to detect *Enterococcus faecalis* (*E. faecalis*) in seawater samples (Scheme 13.1) [98]. In this work sensors were developed with a low LOD (down to ~100 bacteria ml^{-1}) and a high correlation coefficient value (>0.99) in the range of 2×10^4–1×10^8 cfu ml^{-1} (Figure 13.5).

The key advantages like selectivity for *E. faecalis* against *Escherichia coli*, *Bacillus subtilis*, and *Staphylococcus aureus* samples, reusability, and stability of the developed sensor platform reveal the potential applications from microbiome to forensic analysis.

Recently Denizli's group described a novel rapid biomimetic SPR sensor for selective and sensitive detection of histamine in food samples [99]. For this sensor platform, two-phase mini-emulsion polymerization method that provides nanoMIPs having surface-exposed binding sites was used to prepare histamine nanoMIPs and subsequently nanoMIP coated SPR chip was employed for real-time sensing of histamine. SPR sensor exhibited high sensitivity to histamine in the range of 0.001 to 10 μg ml^{-1} with 0.58 ng ml^{-1} LOD by the nanoMIP chip.

Recently, an interesting research on investigation of whether MIP nanogels that undergo analyte-induced deformation can be explored to attain improved analytical performances

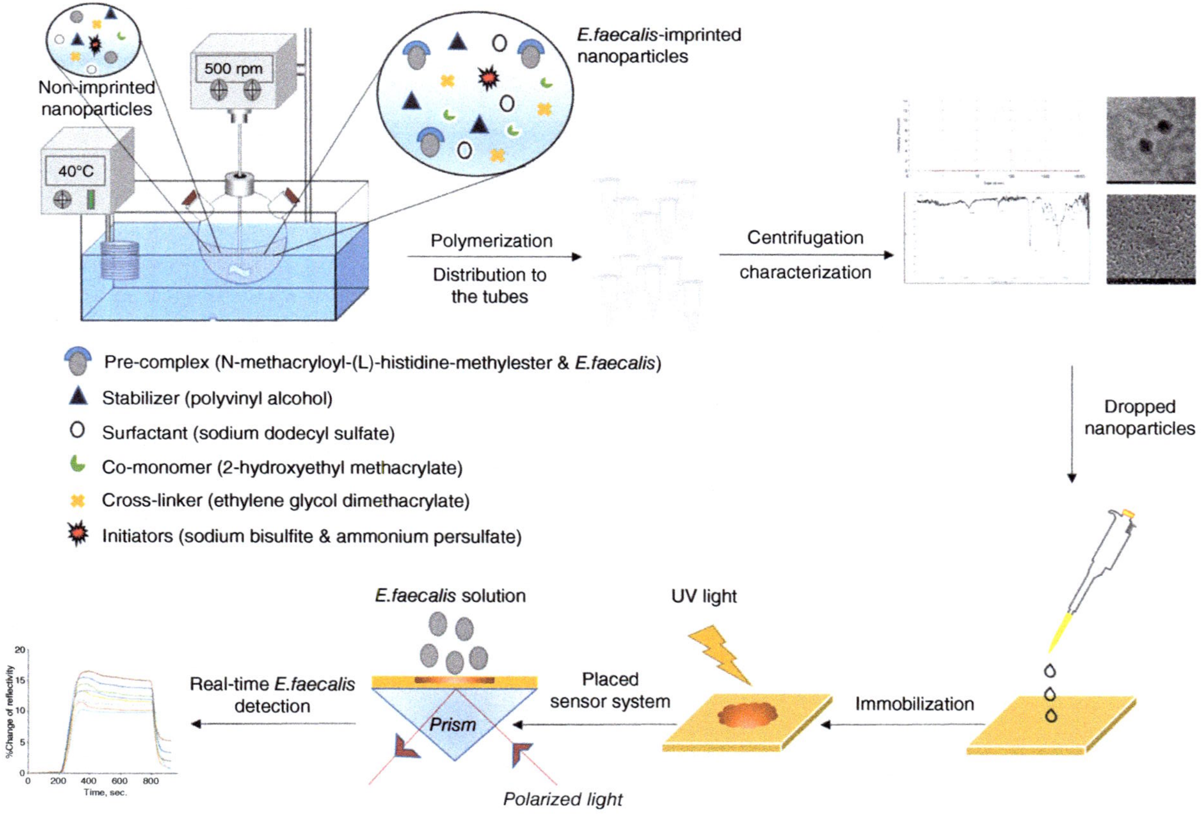

Scheme 13.1 The scheme for the preparation of *E. faecalis*-imprinted plasmonic sensor. Source: Reprinted with permission from reference [98]. Copyright (2019) Elsevier.

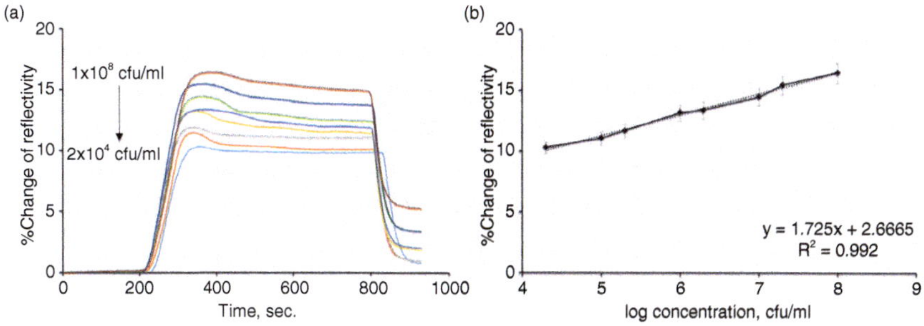

Figure 13.5 (a) The real-time *E. faecalis* detection (b) The relationship between the change of reflectivity and logarithm of *E. faecalis* concentration. Source: Reprinted with permission from reference [98]. Copyright (2019) Elsevier.

in plasmonic sensing was carried out by Cennamo et al. [100]. In this work, nanoMIPs for the recognition of the model protein human serum transferrin (HTR) were prepared by precipitation polymerization using HTR as the template. A plastic optical fiber (POF) was used to produce the D-shaped POF-SPR sensing platform and the selectivity of the sensor was obtained by covalent coupling of nanoMIPs to the POF. The recognition events at the POF-SPR surface permitted to observe whether the nanoMIP network rearrangements contributed to the sensor responses. The nanoMIP-POF sensor showed a linear range of 1.2 fM–1.8 pM response for HTR with the LOD of 1.2 fM corresponding to ~0.7 attomol of HTR on the platform. According to their results, Cennamo et al. anticipate that the binding of HTR to the nanoMIPs triggered conformational modifications in the nanogel structure yielding significant optical shifts. Consequently, soft nanoMIPs act as dynamic amplifiers of the plasmonic spectral responses and stand as a novel interesting class of sensitivity-gain nanomaterials for the design of optical sensing devices.

13.3 Conclusions and Future Prospects

The use of MIPs for sensor purposes has increased exponentially during the last two decades, being nowadays the main application field where molecular imprinting technology focuses more its attention, compared to separation/chromatography. Inside the optical sensing field, the recent integration of SPR and molecular imprinting via the combination of SPR sensors with MIPs is certainly of prime interest. Although SPR sensing is very sensitive, the applications of MIPs in SPR systems are limited due to difficulty in determination very small refractive index change of template molecules with low molecular mass in molecular imprinting process that results decrease in analytical performance. Despite these disadvantages, encouraging studies have been reported showing the potential application of MIPs in SPR sensors as recognition elements in the last decade.

Herein, an overview of MIPs with a special focus on recent advances in MIP-based SPR sensors is presented. As mentioned previously, MIPs are alternatives of the natural receptors with outstanding affinity, selectivity, and stability in SPR sensor platforms. However, technological developments which make their way into society such as smartphones and 3D printing have intensified the development of sensing platforms. MIP-based SPR sensors could also profit from this trend. With the help of this trend, efforts must be devoted to the development and application of MIP-based SPR sensing platform paving the way to on-field or point-of-care performance. The main bottleneck of hindering MIP-based SPR sensing from being commercial can be tackled by the integration of MIPs in SPR sensor platform suitable for commercial production. For this goal, nanoMIPs could be the convenient MIP format to be prepared in large batches which are uniform and monodisperse with high affinity toward their target that can not only amplify the SPR signal but also diminish the mass transfer limitations. The increasing utilization of MIPs in SPR sensors warrants undoubtedly further updates in the forthcoming future.

References

1 Gordon, J.G. II and Ernst, S. (1980). Surface plasmons as a probe of the electrochemical interface. *Surf. Sci.* 101: 499–506.

2 Pockrand, I., Swalen, J.D., Gordon, J.G., and Philpott, M.R. (1978). Surface plasmon spectroscopy of organic monolayer assemblies. *Surf. Sci.* 74: 237–244.

3 Nylander, C., Liedberg, B., and Lind, T. (1982). Gas detection by means of surface plasmons resonance. *Sensors Actuators* 3: 79–88.

4 Liedberg, B., Nylander, C., and Lundström, I. (1983). Surface plasmons resonance for gas detection and biosensing. *Sensors Actuators* 4: 299–304.

5 Liedberg, B., Nylander, C., and Lundström, I. (1995). Biosensing with surface plasmon resonance- how it all started. *Biosens. Bioelectron.* 10: i–ix.

6 Homola, J., Yee, S.S., and Gauglitz, G. (1999). Surface plasmon resonance sensors: review. *Sensor. Actuat. B-Chem.* 54: 3–15.

7 Helmerhorst, E., Chandler, D.J., Nussio, M., and Mamotte, C.D. (2012). Real-time and label-free bio-sensing of molecular interactions by surface plasmon resonance: a laboratory medicine perspective. *Clin. Biochem. Rev.* 33 (4): 161.

8 Kausaite, A., van Dijk, M., Castrop, J. et al. (2007). Surface plasmon resonance label-free monitoring of antibody antigen interactions in real time. *Biochem. Mol. Biol. Educ.* 35 (1): 57–63.

9 Mohammadzadeh-Asl, S., Aghanejad, A., de la Guardia, M. et al. (2021). Surface plasmon resonance signal enhancement based on erlotinib loaded magnetic nanoparticles for evaluation of its interaction with human lung cancer cells. *Opt. Laser Technol.* 133: 106521.

10 Mauriz, E., Garcia-Fernandez, M.C., and Lechuga, L.M. (2016). Towards the design of universal immunosurfaces for SPR-based assays: a review. *Trends Anal. Chem.* 79: 191–198.

11 Otto, A. (1968). Excitation of nonradiative surface plasma waves in silver by the method of frustrated total reflection. *Z. Phys.* 216 (4): 398–410.

12 Kretschmann, E. and Raether, H. (1968). Notizen: radiative decay of non radiative surface plasmons excited by light. *Z. Nat. A* 23: 2135.

13 Daghestani, H.N. and Day, B.W. (2010). Theory and applications of surface plasmon resonance, resonant mirror, resonant waveguide grating, and dual polarization interferometry biosensors. *Sensors* 10 (11): 9630.

14 Chen, Y. and Ming, H. (2012). Review of surface plasmon resonance and localized surface plasmon resonance sensor. *Photonic Sens.* 2 (1): 37–49.

15 Boozer, C., Kim, G., Cong, S. et al. (2006). Looking towards label-free biomolecular interaction analysis in a high-throughput format: a review of new surface plasmon resonance technologies. *Curr. Opin. Biotechnol.* 17: 400–405.

16 Wulff, G. and Sarhan, A. (1972). The use of polymers with enzyme-analogous structures for the resolution of racemates. *Angew. Chem. Int. Ed.* 11: 341.

17 Arshady, R. and Mosbach, K. (1981). Synthesis of substrateselective polymers by host-guest polymerization. *Macromol. Chem. Phys.* 182: 687–692.

18 Haupt, K. and Mosbach, K. (1998). Plastic antibodies: developments and applications. *Trends Biotechnol.* 16: 468–475.

19 Haupt, K. (2003). Molecularly imprinted polymers: the next generation. *Anal. Chem.* 75: 376A–383A.

20 Chen, L., Wang, X., Lu, W. et al. (2016). Molecular imprinting: perspectives and applications. *Chem. Soc. Rev.* 45: 2137–2211.

21 Piletsky, S., Canfarotta, F., Poma, A. et al. (2020). Molecularly imprinted polymers for cell recognition. *Trends Biotechnol.* 38 (4): 368–387.

22 Haupt, K., Medina Rangel, P.X., and Bui, B.T.S. (2020). Molecularly imprinted polymers: antibody mimics for bioimaging and therapy. *Chem. Rev.* 120 (17): 9554–9582.

23 Cheng, Y., Nie, J., Liu, H. et al. (2020). Synthesis and characterization of magnetic molecularly imprinted polymers for effective extraction and determination of kaempferol from apple samples. *J. Chromatogr. A* 1630: 461531.

24 Mehta, R., van Beek, T.A., and Tetala, K.K.R. (2020). A micro-solid phase extraction device to prepare a molecularly imprinted porous monolith in a facile mode for fast protein separation. *J. Chromatogr. A* 1627: 461415.

25 Xing, R., Wen, Y., He, H. et al. (2019). Recent progress in the combination of molecularly imprinted polymer-based affinity extraction and mass spectrometry for targeted proteomic analysis. *Trends Anal. Chem.* 110: 417–428.

26 Shinde, S., Incel, A., Mansour, M. et al. (2020). Urea-based imprinted polymer hosts with switchable anion preference. *J. Am. Chem. Soc.* 142: 11404–11416.

27 Garcia Cruz, A., Haq, I., Cowen, T. et al. (2020). Design and fabrication of a smart sensor using in silico epitope mapping and electro-responsive imprinted polymer nanoparticles for determination of insulin levels in human plasma. *Biosens. Bielectron.* 169: 112536.

28 Ahmad, O.S., Bedwell, T.S., Esen, C. et al. (2019). Molecularly imprinted polymers in electrochemical and optical sensors. *Trends Biotechnol.* 37 (3): 294–309.

29 Chunta, S., Boonsriwong, W., Wattanasin, P. et al. (2021). Direct assessment of very-low-density lipoprotein by mass sensitive sensor with molecularly imprinted polymers. *Talanta* 221: 121549.

30 Saylan, Y., Akgönüllü, S., Yavuz, H. et al. (2019). Molecularly imprinted polymer based sensors for medical applications. *Sensors* 19 (6): 1279.

31 Cao, N., Zhao, F., and Zeng, B. (2020). A novel ratiometric molecularly imprinted electrochemiluminescence sensor T for sensitive and selective detection of sialic acid based on PEI-CdS quantum dots as anodic coreactant and cathodic luminophore. *Sensor. Actuat. B-Chem* 313: 128042.

32 Uzun, L. and Turner, A.P.F. (2016). Molecularly-imprinted polymer sensors: realising their potential. *Biosens. Bielectron.* 76: 131–144.

33 Qin, Y., Feng, Y., Ma, Y. et al. (2020). Tumor-sensitive biodegradable nanoparticles of molecularly imprinted polymer-stabilized fluorescent zeolitic imidazolate framework-8 for targeted imaging and drug delivery. *ACS Appl. Mater. Interfaces* 12: 24585–24598.

34 Sanadgol, N. and Wackerlig, J. (2020). Developments of smart drug-delivery systems based on magnetic molecularly imprinted polymers for targeted cancer therapy: a short review. *Pharmaceutics* 12: 831.

35 Griffete, N., Fresnais, J., Espinosa, A. et al. (2015). Design of magnetic molecularly imprinted polymer nanoparticles for controlled release of doxorubicin under an alternative magnetic field in athermal conditions. *Nanoscale* 7: 18891–18896.

36 Hassanpour, A., Irandoust, M., Soleimani, E., and Zhaleh, H. (2019). Increasing the anticancer activity of azidothymidine toward the breast cancer via rational design of magnetic drug carrier based on molecular imprinting technology. *Mater. Sci. Eng. C* 103: 109771.

37 Kordea, B.A., Mankara, J.S., Phuleb, S., and Krupadam, R.J. (2019). Nanoporous imprinted polymers (nanoMIPs) for controlled release of cancer drug. *Mater. Sci. Eng. C* 99: 222–230.

38 Canfarotta, F., Smolinska Kempisty, K., and Piletsky, S. (2017). Replacement of antibodies in pseudo-ELISAs: molecularly imprinted nanoparticles for vancomycin detection. *Methods Mol. Biol.* 1575: 389–398.

39 Esen, C., Czulak, J., Cowen, T. et al. (2019). Highly efficient abiotic assay formats for methyl parathion: molecularly imprinted polymer nanoparticle assay as an alternative to enzyme-linked immunosorbent assay. *Anal. Chem.* 91: 958–964.

40 Mattsson, L., Xu, J., Preininger, C. et al. (2018). Competitive fluorescent pseudo-immunoassay exploiting molecularly imprinted polymers for the detection of biogenic amines in fish matrix. *Talanta* 181: 190–196.

41 Garcia-Cruz, A., Cowen, T., Voorhaar, A. et al. (2020). Molecularly imprinted nanoparticles-based assay (MINA)- detection of leukotrienes and insulin. *Analyst* 145: 4224–4232.

42 Elbelazi, A., Canfarotta, F., Czulak, J. et al. (2019). Development of a homogenous assay based on fluorescent imprinted nanoparticles for analysis of nitroaromatic compounds. *Nano Res.* 12 (12): 3044–3050.

43 Mahajan, R., Rouhi, M., Shinde, S. et al. (2019). Highly efficient synthesis and assay of protein-imprinted nanogels by using magnetic templates. *Angew. Chem. Int. Ed.* 58: 727–730.

44 Panagiotopoulou, M., Salinas, Y., Beyazit, S. et al. (2016). Molecularly imprinted polymer coated quantum dots for multiplexed cell targeting and imaging. *Angew. Chem. Int. Ed.* 55: 8244––8248.

45 Demir, B., Lemberger, M.M., Panagiotopoulou, M. et al. (2018). Tracking hyaluronan: molecularly imprinted polymer coated carbon dots for cancer cell targeting and imaging. *ACS Appl. Mater. Interfaces* 10: 3305–3313.

46 Wang, S., Yin, D., Wang, W. et al. (2016). Targeting and imaging of cancer cells via monosaccharide-imprinted fluorescent nanoparticles. *Sci. Rep.* 6: 22757.

47 Medina Rangel, P.X., Laclef, S., Xu, J. et al. (2019). Solid-phase synthesis of molecularly imprinted polymer nanolabels: affinity tools for cellular bioimaging of glycans. *Sci. Rep.* 9: 3923.

48 Ekpenyong-Akiba, A.E., Canfarotta, F., Abd, B. et al. (2019). Detecting and targeting senescent cells using molecularly imprinted nanoparticles. *Nanoscale Horiz.* 4: 757––768.

49 Zhang, Y., Li, S., Ma, X.T. et al. (2020). Carbon dots-embedded epitope imprinted polymer for targeted fluorescence imaging of cervical cancer via recognition of epidermal growth factor receptor. *Microchim. Acta* 187: 228.

50 Yin, D., Li, X., Ma, Y., and Liu, Z. (2017). Targeted cancer imaging and photothermal therapy via monosaccharide-imprinted gold nanorods. *Chem. Commun.* 53: 6716–6719.

51 Zhang, Y., Deng, C., Liu, S. et al. (2015). Active targeting of tumors through conformational epitope imprinting. *Angew. Chem. Int. Ed.* 54: 5157–5160.

52 Liu, S., Bi, Q., Long, Y. et al. (2017). Inducible epitope imprinting: 'generating' the required binding site in membrane receptors for targeted drug delivery. *Nanoscale* 9: 5394–5397.

53 Peng, S., Wang, Y., Lia, N., and Li, C. (2017). Enhanced cellular uptake and tumor penetration of nanoparticles by imprinting the "hidden" part of membrane receptors for targeted drug delivery. *Chem. Commun.* 53: 11114–11117.

54 Medina Rangel, P.X., Moroni, E., Merlier, F. et al. (2020). Chemical antibody mimics inhibit cadherin-mediated cell-cell adhesion: a promising strategy for cancer therapy. *Angew. Chem. Int. Ed.* 59: 2816–2822.

55 Luoa, Q., Yua, N., Shib, C. et al. (2016). Surface plasmon resonance sensor for antibiotics detection based on photoinitiated polymerization molecularly imprinted array. *Talanta* 161: 797–803.

56 Muratsugu, S., Shirai, S., and Tada, M. (2020). Recent progress in molecularly imprinted approach for catalysis. *Tetrahedron Lett.* 61: 151603.

57 Li, X. and Zhao, Y. (2019). Chiral gating for size- and shape-selective asymmetric catalysis. *J. Am. Chem. Soc.* 141: 13749–13752.

58 Muratsugu, S., Baba, H., Tanimoto, T. et al. (2018). Chemoselective epoxidation of cholesterol derivatives on a surface-designed molecularly imprinted Ru–porphyrin catalyst. *Chem. Commun.* 54: 5114–5117.

59 Muratsugu, S., Maity, N., Baba, H. et al. (2017). Preparation and catalytic performance of a molecularly imprinted Pd complex catalyst for Suzuki cross-coupling reactions. *Dalton Trans.* 46: 3125–3134.

60 Hu, L., Arifuzzaman, M., and Zhao, Y. (2019). Controlling product inhibition through substrate-specific active sites in nanoparticle-based phosphodiesterase and esterase. *ACS Catal.* 9: 5019–5024.

61 Ahmad, O.S., Bedwell, T.S., Esen, C. et al. (2019). Molecularly imprinted polymers in electrochemical and optical sensors. *Trends Biotechnol.* 37 (3): 294–309.

62 Lai, E.P.C., Fafara, A., VanderNoot, V.A. et al. (1998). Surface plasmon resonance sensors using molecularly imprinted polymers for sorbent assay of theophylline, caffeine, and xanthine. *Can. J. Chem.* 76: 265–273.

63 Safran, V., Göktürk, I., Derazshamshir, A. et al. (2019). Rapid sensing of Cu^{+2} in water and biological samples by sensitive molecularly imprinted based plasmonic biosensor. *Microchem. J.* 148: 141–150.

64 Jalilzadeh, M., Çimen, D., Özgür, E. et al. (2019). Design and preparation of imprinted surface plasmon resonance (SPR) nanosensor for detection of Zn(II) ions. *J. Macromol. Sci. Part A* 56: 877–886.

65 Bakhshpour, M. and Denizli, A. (2020). Highly sensitive detection of Cd(II) ions using ion-imprinted surface plasmon resonance sensors. *Microchem. J.* 105572: 159.

66 Sharma, S. and Gupta, B.D. (2018). An efficient and selective sensing of creatinine based on fiber optic SPR technique exploiting the advantages of molecular imprinting technique. In Optical Sensing and Detection V. International Society for Optics and Photonics, p. 106801T.

67 Göçenoğlu Sarıkaya, A., Osman, B., Çam, T., and Denizli, A. (2017). Molecularly imprinted surface plasmon resonance (SPR) sensor for uric acid determination. *Sensor. Actuat. B-Chem.* 251: 763–772.

68 Ertürk, G., Özen, H., Tümer, M.A. et al. (2016). Microcontact imprinting based surface plasmon resonance (SPR)biosensor for real-time and ultrasensitive detection of prostate specific antigen (PSA) from clinical samples. *Sensor. Actuat. B-Chem.* 224: 823–832.

69 Rebelo, T.S.C.R., Costa, R., Brandao, A.T.S.C. et al. (2019). Molecularly imprinted polymer SPE sensor for analysis of CA-125 on serum. *Anal. Chim. Acta* 1082: 126–135.

70 Altintas, Z., France, B., Ortiz, J.O., and Tothill, I.E. (2016). Computationally modelled receptors for drug monitoring using an optical based biomimetic SPR sensor. *Sensor. Actuat. B-Chem.* 224: 726–737.

71 Sari, E., Üzek, R., Duman, M. et al. (2018). Prism coupler-based sensor system for simultaneous screening of synthetic glucocorticosteroid as doping control agent. *Sensor. Actuat. B-Chem.* 260: 432–444.

72 Shrivastav, A.M., Usha, S.P., and Gupta, B.D. (2017). Highly sensitive and selective erythromycin nanosensor employing fiber optic SPR/ERY imprinted nanostructure: application in milk and honey. *Biosens. Bioelectron.* 90: 516–524.

73 Perçin, I., Idil, N., Bakhshpour, M. et al. (2017). Microcontact imprinted plasmonic nanosensors: powerful tools in the setection of salmonella paratyphi. *Sensors* 17: 1375.

74 Özgür, E., Topçu, A.A., Yılmaz, E., and Denizli, A. (2020). Surface plasmon resonance based biomimetic sensor for urinary tract infections. *Talanta* 212: 120778.

75 Çakır, O. and Baysal, Z. (2019). Pesticide analysis with molecularly imprinted nanofilms using surface plasmon resonance sensor and LC-MS/MS: comparative study for environmental water samples. *Sensor. Actuat. B-Chem.* 297: 126764.

76 Pathak, A., Parveen, S., and Gupta, B.D. (2017). Ultrasensitive, highly selective, and real-time detection of protein using functionalized CNTs as MIP platform for FOSPR-based biosensor. *Nanotechnology* 28: 355503.

77 Rico-Yuste, A. and Carrasco, S. (2019). Molecularly imprinted polymer-based hybrid materials for the development of optical sensors. *Polymers* 11: 1173.

78 Chen, B., Liu, C., Xie, Y. et al. (2016). Localized surface plasmon resonance gas sensor based on molecularly imprinted polymer coated au nano-island films: influence of nanostructure on sensing characteristics. *IEEE Sensors J.* 16: 3532–3540.

79 Luo, Q., Yu, N., Shi, C., and Wang, X. (2016). Jianmin Wu, surface plasmon resonance sensor for antibiotics detection based on photo-initiated polymerization molecularly imprinted array. *Talanta* 161: 797–803.

80 Zhang, L., Zhu, C., Chen, C. et al. (2018). Determination of kanamycin using a molecularly imprinted SPR sensor. *Food Chem.* 266: 170–174.
81 Ertürk, G. and Mattiasson, B. (2016). From imprinting to microcontact imprinting- a new tool to increase selectivity in analytical devices. *J. Chromatogr. B* 1021: 30–44.
82 Ertürk Bergdahl, G., Andersson, T., Allhorn, M. et al. (2019). In vivo detection and absolute quantification of a secreted bacterial factor from skin using molecularly imprinted polymers in a surface plasmon resonance biosensor for improved diagnostic abilities. *ACS Sens.* 4: 717–725.
83 Kidakova, A., Reut, J., Rappich, J. et al. (2018). Preparation of a surface-grafted protein-selective polymer film by combined use of controlled/living radical photopolymerization and microcontact imprinting. *React. Funct. Polym.* 125: 47–56.
84 Walcarius, A. and Collinson, M.M. (2009). Analytical chemistry with silica sol-gels: traditional routes to new materials for chemical analysis. *Annu. Rev. Anal. Chem.* 2: 121–143.
85 Ayankojo, A.G., Reut, J., Öpik, A. et al. (2018). Hybrid molecularly imprinted polymer for amoxicillin detection. *Biosens. Bioelectron.* 118: 102–107.
86 Yao, T., Gu, X., Li, T. et al. (2016). Enhancement of surface plasmon resonance signals using a MIP/GNPs/rGO nano-hybrid film for the rapid detection of ractopamine. *Biosens. Bioelectron.* 75: 96–100.
87 Brule, T., Granger, G., Bukar, N. et al. (2017). A field-deployed surface plasmon resonance (SPR) sensor for RDX quantification in environmental waters. *Analyst* 142: 2161–2168.
88 Pastoriza-Santos, I. and Liz-Marzan, L.M. (2008). *J. Mater. Chem.* 18 (15): 1724–1737.
89 Jiang, X., Zeng, Q., and Yu, A. (2007). *Langmuir* 23 (4): 2218–2223.
90 Cheng, Z., Wang, Z., Gillespie, D.E. et al. (2015). *Anal. Chem.* 87 (3): 1466–1469.
91 Özkan, A., Atar, N., and Yola, M.L. (2019). Enhanced surface plasmon resonance (SPR) signals based on immobilization of core-shell nanoparticles incorporated boron nitride nanosheets: development of molecularly imprinted SPR nanosensor for anticancer drug, etoposide. *Biosens. Bioelectron.* 130: 293–298.
92 Palladino, P., Minunni, M., and Scarano, S. (2018). Cardiac Troponin T capture and detection in real-time via epitope-imprinted polymer and optical biosensing. *Biosens. Bioelectron.* 106: 93–98.
93 Baldoneschi, V., Palladino, P., Banchini, M. et al. (2020). Norepinephrine as new functional monomer for molecular imprinting: an applicative study for the optical sensing of cardiac biomarkers. *Biosens. Bioelectron.* 157: 112161.
94 Sari, E., Üzek, R., Duman, M., and Denizli, A. (2016). Fabrication of surface plasmon resonance nanosensor for the selective determination of erythromycin via molecular imprinted nanoparticles. *Talanta* 150: 607–614.
95 Yılmaz, E., Özgür, E., Bereli, N. et al. (2017). Plastic antibody based surface plasmon resonance nanosensors for selective atrazine detection. *Mater. Sci. Eng. C* 73: 603–610.
96 Altintas, Z. (2018). Surface plasmon resonance based sensor for the detection of glycopeptide antibiotics in milk using rationally designed nanoMIPs. *Sci. Rep.* 8: 11222.
97 Ashley, J., Shukor, Y., D'Aurelio, R. et al. (2018). Synthesis of molecularly imprinted polymer nanoparticles for α-casein detection using surface plasmon resonance as a milk allergen sensor. *ACS Sens.* 3: 418–424.

98 Erdem, Ö., Saylan, Y., Cihangir, N., and Denizli, A. (2019). Molecularly imprinted nanoparticles based plasmonic sensors for real-time enterococcus faecalis detection. *Biosens. Bioelectron.* 126: 608–614.

99 Rahtuvanoğlu, A., Akgönüllü, S., Karacan, S., and Denizli, A. (2020). Biomimetic nanoparticles based surface plasmon resonance biosensors for histamine detection in foods. *Chemistry Select* 5: 5683–5692.

100 Cennamo, N., Maniglio, D., Tatti, R. et al. (2020). Deformable molecularly imprinted nanogels permit sensitivity-gain in plasmonic sensing. *Biosens. Bioelectron.* 156: 112126.

Index

Plasmonic Sensors and their Applications, First Edition. Edited by Adil Denizli.